Human Factors:
Theory and Practice

WILEY SERIES IN HUMAN FACTORS

Edited by David Meister

Human Factors in Quality Assurance

Douglas H. Harris and Frederick B. Chaney

Man-Machine Simulation Models

Arthur I. Siegel and J. Jay Wolf

Human Factors Applications in Teleoperator Design and Operation

Edwin G. Johnsen and William R. Corliss

Human Factors: Theory and Practice

David Meister

Human Factors: Theory and Practice

David Meister
The Bunker-Ramo Corporation

Wiley-Interscience a Division of John Wiley & Sons, Inc.
New York · London · Sydney · Toronto

Library of Congress Catalogue Card Number: 77-148505

ISBN 0-471-59190-4

Printed in the United States of America.

10 9 8 7 6 5 4 3 2

SERIES PREFACE

Technology is effective to the extent that men can operate and maintain the machines they design. Equipment design which consciously takes advantage of human capabilities and constrains itself within human limitations amplifies and increases system output. If it does not, system performance is reduced and the purpose for which the equipment was designed is endangered. This consideration is even more significant today than in the past because the highly complex systems that we develop are pushing human functions more and more to their limits of efficient performance.

How can one ensure that machine and machine operations are actually designed for human use? Behavioral data, principles, and recommendations—in short, the Human Factors discipline—must be translated into meaningful design practices. Concepts like ease of operation or error-free performance must be interpretable in hardware and system terms.

Human Factors is one of the newer engineering disciplines. Perhaps because of this, engineering and human-factors specialists lack a common orientation with which their respective disciplines can communicate. The goal of the Wiley Human Factors Series is to help in the communication process by describing what behavioral principles mean for system design and by suggesting the behavioral research that must be performed to solve design problems. The premise on which the series is based and on which each book is written is that Human Factors has utility only to the degree that it supports engineering development; hence the Series emphasizes the *practical application* to design of human-factors concepts.

Because of the many talents on which Human Factors depends for its implementation (design and systems engineering, industrial and experimental psychology, anthropology, physiology, and operations research, to name only a few), the Series is directed to as wide an audience as possible. Each book is intended to illustrate the usefulness of Human Factors to some significant aspect of system development, such as human factors in design or testing or simulation. Although cookbook answers are not provided, it is hoped that this pragmatic approach will enable the many specialists concerned with problems of equipment design to solve these problems more efficiently.

DAVID MEISTER
Series Editor

PREFACE

To avoid any misconceptions of what he is about to read, it is only fair to warn the reader about what this book is and is not. Summed up as simply as possible, the book tries to describe what the human factors specialist actually does during system development, what he should be doing, and what the factors underlying and impacting on his discipline are.

I have endeavored to make the book both practical and theoretical. Part One deals with the underlying concepts of human factors, Part Two with the realities of the human factors job. In both aspects I have attempted to "tell it as it is," neither overemphasizing nor underplaying the role human factors has in system development. Because of its candor, the book may disturb some readers by its recognition of the inadequacies in both human factors and engineering. Some of the things I have said might even be considered "controversial", although not to the experienced specialist. However, I make no apologies for telling the truth as I see it.

It is my hope that the book can be used as a practical guide to the daily performance of tasks in human factors as well as a general tutorial introduction to the discipline. As a guide, it contains step-by-step procedures and models for performing some of the tasks the specialist will have to face. It is not a design guide, in the sense that specific human engineering data are presented as in Woodson and Conover's *Human Engineering Guide for Equipment Designers* although it discusses methods of designing for the requirements of human factors. Moreover, it is not a text on applied experimental psychology, although it can be used to teach what the human factors job actually is.

The book is adressed to several types of reader:

1. Obviously, the *human factors specialist*. For him the book contains specific instructions on the most effective way of assisting in system development, among which are the following examples: (a) a step-by-step description of how to perform human factors analyses; (b) how to write the human factors part of a proposal, how to solicit contract research, and how to develop a test plan and write design recommendations; (c) how to organize and run a human factors group.

2. The *design engineer* and *engineering manager* hopefully will also be readers especially since I have made the relationship between the engineer and the specialist a cornerstone of my treatment. If this book gives the former an appreciation of what the human factors specialist is trying to accomplish, I will consider myself successful. The book will, in addition, indicate to the manager the services his human factors group should perform, thereby enabling him to make better use of those services; for example, it analyzes in detail MIL-H-46855 and MIL-STD 1472, the governing human factors specification and standard, pointing out what can be done with them and what cannot. An entire chapter is devoted to an experimentally based discussion of the relationship between human factors and engineering and how the specialist views the design process; it may point out a few things about the process that the engineer may not have been aware of.

3. The *student* interested in becoming a human factors specialist will find out what he will be asked to do, the problems he will face, and the training and experience he should have. Previous texts in this field have emphasized the academic background and data it is presumed the specialist should ideally possess. This volume differs from its predecessors by concentrating on what the specialist actually does, hence on what he really needs to know.

4. It will give the *psychologist* a backstage view of the drama of human factors. What I have said about the differences between human factors and psychology and the implications of those differences for research may stimulate a more effective research program in both disciplines.

5. For the *government representative*, whether or not he is a specialist, it will point up the need for exercising more effective control over the human factors aspects of system development.

How well I have succeeded in my aims the reader alone will judge. It remains only for him to turn the page and read.

DAVID MEISTER

Los Angeles, California
January 1971

CONTENTS

HUMAN FACTORS: Theory

Overview of Part One

Chapter One is an introduction to the rest of the book. It explores the various ways in which human factors can be defined, describes the assumptions under which human factors specialists perform, introduces the man-machine system concept and the effect of error and time discrepancies on system performance. The questions the discipline seeks to answer are listed. The chapter concludes with a short description of human factors as a discipline and the background of the human factors specialist.

Chapter Two points out the significance of the error concept to human factors theory and practice. The various types of human error are described in terms of their causes, their consequences, and their occurrence at various stages of the system development cycle. Evidence is provided for the existence of error in system development and operation. The significant effect of error on system functioning is shown. The requirements for demonstrating that human factors can reduce error are pointed out as well as some examples of the usefulness of the discipline in system design.

Chapter Three describes in step-by-step fashion how the human factors specialist analyzes system requirements, allocates and verifies functions, describes and analyzes tasks, and identifies the interface equipment needed to implement man-machine interaction. The data requirements for performing these analyses are indicated as well as the availability of data to initiate these analyses.

Chapter Four describes the human factors research required to supply the data for Chapter Three analyses. Research needs are analyzed in terms of six categories: equipment, personnel, functions/tasks, environment, measures, and methodology. Differences in orientation between human factors and psychology are pointed out together with their implications for research requirements. A list of priority areas for man-machine research is drawn up.

I

HUMAN FACTORS AND THE MAN MACHINE CONCEPT

This chapter is an introduction to the rest of the book. On the assumption that at least a few of our readers may never have heard of the human factors discipline before, this chapter will provide an overview and background for later more detailed discussion. Topics briefly mentioned in this chapter will be discussed in greater detail in later chapters.

DEFINITIONS

The term *human factors* has several connotations. This may lead to confusion unless their different meanings are carefully differentiated. When a human factors specialist uses the term, he may mean one of the following:

1. Human factors are those elements which influence the efficiency with which people can use equipment to accomplish the functions of that equipment. The most important of these elements are the following:

(a) Equipment—the physical characteristics of the equipment to which personnel must respond (e.g., such things as the arrangement of controls and displays).

(b) Environment—the physical surroundings in which the equipment must be operated and maintained (e.g., the physical layout of the room, noise level, temperature, lighting and the arrangement of other equipment in the work area).

(c) Task—the characteristics of the jobs which people must perform in order to accomplish performance goals (e.g., length and complexity of operating procedures).

(d) Personnel—not least, the capabilities and limitations of equipment operators and maintainers themselves (e.g., their intelligence and their visual, auditory, and motor acuity, training, and experience in operation of the equipment).

The engineer and the human factors specialist must consider the same elements during design to ensure that equipment under development is designed for most effective use by its operators. How these elements influence operator performance will be described in greater detail later.

2. A second meaning for, or implication of, the term human factors is to refer to the number and type of personnel selected to run the system and how they function. This includes

(a) the number of operators and maintenance technicians in the system (e.g., six radar mechanics, two tank drivers);
(b) their skill level (e.g., apprentices, slightly skilled, highly skilled);
(c) the functions and tasks they must perform in controlling their equipment (e.g., detect and classify submarines, program a computer);
(d) how they perform their tasks (e.g., as single operators or as a crew; in what interaction with other personnel; shift duration).

3. A third implication of the term human factors is its reference to

(a) the manner in which personnel perform in using the equipment (e.g., their errors, the time they take to complete their tasks) and
(b) the effect of that performance on other system elements (e.g., an error resulting in a malfunctioned equipment) or on over-all system goals (e.g., an operator error or delay in response resulting in failure to accomplish the mission, such as, a gunner firing at a target after it is out of range).

4. A fourth way in which the term human factors can be used is to refer to the effect of the over-all system upon its personnel elements. The system of which the operator is a part may influence his performance (causing him to work harder or slower), his health and his attitudes toward nonwork related activities. In ancient Roman galley ships, slaves pulled harder or slower on their oars, depending on the operational requirements, such as the need to pursue a pirate. In more modern times, working at a physically demanding job like a blast furnace may reduce the worker's life expectancy; working at a job which is intellectually and emotionally demanding may create frustrations which affect the worker's attitudes toward his family, friends and society in general.

5. The final sense in which the term human factors is used is to refer to the *profession* in which human factors specialists function. In this

sense human factors encompasses the specialists, the methods they employ, the data they use, and the work they do. In order to distinguish between human factors as a set of variables influencing equipment and personnel efficiency and human factors as a professional discipline, the latter will be designated by the symbol $\underline{H}$, that is H underlined. When the term human factors is spelled out, it will stand for the complex of elements and parameters about which the $\underline{H}$ specialist and the engineer must be concerned if total system efficiency is to be maximized.

ASSUMPTIONS

As a discipline $\underline{H}$ makes a number of assumptions, for all of which some empirical proof (to be described later) exists:

1. There is a relationship between the efficiency with which people operate and maintain an equipment and the ultimate effectiveness of that equipment's functioning. If this is true, then any reduction in personnel performance should cause the equipment to fail to perform *its* function or to perform that function less effectively than it should. Although a radar may have the capability of detecting aircraft at a distance of 30 miles, it will fail to perform this function if its operator fails to perceive the target signal on its scope because he is unduly fatigued or improperly trained to recognize the signal.

2. Equipment characteristics influence how men operate and maintain that equipment. These act as stimuli (inputs) to which the equipment user must respond (output). If the stimuli require more than the user can supply, he will respond erroneously, out of synchronization (delayed response), or perhaps not at all.

Equipment characteristics demand selective attention of the user; that is, he must pay more attention to some features of the equipment than others. Procedures for operating and maintaining the equipment impose loads on his memory. They require that he be trained to perform his job.

Just as materials such as steel, aluminum, or plastic have certain tolerance limits within which they function effectively, so the capabilities of men are set within certain natural limits within which they function best. At a temperature of 3000 or more degrees certain types of plastic begin to melt; in the same manner, when people are required to operate equipment whose configuration overloads them, their efficiency decreases.

3. Since equipment characteristics (e.g., arrangement of controls and displays, the organization of internal components) function as user stim-

uli, it follows that certain arrangements of these characteristics/stimuli will be more efficiently responded to by personnel than will others. Thus, if equipment characteristics are matched to the capabilities and limitations of men, the performance of the latter should be more efficient. An auditory message will obviously be heard more readily if it is transmitted at 10,000 cps (cycles per second) than if it is transmitted at 25,000 cps because the range of human hearing varies between 50 and 18,000 cps.

4. It is cheaper and easier to adapt equipment to human capacities than it is to modify human capabilities to equipment requirements. It is easier to select different components or to arrange them differently than it is to add more sensitive visual acuity to the human, endow him with more than his native intelligence, or change his physical dimensions to a more suitable size.

Why not select people more discriminatingly for equipment operating jobs? or train them longer and more intensively? Take sonar for example. Because sonar detection requires fairly precise pitch and loudness discrimination, it seemed reasonable in early World War II to test applicants' hearing for special sensory ability and to select them on this basis. The trouble was that too few people measured up to these sensory requirements. Higher selection standards—and more intensive training—are an ideal solution as long as one is designing only for a few highly qualified people, such as astronauts or test pilots. This solution falls apart, however, as soon as masses of people are needed (as they were in World War II) as equipment users. There is a great difference in designing an automobile for a Grand Prix racing driver, on the one hand, and for Mr. Average Driver, on the other.

The technological environment in which we live is above all a *mass* culture and becoming more so every day, even in the underdeveloped countries. Because of his modest store of aptitudes and skills, the infusion of technology is likely to disadvantage the average equipment user.

Moreover, the complexity of our modern technological devices does not respect even the highly selected and trained. Aircraft pilots are both highly selected and intensively trained, and yet many aircraft accidents result from human error (Beaty, 1969). The goal of H is, therefore, to optimize the design of equipment from the standpoint of the equipment user so that his efficiency will be at its greatest. In a nutshell H attempts to "tailor" equipment to the capabilities and limitations of the user.

5. It follows from all this that a primary influence (although not the only one) on the way in which humans operate and maintain machines is the efficiency with which machines are developed. It is also logical that improving the efficiency of system development will result in superior equipment-usage.

ELEMENTS AND CHARACTERISTICS

H can now be more precisely defined as the application of behavioral principles and data (i.e., those describing how people behave) to engineering design to do two things: to maximize the human's contribution to the effectiveness of the system of which he is a part and to reduce the impact of that over-all system on him. The basic concept which underlies the application of H to equipment design is that of the *man-machine system* (see Figure 1-1). This concept says in effect that there is a closed-loop relationship between the human and his equipment. The operator receives information (1) from the equipment via displays (2), makes certain decisions (3) involving that information, and operates controls (4) to affect equipment status (5). The equipment in turn provides information about its changed status to the operator (1). And so the cycle continues until one turns either the machine or the man off. The interaction between man and his machine creates the system relationship.

The major elements in this system are the same human factors referred to at the start of this chapter—equipment, environment, tasks, and personnel. Each of these consists of many subelements, each of which may influence the efficiency of the man-machine system.

Equipment	Environment	Tasks	Personnel
Controls	Temperature	Content (procedures)	Intelligence
Displays	Illumination	Duration	Sensory capability
Equipment dimensions	Vibration	Feedback	Motor capability
* Type and arrangement of internal components	Noise	Response frequency requirements	Training
* Test points	Ventilation	Accuracy requirements	Experience
		Speed requirements	Motivation

* Primarily for maintenance men.

Organizing all of these into a meaningful whole is the *system* which sets the requirements for this organization in terms of performance criteria or goals: number of outputs per unit time (e.g., messages transmitted each hour) or precision of response (e.g., allowable bomb miss distance of 50 yards) or speed of response (e.g., 1600 mph in a fighter).

The man-machine system (MMS) can be defined as an organization of machines and men plus the processes by which they interact within an environment to produce some desired system output. The goal of system development, with which H (along with engineering) is primarily con-

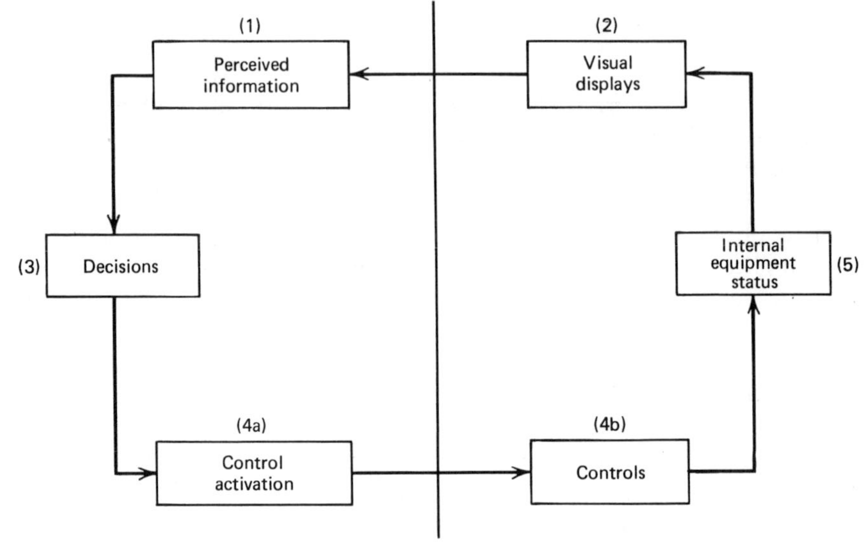

Figure 1-1 The man-machine system loop.

cerned, is to organize equipment, environment, tasks, and personnel variables in such a way that the whole meets these system goals in a manner which is satisfactory both to the system and to the operator.

The man-machine system has certain characteristics which are described graphically in Figure 1-2.

1. Performance is directed to achieve specified goals. The system is created by men to perform certain functions. The machine which the operator feeds has a certain rated output. System adequacy is evaluated in terms of how well the system achieves that design capacity.

2. Men and machines are both required in the system; they interact with each other. Although systems vary in the degree of their automaticity, even industrial process control systems (e.g., as in petrochemicals) which are almost completely automated have one or two men to control the system and many more to maintain it.

3. The system receives inputs and makes outputs which must meet system goals; for example the U.S. Post Office, which is a transmission system, receives millions of letters and parcels per day from patrons; it must deliver these to their recipients within a specified time period.

4. The system has an environment which impacts upon it. Within the factory, for example, this environment consists of the physical room structure, loading docks, lighting, etc.

5. The system is a closed loop which involves feedback. Managers

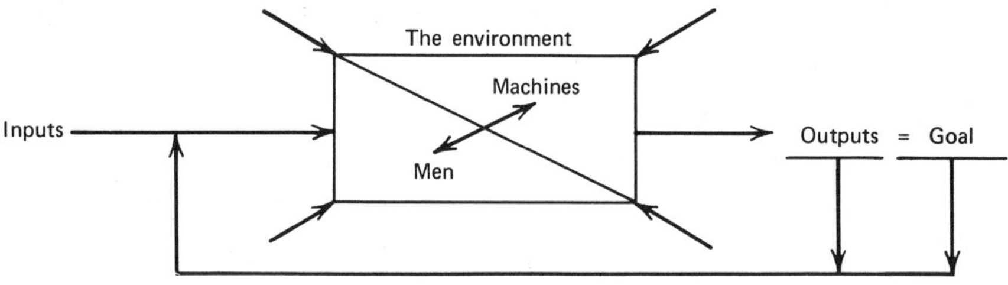

Figure 1-2 Man-machine system interactions.

compare outputs with the design goal to determine system efficiency; input rate may be adjusted to compensate for changes in output or more equipment and personnel may be added to increase output.

In order to design the system effectively one must consider all the factors which might influence the ultimate performance of the system: men, machines, environment, inputs/outputs, and goals.

The essence of a closed loop system like the MMS is that a change in one element produces a change in another. Increase task accuracy requirements (e.g., the speed with which the operator must activate his controls) and personnel elements are affected; motor capability must increase, and training may have to be intensified. Increase the complexity of displays (e.g., substitute a multiscale meter for a simple indicator), and lighting which was adequate for the indicator must be increased for the meter. Increase system performance criteria, and all MMS elements must adjust accordingly.

The extent of the involvement between the man and the machine obviously varies with particular types of equipment. The attribute most crucial for this involvement is *automation* which simplifies the operator's task while complicating the maintenance man's. In highly automated equipment the operator's task may be simply to push a button to activate the equipment and to monitor indicators which enable him to shut the equipment down if an emergency arises. (With increased automation, however, equipment maintenance becomes correspondingly more complex. More attention must then be paid to the maintenance man's job.)

Obviously, the role of the human in the man-machine system will vary depending on the nature of the equipment. His simplest role is, as we have seen, that of equipment activator and monitor; the equipment does most of the complex work. An example is a computer-controlled milling machine which automatically measures, cuts, and finishes steel bars when a control is depressed.

Few equipments are so automatic, however. The human's most important role is that in which he is an equipment "transducer." In this situation the equipment senses, collects, and displays information, and the operator's role is to integrate, interpret, and decide upon that information. In such a distribution of labor, typically represented by information-processing equipment such as radar or sonar, the efficiency of the man-machine combination is critically dependent on the operator's effectiveness.

Most man-machine systems vary between these two extremes. Each situation must be analyzed individually and in detail to determine the degree of human involvement. The amount of that involvement determines how much and what type of work the H specialist will be called on to perform. The more interaction there is between the man and his machine, the more human factors aspects the specialist must consider.

The man-machine system we have been talking about can be viewed at various levels. Within any system there are a large number of subsystems and sub-subsystems, each of which fits the definition of the MMS at a lower level like a cell within a body. At the most molar level the U.S. Steel Corporation can be considered a MMS. U.S. Steel, however, consists of a number of divisions, each of which in turn is a man-machine subsystem. Within each factory there are individual departments such as receiving, inspection, processing, etc., which are also MMS. Within each factory department one finds individual workers at their machines; each of these man-machine combinations, viewed as a closed loop unit (i.e., ignoring its interactions with all other units), is a MMS in miniature.

HUMAN FACTORS AND SYSTEM DESIGN

In the process of developing the MMS the designer and the H specialist must ask the following questions:

1. What inputs and outputs must be provided to satisfy system goals?
2. What operations are required to produce system outputs? (What must the MMS do?)
3. What should be the assignment of system responsibilities (i.e., what functions should the man perform within the system)?
4. What kinds of men are required to perform these functions?
5. What tasks should they perform? Can they respond to system inputs and produce the necessary outputs?
6. What equipment interfaces does the man need to perform his tasks (e.g., controls/displays, test points, diagnostic information, procedures)?

7. What is the effect of the machine on the man; the man on the machine? (Since there is an interaction between the two, the machine may demand more of the man than he can produce; likewise, the man may hinder the functioning of the machine.)

Human factors are, of course, only one parameter affecting design of the MMS. It is unreasonable to expect that the engineer will be concerned only about human factors in his design.

Other parameters which must be considered are performance requirements, cost, reliability, producibility, maintainability, and safety. However, human factors considerations interact with all of these; we cannot consider these parameters correctly without involving human factors.

For any system design requirements there are usually several alternative ways of satisfying that requirement, that is, several different design concepts, all of which may satisfy requirements but some of which may be better than others, depending on the criteria applied; for example, a preference for a machine solution to the design problem may lead to

(a) higher degree of automation,
(b) higher reliability in operations,
(c) higher cost,
(d) lowered producibility,
(e) lowered maintainability.

Alternatively, a preference for operator performance solutions may lead to

(a) lowered degree of automation,
(b) lowered reliability in operation,
(c) lower cost,
(d) higher producibility,
(e) higher maintainability.

No one design solution is good or bad; it depends on trading off all the design parameters mentioned, *including human factors.*

The starting point for our consideration of H̲ as a discipline is obviously system development. Conceivably we could confine our interest merely to the *use* of equipment after that equipment had been designed and produced. In that event, however, we would merely be describing how people use the equipment, which is interesting but not crucial. If we wish to influence user behavior (control being one of the goals of science), we can do so only by going back historically to the genesis of the equipment and modifying the development concepts which ultimately led to the user behavior. Above all else, H̲ is a pragmatic discipline; data on equipment usage serve primarily as feedback to enable the H specialist to improve methods of developing future systems.

We consider therefore that H is a system development discipline and that the ultimate goal of human factors theory, research and practice is to influence system development.

THE EFFECT OF HUMAN PERFORMANCE ON SYSTEM EFFECTIVENESS

Because system functioning is often critically dependent upon human activity, the effect of the human on the performance of the equipment he is operating and maintaining is highly significant. Any characteristic of the system which makes it difficult for operators and maintenance men to do their job reduces the efficiency of equipment functioning. This follows because the man and his machine are interdependent.

The way in which human inefficiency is indicated is through *errors* and *time* (delays). Chapter Two defines these indices more completely (with special emphasis on error) and demonstrates just how significant they are to system performance. For the moment it is enough to consider an error as any deviation from the performance required of the operator to accomplish the system function. The deviation must, of course, impair system functioning if it is to be considered a meaningful error.

The error may reveal itself as

(a) a failure to perform a required action—that is, an error of omission;
(b) the performance of that action in an incorrect manner—that is, an error of commission; or
(c) its performance out of sequence or at an incorrect time.

Time as an index of performance refers to the operator's failure to complete an equipment task *when* required by the system mission, or his failure to respond quickly enough to some signal or cue requiring an action. The time measure is an error also, expressed, however, in time.

Thus the operator's failure to perform correctly often degrades the performance of his equipment. For example, in the U.S. Post Office the average throughput of parcel sorting machines is 42% of design capacity, of letter sorting machines, 40 to 77% of design capacity (Communications & Systems, 1969). The major factor influencing that throughput is the operator input of parcels and letters.

The effect of the operator's failure can be scaled in terms of the seriousness of its consequences to the system goal, to the equipment, and to the operator himself.

The least serious effect is that accomplishment of the equipment function or goal is delayed (where the delay is not crucial to that goal). For example, a railroad stop signal is hoisted erroneously just outside of Philadelphia and delays the New York to Washington Limited. The train, which ordinarily takes 3½ hours, may reach Washington 15 minutes late. Passengers may be inconvenienced, but the goal or function of the train is ultimately accomplished. In the case of the U.S. Post Office, mail which is misdirected is usually not lost but is delayed in getting to its recipients.

More serious, however, is that the goal of equipment functioning cannot be accomplished because of the operator's failure. In the radar system referred to earlier, the operator's failure to see the target signal on his scope means that, although the equipment continues to function, it does not perform its detection mission.

Still more serious, an operator error may lead to equipment malfunction so that not only is the equipment's mission not accomplished but the means of accomplishing the mission is at least temporarily lost; for example, in a recent study performed by the author and his colleagues (Meister et al., 1970) it was found that 13% of all avionics equipment failures reported in B-52's at two airbases were caused directly by operator error.

Most serious is that the effect of an operator's error may be to hazard his own or others' safety; for example, when the pilot of an aircraft makes a mistake, his crew and passengers may be injured or killed.

Everyone accepts the possibility of human error. H assumes, however, that error is not inevitable, is not the consequence of human "fallibility" (something like a flaw in the human material). Rather, the assumption is made that error occurs only when the conditions which predispose the human to make the error exist. In man-machine systems equipment characteristics and the nature of the tasks required to use the equipment may lead to a higher human error rate than one would normally expect on the basis of chance alone.

The role of H in system development is, among other things, to prevent the design of equipment whose characteristics may predispose to operator error. In that sense the analysis of equipment from the standpoint of those factors which could lead to error is an integral part of good design practice.

However, "good" design practice by which engineers mean the electronic, mechanical, etc. principles which they ordinarily apply to solve design problems will not be sufficient to deal with most human factors design problems. Solutions to these problems require a special way of analyzing equipment—in terms of the effect of equipment upon the operator as well as the effect of the operator on the equipment—and special techniques. This is why a distinctive discipline devoted to solving problems of this sort has developed.

This does not mean, however, that an engineer cannot with the proper training learn to become his own H specialist or, even if he confines himself to the equipment aspects of design, that he cannot learn enough to know when to ask assistance of the H specialist or to help the latter do his job better. One of the goals of this book is to help the engineer learn enough about human factors and H techniques to do a good deal of his own H design.

BACKGROUND OF THE H DISCIPLINE

Although the major impetus to H began during World War II, its origins go much further back. Two major disciplines, industrial engineering (concerned with the measurement of work output) and psychology (interested in the factors responsible for learning, performance, and personnel selection), contributed to the development of H.

H in World War II was largely research oriented but with a very strong applications-orientation. Interest was centered on relatively new systems, such as aircraft, radar, sonar, which imposed special demands on personnel. The investigation of personnel performance in these systems (e.g., the study of pilot errors in reading altimeters (Fitts and Jones, 1947) obviously suggested ways of improving the design of these systems.

The research psychologist (he was not yet an H specialist) was thrown willy-nilly into intimate contact with engineers. To apply his research results concretely he had to talk and work within the context of actual system development. We cannot apply human factors research to design as "on-call" consultants, who merely refer the engineer for solutions to specialized journal reports or handbooks. As a consequence, direct communication concerning design problems was required between the engineer and the man who now became an H specialist if the former was to apply the latter's findings and recommendations to specific problems.

This direct, continuing communication has become all the more necessary because design problems and solutions tend to change, sometimes markedly, as system development progresses. Design is not static, nor does equipment spring into existence full blown. Design is progressively elaborated and refined from its original rather gross concept. The products of development, such as functional flow diagrams, equipment descriptions, control panel drawings, etc. are ordinarily developed sequentially and iteratively. This means that new information and new hypotheses must constantly be brought forward during system development. For this reason H cannot be a "one-shot"affair.

In addition, system development has become progressively a team-oriented affair. The sheer volume of work forces a division of labor among engineers. The project team—on which the H specialist must play his role—has become a commonplace in system development.

Because the various products of system development may present individual human factors problems, the H specialist must be intimately involved in many aspects of system development. The man-machine system concept described earlier means that many system aspects have human factors implications which require analysis. The kinds of human factors problems found in the individual developmental phases are described in detail later.

The point is that because these problems arise throughout system development and can only be solved by following them during that development by interaction with the engineer, the H specialist, to be effective, has had to leave his research laboratory and become an intimate part of the design project team.

THE H SPECIALIST

Since H analyzes the interrelationship of man and machine, it follows that it is almost exclusively a product of Western technological development. In addition to the United States, we find H specialists in Britain, France, Holland, Germany, Russia, Japan, and Italy (Bertone, 1969). In so-called underdeveloped countries depending largely on the West for their technology, such as the Arab nations, Indonesia, and the African republics, we will find few, if any, H specialists.

The number of H practitioners in this country and overseas is difficult to specify largely because many engineers (and nonengineers) may perform H functions without thinking of themselves as, or indeed being, H specialists. Within this country there are probably about 3000 people specializing in this discipline (based on 1500 members of the *Human Factors Society*, multiplied by 2 to include nonmembers). In Europe one finds approximately 500 ergonomists (who are essentially the same as H specialists).

The H specialists referred to in this book are concentrated largely in industries serving military defense. There are several reasons for this. The discipline was first developed under military sponsorship and still continues in a client relationship to the Department of Defense. The military are the most concerned about the effectiveness of the man-machine relationship because the systems they develop are the most complex and demanding of human personnel and because they cannot, as civilian industry often does, accept inefficient performance with their systems. Human factors are most important when advanced systems are being developed; most of these are developed by the military. (It is a cliché to remark that the basic technology of the automobile has not changed significantly since the introduction of the model T automobile.)

Military systems are usually operated under greater stress than are civilian systems; this imposes greater burdens on the operator and makes the proper man-machine relationship more essential in their design.

Unfortunately, commercial industry has not shown similar appreciation of the *need* for incorporating human factors in design. Examples of this need in civilian life are apparent: few activities are as stressful as daily driving. Commercial industry is, however, less concerned than the military with the efficiency with which performance goals are accomplished since the major criteria of commercial system success are sales and profit.

The H̲ specialist regrets this since there is substantial evidence of need for his services in civilian industry (discussed later). The number of H̲ specialists in civilian-oriented industry, such as automobiles, hospitals, etc., is small, but it is increasing. This area of activity will be discussed in greater detail in Chapter Ten.

There are four major areas within which the H̲ specialist may work. The most important from the standpoint of sheer numbers is *industry*, both defense and civilian. In industry the specialist performs as a member of the project team designing, developing, and testing new equipment. This book is concerned primarily with the industrial human engineer. Within *government* a small number of H̲ specialists are involved in two types of activity: (a) monitoring the development of government-sponsored equipment to ensure its adequacy from a man-machine standpoint; (b) performing research themselves or monitoring those who perform human factors research for the government.

In contrast to the government specialist, one finds the *consultant* whose major job it is to perform studies and experiments to supply the data needed by government and industrial H̲ specialists to do their system development job better.

Finally, within *universities* one finds a small number of H̲ people performing research and teaching others who in time will become H̲ specialists themselves.

Because of the varied nature of the jobs the H̲ specialist may be called upon to perform, people of varied backgrounds are to be found in the discipline. Chapter Eight describes the academic background of these people (Kraft, 1969). The largest contingent has had training in the behavioral sciences, followed by engineers, with a substantial minority made up of a great variety of academic specialities. The predominance of those with a background in psychology has significant implications for communication with engineers, which will be discussed in Chapter

Seven. In almost all cases the minimum academic degree is the BA or BS, but a substantial percentage have the MA and the PhD.

The range of problems the H specialist is likely to encounter at any time suggests that to solve them effectively he ought to have a very broad, cross-disciplinary education. The ideal specialist should have his doctorate in psychology, a bachelor's degree in some specialized area of engineering, a medical or other advanced degree in physiology, a degree in mathematics or statistics, etc. Obviously, such a paragon does not exist. It is inevitable, therefore, that the average H specialist will at one time or another be less than completely successful in solving the problems he encounters. This failure merely demonstrates that he is like anyone else who works in system development.

REFERENCES

Beaty, D., 1969. *The Human Factor in Aircraft Accidents.* London: Becker and Warburg.

Bertone, C. M., 1969. Special Section on Soviet and European Human Factors, *Human Factors,* **11** (1), 65–94.

Communications and Systems, Inc., 1969. *System Description Report,* System Engineering Program, Report 105–68–1–1. Washington, D.C.: U.S. Post Office Department, 24 March.

Fitts, P. M., and R. E. Jones, 1947. *Psychological Aspects of Instrument Display. I: Analysis of 270 "Pilot-Error" Experiences in Reading and Interpreting Aircraft Instruments,* Report No. TSEAA-694-12A, Aero Medical Laboratory, Air Materiel Command, Dayton, Ohio, 1 October.

Kraft, J. A., 1969. *Human Factors and Biotechnology*—A Status Survey for 1968—1969, Report LMSC-687154, Lockheed Missiles & Space Company, Sunnyvale, California, April.

Meister, D., et al., 1970. *The Effect of Operator Performance Variables on Airborne Equipment Reliability,* Final Report, RADC-TR-70-140, Rome Air Development Center, Griffiss AFB, New York, July.

II

HUMAN ERROR AND HUMAN FACTORS

THE IMPORTANCE OF HUMAN ERROR

The concept of error, considered as a deviation from required performance (which includes also delays in response), is central to human factors. H seeks to reduce error to a minimum; and when, as is inevitable, error occurs, it seeks to understand and to eliminate that error by modifying the human factors which produce it.

Error occurs not only in the operation and maintenance (use) of equipment but also in the development of that equipment. Indeed, it was assumed in Chapter One that a major cause of user error is the error "built into" the system during its development by inappropriate design practices. Since this text deals largely with the human factors to be considered in system development, it is necessary to explore the nature of the errors that may occur during that development.

Another reason for discussing error at some length is to provide evidence for the necessity of including human factors principles in design. If errors do not occur, or if they are trifling, or if they have little impact on system functioning, then it appears hardly necessary to be concerned about H as a system development discipline. But if error does affect system development and performance significantly, and if the inclusion of human factors principles in design does reduce the incidence of that error, the discipline must be taken seriously.

The purpose of this chapter is, therefore, to provide not only a logical argument, but the evidence, for the inclusion of human factors principles in design; it does so by exploring the nature of error. It is usually said that since equipment is always operated and maintained by men, the design engineer *must* consider the man in his design. This is not a convincing argument; stated in this way, the proposition is only an article of faith.

Indeed, without evidence, the proposition may be disputable. The argument suggests that because men, who are fallible, operate and maintain equipment, they make errors in that operation and maintenance. These errors exercise a significant negative effect on equipment functioning. Therefore, something (i.e., consideration of human factors) must be done during equipment development to prevent these errors.

The following counter-arguments may be advanced:

1. If all the data were available, it might show that few errors are made.
2. Even if many errors do occur, they may have only a minor effect on equipment functioning.
3. In any event, "considering human factors in design" does not provide any meaningful guidance about how to avoid these errors.

This book would not have been written unless the author believed in the importance of human factors principles. It is unreasonable, however, to expect the nonspecialist to accept this proposition on faith alone. To prove the proposition requires that four points be demonstrated:

1. That human error or other operator [1] inadequacy occurs in significant numbers in system operation;
2. That this human error results from deficiencies in the design, production, testing, or operation of that system (in addition to other factors, of course, like task demands or lack of personnel skill);
3. That this human error significantly affects the performance of the system;
4. That H (as a set of principles and techniques), if applied to design at appropriate times, will reduce the potential for human error and improve system performance.

The rationale for these four points should be quite clear; only if human error is sizeable (i.e., is not an isolated, unusual phenomenon), need it be considered. That error must have been derived from modifiable equipment or system characteristics; otherwise the system design cannot be held accountable. But, even if the error were system-determined and amounted to sizeable proportions, we would not be concerned about it unless it adversely affected the system's operation. Finally, even if errors were extensive, resulted from system deficiencies, and significantly affected system operation, we would be entitled to ask whether the application of H principles can, in fact, help to avoid or reduce these error effects.

None of these points is easy to demonstrate, even though, as users of equipment, we make errors every day and experience their effects. As he

[1] The term *operator* when used from now on includes the maintenance man.

reads the remainder of this chapter the reader should ask himself, "Is the error reported substantial; does it influence system performance; will H reduce that error?"

TYPES OF ERROR

Error was defined in Chapter One as any deviation from a procedure required to operate or maintain an equipment. One should not assume from this, however, that all errors are alike in terms of their causes or their effects on the system.

Errors may be classified in various ways:

1. In terms of what caused the error.
2. In terms of what the error consequences are.
3. In terms of the stage of system development in which the errors occurred.

ERROR CAUSES

A distinction must first be made among what can be termed "system-induced error," "design-induced error," and "operator-induced error."

To design a system requires that one specify not only the individual items of equipment but also the number and types of personnel using the equipment; their background and training; appropriate data resources (e.g., technical manuals, instruction material, blueprints, etc.); logistics (e.g., correct number and type of spares and tools, properly stockpiled); and maintenance programs (e.g., preventive maintenance schedules, methods of malfunction detection and correction).

Errors may arise not only from inadequacies in the design of the individual equipment but also from inadequacies in the "software" features mentioned in the previous paragraph; for example, specifying too few operational personnel or personnel with too little skill to perform required jobs will lead to error. Providing too few or inadequate tools or spares will also predispose to error. Such problems, which we term system error, cannot be blamed on the design of the individual equipment or on the operator himself; they reflect deficiencies in the manner in which the total system was planned. System-induced error, therefore, describes errors made by personnel which result from the inadequate design of the total system.

Design-induced error results from inadequacies in the design of the individual equipment. The resulting equipment characteristics create special difficulties for the operator which substantially increase the potentiality for error. Examples of improperly human engineered equipment are very common.

A few years ago the author bought a color television set. The color adjustment controls are recessed at the bottom of the set (undesirable because he has to stoop to make the adjustment) inside a panel set flush with the console surface. The problem is that the panel can be opened only by pulling on a rectangular bar raised slightly (0.15 inch) from the console surface. Since the grasping surface is so small and, since the panel, which has no handle, is snugly recessed, it sticks and requires great exertion to pry open. Once the panel is open, the controls are revealed as screw stems also without a firm grasping surface (no knob). If the control sticks (as it often does), there is nothing for the fingers to grasp to turn it.

Other examples of human factors design deficiencies in everyday life which are high error potential sources involve the ubiquitous automobile. The following examples are abstracted from the files of *Consumer Reports* (CR).[2]

"The ________'s . . . windshield wiper and headlights are operated by pulling knobs that are identical in shape and are placed side by side to the left of the steering column. Though they are labeled and lighted, you can easily pull the wrong one inadvertently." (*CR*, January 1967, p. 34)

"The ______ . . . (has) a diabolical arrangement of three identical knobs in a vertical row on the extreme left of the instrument panel . . . None of the three controls is lighted and the vent pull is unlabeled. It would be quite easy for the ______ driver to put out his headlights when he intended to close the fresh air vent or to cover his windshield with water when he needed lights." (*CR*, January 1967, p. 35)

"One drawback of ______'s . . . styling is limited rear visibility. Its rear deck slopes too sharply to be seen from the driver's seat, making it difficult to judge distance when backing up . . . In all . . . of the specialty cars . . . wide rear roof pillars create blind spots that we consider dangerous." (*CR*, July 1967, p. 356)

"Credit ______ with consistency . . . Directional signals, for example, are operated by a lever on the steering column. But it's on the righthand side, not the left. On the left side is a very similar lever. *It* controls the lights. The chance that a driver will turn off his headlights when he intended only to signal a left turn is dangerously real . . ." (*CR*, August 1967, p. 441)

"______ has recessed its controls so far into the instrument panel that operating them is difficult with a gloved hand, and the ignition key is difficult to turn bare-handed." (*CR*, January 1968, p. 28)

[2] Reprinted with permission from *Consumer Reports*.

". . . The speedometer's index lines were printed on the inner instrument face, but the indicating numbers were printed on a clear plastic lens at considerable distance out from the index lines, making the speedometer particularly difficult to interpret accurately." (*CR*, February 1968, p. 88)

Operator-induced errors can be traced directly to an inadequacy on the part of the individual who makes that error. Errors resulting from lack of capability (e.g., deficient vision), training, skill, motivation, or fatigue would be categorized as operator-induced errors. Driving the wrong way on a properly marked one-way street is an example of operator error that cannot be referred either to the system or the individual equipment.

In examining an error, therefore, we are faced with a number of possible causes. Assume, for example, that a circuit is miswired during production. The error may have been caused by inattention or lack of skill on the part of the worker; by an incorrect blueprint, by inadequate lighting, by an improperly designed tool, etc.

A word of caution. It is often difficult to make an unequivocal assignment of error cause. Lack of training which produces an operator error may result in part from a deficiency in planning the system. This deficiency may have resulted in providing the operator with inadequate skill. Is this a system or an operator error? Hence all causal categories must be viewed with some scepticism. Moreover, to assign a causal classification does not mean that one has understood the error source or mechanism. The categories supplied above are merely a handy means of sorting errors into smaller "bins."

ERROR CONSEQUENCES

The effect of an error is also a variable. Imagine that a timing circuit has been miswired. Depending on the function performed by the circuit and the system of which it is a part, the miswiring may simply cause the speedometer of an automobile to read 5 mph less than actual speed, or it may cause an autopilot to supply incorrect steering directions to an aircraft, leading to the latter's destruction.

Table 2-1 presents, among other things, a classification of potential error consequences, ranging from the relatively minor (delay in system performance) to the most severe (loss of life). Note that these consequences are affected by the stage of system development at which they occur: errors occurring in operation are more severe than those occurring in production and test, because in the former they are more likely to affect the equipment user.

Table 2-1 Error Classification Categories

	Stage of System Development		
Design	Production	Test	Operation
Type of Error			
Design error	Fabrication error Inspection error	Operating error Installation error Maintenance error	Operating error Installation error Maintenance error
Causal Factors			
Inappropriate function allocation Failure to implement requirements Poor human engineering design	Incorrect blueprints Incorrect instructions Inadequate tools Inadequate environment Inadequate training/skill Poor human engineering design of equipment Poor workplace layout	Inadequate/incomplete technical data Inadequate logistics Poor human engineering design Poor workplace layout	Inadequate/incomplete technical data Inadequate logistics Inadequate training/skill Inadequate motivation Poor human engineering design Poor workplace layout Inadequate environment Overload conditions Task complexity Poor personnel selection
Error Consequences			
Inadequately designed equipment	Scrapped/reworked equipment Production delays Higher cost Malfunctioning equipment classified as functioning Good equipment rejected	Delay in system operations Human-initiated malfunctions System breakdown Failure to accomplish test Degradation in system performance Possible danger and loss of life	Delay in system operations Human-initiated malfunctions System breakdown Failure to accomplish mission Degradation in system performance Possible danger and loss of life

Table 2-2 System Failure as a Function of Human Error

Type of System	Type of Failure	Number of Operations or Time Period	Failures Resulting from Human Error (%)	Reference
"Defense"	System failures	4½ years	40	Cornell (1968)
X-15 Aircraft	Unsuccessful flights	164	12	Wilson & Gaffney (ND)
Missiles	Accidents	3 years	65	Willis (1962b)
Missiles/aircraft	Unspecified problems	Unknown	40	Willis (1962a)
Missiles	Failure reports	548 failure reports	46	Willis (1962b)
Ships	Collisions, floodings, groundings	4 years	63.6	Willis (1962a)
Aircraft	Major accidents	47 accidents	51	Willis (1962a)
Missiles/aircraft	Equipment failures	Unknown	26–50	Willis (1962a)
Missiles	Human-initiated malfunctions	3829 failure reports	20–53	Shapero & Bates (1960)
Rocket engines	Human-initiated malfunctions	600 failure reports	35	Majesty (1962)
Missiles	Holds, postponements, aborts	Unknown	40	Majesty (1962)
Aircraft	Equipment failures	1642 failure reports	37	Meister (personal data)
Missiles	Failure reports	122 major system failures	35	Meister (1967)
Nuclear weapons	Production defects	23,000 defects	82	Rook (1962)
Three electronic systems	Human-initiated malfunctions	1820 failure reports	23–45	Levan (1960)
Missiles	Equipment failures	1425 failure reports	20	Meister (1961)
Missiles	Human-initiated malfunctions	35,000 failure reports	20–30	Meister (1965)
Various	Engineering design errors	Unknown	2–43	Rigby & Cooper (1961)
Aircraft	Accidents	Unknown	60	Rigby & Cooper (1961)
B-52 aircraft	Human-initiated malfunctions and other human-related failures	552 failures	36	Meister et al. (1970)
Air defense	Maintenance-induced malfunctions	213	11.7 (wholly) 14.1 (partially)	Robinson et al. (1970)

A major distinction is between what we term "human-initiated failures" (HIF) and human error. HIF is an error which results in an equipment malfunction. Hence, it is merely one subcategory of errors in general, many of which do not result in equipment failure. Because an equipment failure is a rather striking event, data on HIF are somewhat easier to collect than on errors which do not result in equipment breakdown. The data in Table 2-2 all deal with errors which resulted in one way or another in equipment failures.

There are other ways of differentiating error consequences. The effect of some errors is immediate, while that of others is delayed. The automobile driver who turns his steering wheel the wrong way in a skid experiences the immediate effects of that error. The error of a production worker who makes a poor solder connection has no consequences until the connection ultimately breaks and fails the equipment.

Because of the relative immediacy of error effects, some errors are more visible than others. Many workmanship errors are difficult to discover even during inspection. Errors performed in equipment operation are usually quite visible because they affect the equipment or the equipment user.

Error visibility may also produce different error consequences. If the error is apparent to the man who made it, the chances of its being rectified are increased. This is one reason why errors made with discrete tasks (e.g., flipping the wrong switch) are more quickly corrected than those made with continuous tasks (e.g., tracking a moving signal).

Some errors are frequent, but minor in their effect; others are infrequent, but critical to system performance. In part, this is because error consequences vary as a function of the system in which they occur. Errors in firing a rifle at a target are frequent but will not lead (except in combat, and often not even then) to disaster. The pilot's failure to release his landing gear while landing an aircraft is infrequent, but invariably results in equipment damage.

Making an error need not necessarily lead to disastrous consequences. Many equipments are so designed that even if one makes an error, that error, once noted, can be rectified by performing the operation again. This is actually characteristic of much, if not most, of our equipment. If I turn my television set to the wrong channel, I still have the option of turning it to the right one. Error is, therefore, not necessarily synonymous with failure to accomplish the equipment's function. On the other hand, if I miscalculate my braking distance on a slippery street, my chances of rectifying the error (in time) are slight.

A great deal also depends on how one defines an error. In many military systems mission accomplishment requires such precise operation that error *may* be synonymous with system failure. If a bomb must be dropped no more than 100 yards from a target, then dropping it 105 yards from the target constitutes a system failure. On the other hand, as we have seen, many nondefense systems are quite forgiving of error. It is therefore necessary to determine the significance of the individual error and the means by which it contributes to mission failure before the piling up of error frequencies means a great deal.

ERROR IN SYSTEM DEVELOPMENT

As Table 2-1 suggests, errors may also be classified in terms of the stage of system development in which they occur. Traditionally, there are four stages of system development: *design, production, test,* and *operation.* Although the following simple definition fails to do justice to the complexities of each stage, it is sufficient to say that equipment is designed in the design stage, turned into hardware during production, tested to ensure that it meets specification during test, and operated by its users during operations. The phasing of these stages overlaps so that, in development of major systems, design of some assemblies or equipments may be going on while others are being produced and tested.

Errors of improper design can be termed "design errors." This type of error, illustrated earlier, occurs during design of the equipment, but its effects are experienced only in operation of the equipment as an operator error.

Errors made in fabricating the designed equipment can be termed "production" or "workmanship" errors. Like design errors, the effects of production errors are delayed; they occur during fabrication but manifest their primary effect during test and operation.

Errors occurring during installation and maintenance of the equipment (which may occur either during test or operations) are "installation" or "maintenance" errors. These too have delayed effects.

Errors performed during testing are either operating, installation, or maintenance errors. Operating and maintenance errors recur, of course, throughout the period of system use.

The relative frequency of errors during system development is largely an unknown primarily because there are few if any efforts to collect error data by the contractor and the user (who might be embarrassed by such data). The lack of systematic efforts—either by the contractor or the government—to collect error data in terms of frequency, causal factors, and criticality is a serious deficiency. Without such data it is difficult to

(a) assess the importance of error in influencing equipment operations;

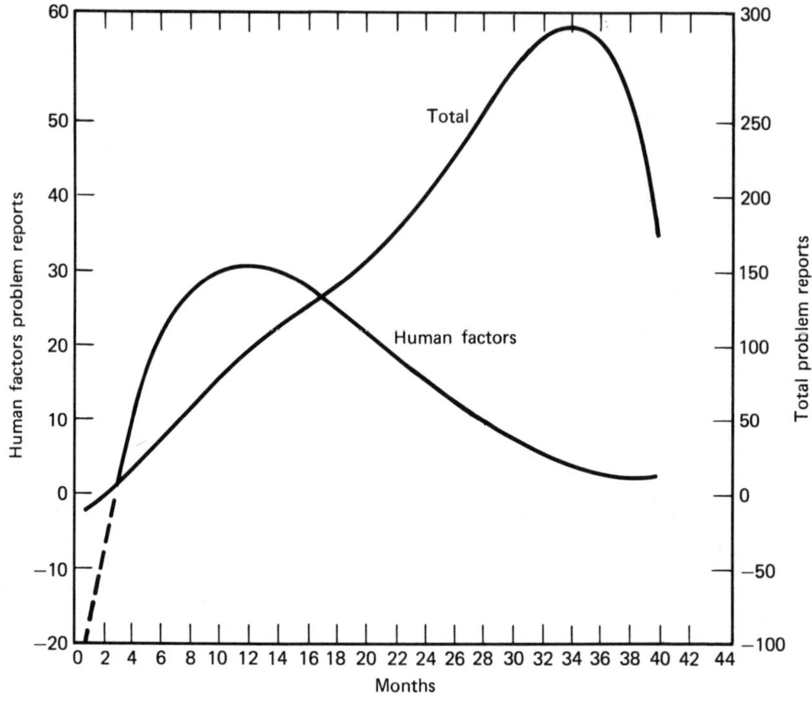

Figure 2-1 Problem reports by month, smoothed curves.

(b) pinpoint factors in system development which could be improved;
(c) predict the operational performance efficiency one can expect of a system, where that efficiency is influenced by personnel errors.

One piece of data does exist. Van Buskirk and Huebner (1962) have analyzed the relative frequency of human-initiated failures (HIF) as a function of all the equipment problems reported in one missile system development program; their results are shown in Figure 2-1. Figure 2-2 indicates what one analyst considers to be changes in error frequency following acceptance of the equipment for operational use. Note that there is an initial build-up in error, followed by a progressive decline until a relatively stable frequency is achieved over the period of system use.

ERROR IMPORTANCE

Despite the relative lack of data, there is substantial evidence to indicate that error is a crucial factor in system performance. This is shown

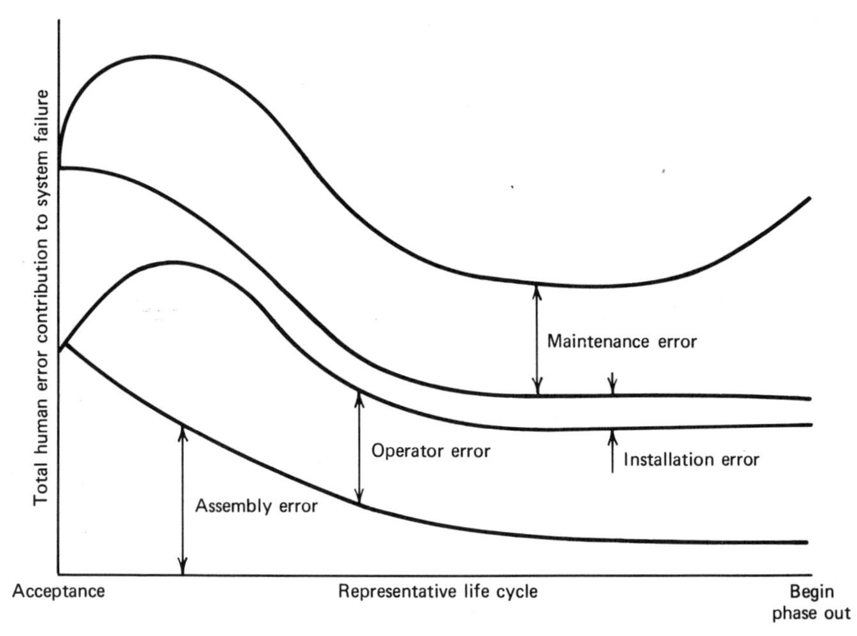

Figure 2-2 Proportional contribution of the different species of human error to system failure (taken from Rigby, 1967).

by Table 2-2 which lists a representative sample of studies describing the frequency of error and its effect on equipment. Data on the specific systems or the time period described in the reports are sometimes lacking. The errors described are those which resulted in failure of the equipment to function (IIIF), the scrapping of equipment components, or accidents. Obviously, this is mixing apples and oranges; scrapping components during production is not equivalent to losing an aircraft and crew. For this very reason, however, Table 2-2 presents a representative cross–section of the types of situations in which errors occur and their consequences.

The mean percentage of system failures (i.e., equipment disabled or prevented from accomplishing its function) resulting from these errors is about 40%. Presumably, the remainder of these failures resulted from normal equipment wearout, nonoperator–related design deficiencies, or other causes which can be ascribed only remotely to personnel.

The irreducible minimum of error to be expected even under optimal conditions—if there is such a minimum—is not known. However, an average of 40% equipment failures resulting from human error, together with their effects, suggests that error is not only frequent, but significant for system performance—hence worth considering as a factor in equipment development. The study by Meister et al. (1970) indicates that HIF significantly reduced the mean time between failure (MTBF) of the equipments involved.

DESIGN ERROR

Design errors are manifested in improperly designed equipment [as determined by H standards in MIL-STD 1472 (United States, 1969)], in failure to assign effective roles to equipment and personnel, and in failure to meet system requirements. We exclude from this definition those preliminary design efforts which are eliminated before final drawings or specifications are released. Obviously, many design errors are made while the engineer is considering a number of alternatives. Only those design errors which are *not* caught and which, appearing in the final design, remain to influence the operational usage of the system are within the definition of the term.

Functional Allocation Error

The first type of design error is to assign the man or the machine to a function for which it is not suited. This means either of two possibilities:

CASE I. The human is assigned to a function which he cannot perform or which could more effectively be performed by a machine component; for example, if a man were asked to time the speed of a 100 yard dash by counting his pulse–beats during the race, we would say that the function could better be performed by an electronic timer.

CASE II. A function is allocated to a machine component which might better be performed by the human. For example, a computer might be designed to make probabilistic judgments of the aesthetic value of paintings.

In the allocation of functions we assume that an error occurs when a function is assigned to a machine component when there is no significant advantage in having that function performed by the machine. In other words, the simplest and best design is always a manual one, provided that there are no significant advantages (e.g., performance capability, reliability, maintainability, and cost or convenience for the user) for automatic design.

The engineer may view this assumption as being unwarranted. However, the reason for selecting a human to perform a function over a machine, when all factors are essentially equal, is that mechanization has more complex consequences than does manual operation. Maintenance and logistics requirements, for example, are substantially increased by automatization. Unwarranted automation leads to unbridled technological proliferation. We recognize that this choice–principle may conflict with the engineer's design tendencies. Because of the engineer's training (and the average user's laziness), it is more natural for him to automatize whatever functions he can. The long–range effects of undisciplined technology (both ecological and psychological) are, however, beginning to become more painful to the user.

After system functions are defined, they must be allocated among machine or crew components (literally, who is to do what?). At the gross system level involving major functions, allocation is in part constrained by the nature of the function. For example, we would not ask the pilot to act as a jet engine, or demand of the operator that he act as an air defense warning station without radar. Thus, at the most molar function levels, the decision as to whether the man or the machine should perform a particular function is usually immediately apparent.

At more detailed function levels (corresponding perhaps to the subsystem and assembly level in equipment), the tradeoffs between man and machine are more molecular and less obvious. Often both the man and the machine can do the job; the question becomes, which can do it better? The analyst is concerned not with an entire system function but with such things as whether a series of manually timed operations (requiring a human to switch on a series of timing mechanisms) will be as effective as an automatic timing circuit which requires no human intervention. Naturally, multiple criteria are involved in making such decisions: cost, safety, performance reliability, maintainability, etc.

Because most function allocation problems are meaningful for H consideration only at relatively molecular levels, the lists of characteristics which differentiate men from machines (Fitts, 1951) (e.g., "men are better at inductive reasoning, machines are better at monitoring") are of little practical value. These general characteristics apply primarily to top level system allocations which, as we have seen, are usually made on the basis of common sense or constrained severely by over-all design concepts.

Moreover, these "men are better at . . . , machines are better at . . ." characteristics have other deficiencies: they are black or white and never any shade of gray; and they are nonquantitative, when the engineer making a tradeoff wants to know, "better by how much?"

Tradeoffs at molecular decision points involve many parameters, while those at the top level usually involve only one overriding consideration; this is why top level function allocations are easier to make. The tradeoffs between the manually operated switch and the timing circuit include such criteria as performance requirements (e.g., how precise must the timing be?); cost (which costs more?); reliability (probability of error vs. probability of circuit failure).

Since the decision involves a number of criteria, the proper combination of these criteria is quite complex. Because of this, allocation tradeoffs are ultimately based almost invariably on purely intuitive methods. There are formal methods of making such allocation tradeoffs (Teeple, 1961), but their validity is dubious and in actual practice they are difficult to apply. Usually they require more data than anyone could possibly have at the time the tradeoff decision is required. A typical example is Barton's (1964) queuing technique for determining manpower in early system development, a technique that requires a minimum of eight data inputs, some of which cannot be determined until after the system is operational.

The H specialist's concern in such allocation tradeoffs is to ensure that the human component gets a fair shake in the tradeoff. The engineer's very natural tendency to select an automatic means of accomplishing a function in preference to a manual means (this is quite understandable; after all, it's his business to create hardware mechanisms to replace human labor) makes it possible that he will not pay full attention to the human side of the picture. Even when the engineer's allocation is justified, the H specialist wishes to ensure that a full hearing has been given to the possibility that a human component could or should do the job. In a sense, it is like being a lawyer for the human plaintiff; once the jury issues a verdict, the question is settled, but the defendant has a right to a full and complete hearing.

Once an allocation has been made, it is very difficult to determine whether it was or was not made correctly. Even when a machine component has been erroneously selected over one which is human, the machine component will probably function and may even perform relatively well; the cost the user pays because machine function is not as efficient as human function is often invisible. The reverse is also true, and the H specialist is just as interested in avoiding Case I errors as those of Case II. Deciding to include an unnecessary human function which requires an additional operator may cost a fantastic sum over the life span of a system. One estimate of over-all human cost the writer encountered was $8.50 an hour per man. Thus over a five-year system life (2000 man hours per year) one unneeded man would add approximately another $85,000 to the cost.

Failure to Implement Human Requirements

A second design error is failure to include in the design certain functions which are required if the operator is to perform in accordance with system requirements or at his maximum efficiency. For example, a display needed to warn the operator of a potential emergency condition may not be included in the design.

This type of error occurs most often when the function is only implicit in system requirements, and the designer fails to include the controls and displays needed to perform the function. Where the requirement is explicitly cited in the engineer's statement of work, he rarely ignores it. An example of an implicit requirement might be the need to provide a display to warn of an impending emergency when the emergency is only a potential one. Most statements of work (SOW) describe operating requirements rather completely but leave "conditional" requirements (those dealing with situations which *might*, but do not necessarily *have*, to occur) to the imagination of the equipment designer. Such requirements may well become obvious if the system is analyzed in depth. However, design, at least in its initial conceptual stages, proceeds very rapidly; by the time the implicit requirement is uncovered, the basic design concept may be so formalized that it is difficult to accommodate it.

Because of the way in which many SOW's are developed, with their concentration on major hardware requirements, the implicit human factors requirements are among those which tend to be overlooked. This does not mean that the system will necessarily fail because such implicit human factors requirements are late in being discovered. After all, the major system needs have usually been taken care of. However, the failure to implement implicit requirements means that the system contains a certain weakness which under great stress may reveal itself.

One example of such a design error is the following, reported by Gordon, 1964:

"During a silo missile accident, approximately 18,000 gallons of diesel fuel were gravity fed into the silo and became the main source of fuel for a follow–on 18 hour fire. There was no way to shut off the main feed line, since no one had provided an emergency cut-off valve for this auxiliary system."

Obviously, no one had anticipated an operational problem in which an auxiliary fuel shut-off valve would have been necessary.

Human Engineering Design Errors

A third very common design error is to fail to ensure that man and machine components interact effectively (inadequate human engineering of equipment). Specifically,

(a) failure to select the most appropriate control–display component for the particular operator function to be performed (e.g., selecting a bank of toggle switches to activate multiple step functions when a rotary control would be more efficient);

(b) failure to arrange controls/displays or other components (e.g., handles or access spaces) in a manner which will permit personnel to use them as rapidly and accurately as possible (e.g., placing a display and the control which is to be used in conjunction with it in widely separated locations);

(c) requiring the operator to use information or perform operations he does not need to accomplish his job (e.g., giving him a display for information on which he cannot act because the equipment for implementing that action has not been provided). Or the reverse—failing to give the operator enough information to do his job, an error often found in procedure development.

Because these are errors of detail, it is often difficult for the engineer to recognize their potentially damaging consequences. Why should the misplacement of a control or the selection of an improper display scale make that much difference to performance? The requirement to human engineer an equipment is even more conditional than the requirements discussed previously. In this third class of design error the engineer deals with a contingent, hence somewhat nebulous possibility; the effect of an improper design characteristic may or may not affect the operator, and even if it affects the operator's performance, the latter may or may not influence over-all equipment functioning.

Examples of possible errors resulting from human engineering design deficiencies were given in the automobile examples cited earlier (see pages 24–25). These design deficiencies may lead to accidents, but, on the other hand, they may not. Inadequate rear visibility could lead to a backup accident; but maybe not. Misinterpretation of an inadequate speedometer scale may lead the driver to drive faster or slower than he intended, and that error could result in an accident, but only with a certain probability, never a certainty.

Thus, there is a probability (p_i) not only for an error to occur as a result of a design deficiency but also for certain consequences to occur as a result of the error (f_i) (Rook, 1962). Even if that probability is phrased in quantitative terms (which it almost never is), what level of probability consititutes a hazard significant enough to warrant special design consideration? Is an error probability of .60 high enough, or need we go to .90? We have no precise ground rules to assist the engineer in determining that significance level.

Are design errors frequent and significant enough that the reader is convinced that this is a factor to which he must pay attention in his design? Do such errors in fact influence equipment/system performance negatively? The evidence is less satisfactory than we would wish. Table 2-2 describes mostly workmanship errors because these are relatively easy to detect and describe.

It is much more difficult to tease out design errors in function allocation or in failure to include human requirements in equipment design. When these lead to failures or accidents, their causal relationship to the failure or accident may be difficult to establish. It must be emphasized that systematic studies of design error are the exception rather than the rule. Industries do not publicize their design errors; when they become aware of them, they attempt to conceal them. It is, moreover, very difficult to assess the significance of such errors. Obviously, no designer is perfect, and to that extent errors of judgment will inevitably occur. Design error becomes apparent only when there are major violations of design principles which lead to catastrophic consequences.

The factors which predispose to design error are the following:

1. Incomplete analysis of the requirements of the equipment and the total system.
2. Too hasty a design effort.
3. Predetermined attitudes or biases in the designer to one or other mode of design solution.
4. Excessive reliance by the engineer on his design experience to the detriment of design analysis ("shooting from the hip," so to speak).

All of these factors are discussed in much greater detail in Chapter Seven.

PRODUCTION ERROR

Many of the statistics reported in Table 2-2 refer to errors made in fabricating hardware: production errors. The great majority of HIF which comprise much of our statistical data on human error in system development involve production error. Of the 122 major system failures analyzed by Meister (1967), the great majority resulted from production defects.

There are several reasons for the preponderance of production defects in our statistics. In contrast to records kept of human error in operational use of an equipment, records of components scrapped or repaired during production are usually very complete. Also, operator–induced HIF (resulting from human engineering design errors) are usually substantially less than HIF resulting from installation and maintenance. For example, only one HIF was reported in a two–month sampling of error data taken in missile test operations (Ehrlich and Horner, 1962).

A production error is actually a *workmanship* error, that is, an error made by the individual production worker which results directly in a failure of the manufactured article to meet a specified standard (usually blueprint). The "visibility" of the production error arises from two sources: (1) to the extent that the standard is unambiguous, as in a blueprint, it is relatively simple to determine when a deviation from that standard exists; (2) a major part of manufacturing (inspection) has a specific responsibility for looking for and discovering such deviations.

Types of Production Error

In view of the variety of things that a production worker does in the course of fabricating equipment, he has many opportunities to make different kinds of errors. Table 2-3 lists a sample of workmanship errors reported by Rook (1962).

Table 2-3 Characteristic Errors Observed in Production

Definition of Error	Error Rate
Soldering operation results in solder splash.	0.001
Soldering operation results in excess solder.	0.0005
Soldering operation results in insufficient solder.	0.002
Soldering operation results in hole in solder.	0.07
Component is damaged by burn from soldering iron.	0.001
Two wires which can be transposed are transposed.	0.0006
A polarized component (diode, etc.) is wired backwards.	0.001
A capacitor with preferred polarity is wired backwards.	0.001
A solder joint is omitted.	0.00005
A component is omitted.	0.00003
A component of wrong value is used.	0.0002
A lead is left unclipped.	0.00003
Staking is omitted on fastener.	0.00003
Staking omitted on adjustment.	0.00003
Small item such as lockwasher is omitted.	0.00003

We may wonder how the small error rates reported in Table 2-3 can result in serious scrap/reject frequencies. There are two reasons: (a) several errors may be made in assembling an individual component; (b) considering the thousands of individual components assembled during a production run, the cumulative volume of these failures becomes impressive.

Other errors which have been observed, but for which error rate data are presently unavailable, include the following:

1. Forcing parts together.
2. Cutting material to wrong dimensions.
3. Improper calibration of check equipment.
4. Holes mislocated, wrong size, or elongated.
5. Mismatched (e.g., incorrect cable connection).
6. Improper processing (heat treatment, aging corrosion treatment).
7. Cracked, torn, ripped, cut, warped, wrinkled, and anything else you can think of in terms of bad treatment.
8. Defective potting.
9. Incorrect blueprints.
10. Wire not grounded or missing.
11. Debris found in equipment.
12. Wire pulled loose.
13. Excessive lubricant.

Causes of Production Error

Although some workmanship defects may be caused by worker deficiencies like lack of training or motivation, many are caused by situations external to the worker which predispose him to make the error (Meister and Rabideau, 1965):

Inadequate Work Space and Poor Work Layout. Highly precise motor manipulations require adequate work space and proper layout. Where containers for parts are not arranged, for example, in accordance with assembly procedures, the probability of selecting an incorrect part increases.

Poor Environmental Conditions (e.g., inadequate lighting, high temperature, and high noise level). Inadequate lighting increases the difficulty of positioning and wiring small components properly; high temperature and noise level reduce work effort.

Inadequate Human Engineering Design (machinery, handtools, and checkout equipment). This factor affects production equipment just as it does operational equipment; for example, in one equipment used to check out autopilot amplifiers in the factory, investigators found test accessories which took up most of the working area, difficulty in hooking up the unit, and poorly laid out control panels (Urmston and Cutchshaw, 1960).

Inadequate Methods of Handling, Transporting, Storing, or Inspecting Equipment. The author recalls an instance of one production department that had an exceptional failure record for one type of highly expensive electronics component until it was discovered that the components were being transported in carts which permitted them to slip off on to the floor. Redesigning the cart cut the failure rate to an acceptable level.

Inadequate Job Planning Information (inadequate or unavailable operating instructions or blueprints). It is not unheard of to find components being fabricated to out-of-date instructions because the information has been delayed in reaching the worker.

Poor Supervision. One production department had a very high defect rate until it was discovered that the supervisor refused to allow his people to sit at the benches at which they worked.

Another factor which cannot be ignored is the complexity of the equipment design which the worker must fabricate. Where design is complex, the difficulty of fabrication presumably increases. [Experimental data on this factor is unavailable, except in the case of inspection processes, where complexity has been shown to be a significant factor affecting inspection accuracy (Harris and Chaney, 1969)]. Designs which are unnecessarily complex represent a distinct form of design error. One of the functions of liaison engineers attached to manufacturing is to make the difficulties of fabricating equipment known to the design engineer. Although it is impossible to ascertain quantitatively the contribution of design complexity to production error, it is probably substantial.

The effect of these factors is to create a work situation favorable to the commission of production errors; that is, the probability of error increases as a function of inadequate production characteristics. Obviously, if one had a factory area with poor lighting, incorrect blueprints, and poorly human engineered handtools, the result would be a serious defect rate.

Note that only one small part of this list of factors involves motivation. There is a general tendency to blame fabrication errors on inadequate "worker motivation." In part, this is the reason for the widespread popularity of the "Zero Defects" program and other production improvement programs. Presumably, if one could increase the worker's motivation, his error rate would decrease correspondingly. However, as Rook (1965) says, ". . . Most errors result from SCE (situation-caused errors) rather than from HCE's (human-caused errors). People who continually make goofs tend to be eliminated."

If one assumes that the factory is a system like any other, in the sense of having a goal (mission), equipment and personnel components, and logistics, maintenance, and communication functions, then the same requirement exists to tailor the factory to the capabilities and limitations of its personnel in order to accomplish production goals.

Historically production has not utilized human factors techniques, nor, for that matter, has H̲ been much concerned with improving the production line. There are several reasons for this: (a) the predominance of the work/motion study or industrial engineering discipline within the factory; (b) the factory's traditional emphasis on rather simple, discrete elements (hence work/motion) rather than on consideration of production as a system, which is the way H̲ specialists tend to view it; (c) the traditional concentration of H̲ on prime equipment development.

The effect of production error can be extremely serious not only in the cost of rework and scrapping but in terms of failures of the operational system. Since the inspection function is only partially successful in screening for defects, inadequate components fail when they reach operational use.

In view of the very large percentage of system failures resulting from human-initiated production defects, this area appears to be a most promising one for H̲ assistance and certainly one requiring that assistance. The question is whether the H̲ discipline possesses the techniques needed to resolve the problem. The work performed by Harris, Chaney, and their co-workers (Harris and Chaney, 1969) in inspection suggests strongly that a good deal can be done in the factory area.

Theoretically it should be simpler to reduce production error than design error. The worker's behavior involves primarily perception and motor coordination with many fewer complex decisions required than in design. It is all the more inexplicable then that greater attention has not been paid to production error.

INSTALLATION/MAINTENANCE ERROR

Installation and maintenance error both involve behavioral and causal factors similar to those found in production error. The kinds of error situations that arise are exemplified by the following extracts from records (Ehrlich and Horner, 1961):

"The quick disconnect covers on the 'mixture ratio control fuel' side and the 'fuel and lube flowmeter' outlet fittings were reversed. 510–570 psi was trapped in the flowmeter fuel outlet line."

"An 8" section of tubing was welded on the assembly. No weld is called out for this assembly."

"Tube assembly installed in four sections instead of three as required."

"Panel is located approximately 14″ lower on wall than called out on blueprint."

Installation errors are short term errors; that is, they occur primarily when a new system is being installed in a fixed site. Once the system has been installed and bugs worked out, the incidence of such errors should decrease markedly. They are, however, costly until they are removed; 42.9% of all HIF's reported over a 12-month period in the early part of a missile test program (Meister, 1961) consisted of assembly and installation errors.

In contrast, maintenance errors persist throughout system life and may in fact increase as the system wears out (see Figure 2-2), thus increasing the opportunity for maintenance (and consequently error) to occur.

INSPECTION ERROR

Another category of error related to manufacturing, but also to be found in lesser degree in test and operations, is inspection error. The inspector's task is to uncover and prevent from entering into operational use any equipment which is defective, that is, does not meet the standards required by blueprints or instructions. Most inspection is performed during production; however, new components and equipment arriving at a test or operational site are often inspected again prior to their being placed in use.

Theoretically, if inspection were completely effective, there would be no need for anyone to be concerned about production error affecting operational system performance; all defective items would be caught and eliminated. The problem then would be only the reduction of the expense involved in scrapping or repairing defective equipment.

Unfortunately, inspection is not 100% accurate. McCornack (1961) reported an average inspection effectiveness of about 85%. However, Harris and Chaney (1969) report that on the basis of their data "variations due to product complexity alone can cause a range of average inspection accuracies from 20 to 80%." They point out "it is seldom that we find over 50 to 60% of the defects being detected at any point in time by a single inspector."

The same factors that were considered responsible for production error are also responsible for inspection error. The inspection process makes great demands on behavioral processes. The very name, inspection, implies visual examination and judgment. The inspector scrutinizes the physical characteristics of the equipment; he measures its performance; he compares equipment-induced values with numerical standards. The accuracy of the inspection process can, therefore, be improved by methods of increasing the discriminability of defective equipment. Some of the improvements in inspection efficiency produced by human factors methods are described later.

OPERATOR ERROR

The kind of error we are all familiar with is that made by the equipment user in the course of a programmed operation. Although a great deal of data is not available, the following trend seems reasonable. Operator error is first noted in the test phase when equipment is first exercised; it gradually builds up and reaches a peak in the initial phase of user operations (see Figure 2-2) when the engineers who designed the system are replaced by its intended users. As user personnel become more familiar with the equipment, the curve of operator error decreases and then levels off. This asymptote (level of continuing error) depends on many factors including the inherent difficulty level of operation and personnel turnover (new operators). However, it appears not to be related to operator experience level once the new personnel entering the system have reached a minimum level of skill. The study by Meister et al. (1970) on operator-induced HIF found no significant relationship between the frequency of that error and experience level.

The frequency of error is very difficult to ascertain in an operational system. Military systems record reliability and maintainability data (e.g., frequency of failure, repair time), but comparable data describing human performance are almost never recorded. The same is even more true of commercial systems.

Generalization from the reader's own consumer-oriented activity is dangerous because consumer errors, except in the case of automobiles or other heavy equipment, rarely stand out in terms of severity. Although consumer errors are frequent, most of them have few serious consequences (except in terms of accidents); for example, I am in a hurry to make an appointment, and I push the wrong elevator button. As soon as I do, I recognize it and push the correct one. My error has cost me a fraction of a moment, at most a slight delay if I get off at the wrong floor.

With major equipment systems such as those ordinarily developed for the military, errors are less frequent; system personnel are specifically trained for their jobs, and many fail-safe precautions are included in design and procedures. However, when errors are made, they are inherently more serious. Military combat systems tolerate much less error than do civilian systems. Their goals demand more precise accomplishment. It may be insignificant to push the wrong button in an elevator but disastrous to do so in a missile launch. The following examples (Meister, 1956) are characteristic:

"In the early days, during a missile flight a missile guidance operator turned the override switch on his equipment inadvertently. The switch was to be used only when the central circuit in the equipment went out. In turning this switch, he failed to make the necessary compensatory manual adjustments, so the missile fell miles short of its target.

. . . "A missile was being landed under automatic control (this was before the development of inertially guided missiles). It appeared to the observation pilot to be coming in too fast. The pilot radioed to the telemetering trailer that the missile should be switched to ground control. In the excitement the control operator missed the correct sequence for transferring control and the missile remained in automatic, except that now it became erratic and nearly crashed into a populated area.

. . . "In one case of a missile misfire, the operator on the tracking panel turned his control to the local (manual) control to stop automatic tracking. The launcher immediately slewed to zero elevation angle, damaging the launcher and endangering the ship and its crew."

The causes of operator error fall into two categories: *idiosyncratic* (operator-determined, e.g., aptitude and motivation) and *situational,* produced by system inadequacies, such as inappropriate procedures or poor training. The operator is directly responsible for the former; system developers are responsible for the latter. Of the two categories, the situational is obviously easier to modify; inadequate training or environment can be improved; poor procedures, revised.

Any individual error is probably multidetermined; that is, more than one factor predisposes that error to occur. This is the reason why, when we backtrack from the error to determine its causes, it is sometimes difficult to pinpoint these causal factors.

It is often assumed that there is an irreducible component of error which will inevitably manifest itself in humans. The author cannot subscribe to that position; H assumes that errors (or any other performance deviation) are caused by a "mismatch" between the capabilities of the operator (idiosyncratic factors) and the demands of the job (situational factors). The human who is required to perform a job brings to that job certain aptitudes and skills for the job (assuming he has been trained). The job itself imposes certain demands upon those skills. An error will occur whenever the balance between demands and capabilities is disturbed. That balance is disturbed whenever one or more of the predisposing error factors listed in Table 2-4 exist.

Table 2-4 Predisposing Operator Error Factors

Situational	Idiosyncratic
Poor human engineering/workplace design	Lack of motivation
Inadequate environmental conditions	Lack of skill/training
Overload conditions	Lack of capability
Improper personnel selection	
Task complexity	
Inadequate technical data and logistics	

In other words, there is an *error potential* in man which is not realized until a predisposing condition, creating a mismatch (e.g., poor human engineering design), permits the error to occur. The predisposing condition is the catalytic agent which translates a potential into an actual error. From this standpoint, there is nothing inevitable about error.

Theoretically, error can be prevented completely because if we eliminate the predisposing factor the potential error will never be realized. This, of course, remains only an ideal, a "design goal."

The number of predisposing situational factors in Table 2-4 and the errors that may result (Table 2-5) suggest that we cannot rely on training alone to overcome inadequate situational factors. Since training is directed at modifying the individual, it only indirectly reduces the impact of situational demands by enabling the operator to cope with them more efficiently. Hence additional training or better personnel selection will never completely catch up with the situational demands. It is possible by training to mitigate the negative effects of poor human engineering or excessive task complexity, but it is impossible to eliminate these effects completely. Although we would never suggest reducing training (which is, in any case, needed to perform the job), it is apparent that only by reducing job/equipment demands (e.g., simplifying design) can the balance between situational demands and personnel responses to these demands be accomplished.

Table 2-5 Resultant Error Behaviors *

Functions	Potential Errors
Sensing, detecting, identifying coding, and classifying	Failure to monitor. Failure to record or report a signal change. Recording or reporting a signal change when none has occurred. Recording or reporting a signal change in the wrong direction. Failure to record or report the appearance of a target. Recording or reporting a target when none is in the field. Assignment of a target to the wrong class.
Sequencing	Making a below-standard response. Omitting a procedural step. Inserting an unnecessary procedural step. Mis-ordering procedural steps.
Estimating and tracking	Failure to respond to an obvious target change. Premature response to a target change. Late response to a target change. Inadequate magnitude of control action. Excessive magnitude of control action. Inadequate continuance of control action. Excessive continuance of control action. Wrong direction of control action.
Decision making	Incorrect weighting of responses to a contingency. Failure to apply an available rule. Application of a correct, but inappropriate, rule. Application of a fallacious rule. Failure to obtain or apply all relevant decision information. Failure to identify all reasonable alternatives. Making an unnecessary or premature decision. Delaying a decision beyond the time it is required.
Problem solving	Formulating erroneous rules or guiding principles. Failure to use available information to derive needed solution. Acceptance of inadequate solution as final.

* Taken from Altman, 1966

H AS A SOLUTION

It should be obvious by now that human error and other inadequacies in human performance are sufficiently frequent and significant to pose a serious problem to system design and operation. The question now becomes: can the application of $\underline{H}$ principles and techniques help to avoid or solve these problems sufficiently so that it is worth the government's and engineering management's money and time to apply these principles and techniques?

The argument can be (and has been) advanced that systems were successfully designed and operated before $\underline{H}$ came on the scene. This is true. Undoubtedly, equipment was designed and operated adequately before $\underline{H}$ existed. However, prior to World War II, equipment was less advanced than it is now; it made fewer demands on the user. As far as military systems are concerned, World War II was the great breakthrough for sophisticated man-machine systems.

Even for earlier systems, however, one wonders at what cost in inefficiency these systems functioned. How many inexplicable system breakdowns? How many operator errors or near-errors? If the criterion of system efficiency is not very stringent, even an inefficient system can perform satisfactorily. The most obvious example of this is the automobile, which is badly designed from a user standpoint and which contributes to approximately 50,000 deaths in the United States alone.

An important philosophical question is suggested by the argument that earlier systems "did without" $\underline{H}$. Should design direct itself to the elimination of all possible error? Would not such an effort, even if it were possible, cost so much as to be unfeasible? All system developers accept reliability goals which are significantly less than 1.0 (e.g., .9950). This recognizes explicitly that a certain amount of equipment unreliability will exist even after development is completed. Certainly it is within the power of equipment developers to reduce this residual unreliability almost to zero; but in the tradeoff between cost and performance, the decision is usually made (except for certain specially hazardous systems, such as manned space vehicles and submarines) that residual unreliability is supportable.

Even if we accept the premise that as much error potential as possible should be eliminated, how much residual error is acceptable? Unfortunately, the precision with which human reliability is measured and predicted is sufficiently poor that one cannot easily make the necessary tradeoffs. We cannot say, "I will accept this much potential human error, but no more than that." The problem is that we cannot be sure how much error we are talking about. Because of this, it becomes necessary to urge that all possible predisposing causes of error be eliminated no matter how much effort is required. In a sense this is overkill, but we must have the entire effort—because less than this may well prove entirely useless.

Some of the available evidence suggests that the magnitude of human error may be much greater than that of equipment unreliability. Although the comparison of human error with equipment unreliability is a little like comparing apples and oranges and although methods of predicting human performance are very crude, predictions of human performance reliability often turn out to be substantially lower than those for equipment performance. Consequently, to ensure a level of human performance not much less effective than equipment performance, it is possible that we must apply a greater effort than would be required to achieve equivalent equipment performance.

Does the H̲ effort pay off? The question is difficult to answer (although there is some concrete evidence—not as much as one would want—that it does). Certainly the evidence is less overwhelming than that which exists for the severity of the human error problem. The reason is that the elimination of human factors deficiencies is not ordinarily recorded during system development. The end product either contains the human engineering deficiency or it does not; but systematic tests of performance improvement are ordinarily not made.

Formal experimental studies to test the efficiency of designing in accordance with human factors principles are few and far between. However, a number of studies have described how H̲ (or ergonomics) has been utilized effectively in the design of equipment and systems. These include the design of a computer console (Shackel, 1962), the improvement of office work (Koskela, 1962), armored tank design (Christian, 1962), a tea blending plant (Lacy, 1967), subway passenger handling (Raffle and Sell, 1969), and automated meat handling (Shackel, 1969).

An even more outstanding example of a system developed as a direct consequence of H̲ design is "Puff the Magic Dragon" or the C-47 gunship utilized by American forces in Southeast Asia. This system was specifically suggested by H̲ specialists (Ralph Flexman and John C. Simons); they took advantage of human factors research on lateral firing from aircraft to develop one of the most effective weapon systems available to the Air Force.

The effect of H̲ recommendations on performance has also been shown by Harris and Chaney (1969) in their studies of the inspection process:

"A new method was developed for inspecting complex electronic equipment. It was found to increase inspection accuracy by 90 percent. . . . The job of the quality engineer was analyzed and redesigned. As a result, engineers were found to spend 34 percent more time on tasks which required their skill and knowledge and correspondingly less time on tasks which did not. . . ." "Techniques were developed to improve the precision measurement performance of machined parts inspectors. They were found to increase inspection accuracy by over 70 percent. . . ." "The use of systematically prepared photographs to illustrate quality standards significantly increased agreement among inspectors on borderline quality cases. . . ."

Similar data come from Teel et al. (1968).

"Two experiments were performed to determine how human performance in the assembly and inspection of microelectronic devices might be improved. The first revealed that use of tools specifically designed to minimize the probability of human error resulted in significant reductions in errors in the handling of silicon wafers. The second indicated that providing the operator with a visual frame of reference resulted in detection of a significantly greater percentage of the defects present in photo masks of the type used in production of microelectronic circuits." (p. 217)

Another line of proof is this. If equipment developers accept recommendations for incorporation of human factors principles into design, it would seem to suggest two things: (a) that a human factors deficiency was considered important enough to warrant some effort at modification of the system; (b) a specific H recommendation was considered to be effective as a solution of the design problem.

Unfortunately, as was pointed out previously, records of design solutions during the time equipment is being developed in drawing form are few and far between. Some evidence is available, however, from test records; here the equipment was tested and found wanting for some specific human factors reason.

Peters and Hall (1963) reported 44 human factors problems during the testing of the MA-3 Rocketdyne engine for the Atlas missile; 20 of the recommendations made for remedying the problem were accepted by the Air Force for implementation. In the same vein, McAbee (1969) reported that, of 980 problems found during the development of the C–5A aircraft, 91 percent were successfully resolved. Such "demonstrations" are indicative only. Qualifying factors are (a) the problems were raised by H personnel and, in many cases, required intensive effort to secure consideration; (b) many of the problems were relatively minor and, therefore, required no extensive redesign; (c) full implementation of these recommendations was not accomplished. Nevertheless, it is suggestive. Chase (1969) presents a number of examples based on a survey of 56 companies, in which H recommendations resulted in substantial improvements and cost savings.

It is almost impossible to demonstrate experimentally the importance of human factors principles with actual operational hardware. By "demonstrate experimentally" we mean to compare performance of two otherwise identical equipments, one with and one without human engineering principles. The expense would in any event be excessive; and the customer for an operational equipment is not interested in making such tests—he simply wants the best equipment for his money.

When operational data can be used for such a demonstration, it is in situations such as that cited by Peters and Hall, in which a human factors problem arises with equipment operation, and we can note whether the human factors principle, if accepted, remedies the problem.

It is especially difficult to demonstrate the importance of good human engineering when one is dealing with a large scale system involving several interactive equipments. An error made by the operator of one equipment may or may not have an effect on other equipments, depending on such factors as error feedback or the number of interlocks provided. The system output may not be a simple product of the individual equipment outputs but may involve feedback loops, redundancies, and parallel operations. Hence the error effects of a poorly human engineering equipment may be attenuated or cancelled out by the other equipments.

The H specialist, therefore, has to rely for proof of the importance of human factors largely on negative examples in which by inadvertance an equipment has been poorly human engineered so that errors are made in the operation of that equipment. To establish that proof he has to trace a direct relationship between a particular poorly designed equipment and repetitive errors (or other performance discrepancies) involving that equipment. If this can be done, he can then point to this as an example of what happens when human engineering principles are ignored. As a result, the need for, and the value of, human factors engineering is most apparent when it is *not* included in design.

It is much more difficult to find equipment examples of what we would consider "good" human engineering. If human factors principles are correctly included in design, it is often impossible to distinguish the effect of these principles from that of other aspects of the equipment design. This is as it should be, no doubt, but it makes it more difficult to put up a "good case" for the special contributions of H.

We may ask why it should be necessary to demonstrate the value of H. After all, the prevalence of human error in ordinary equipment usage is recognized by most people. That equipment deficiencies are responsible for most, if not all, of these errors is also quite clear to the observant. Why then the necessity for such a demonstration? Granted that error and error-causal factors are recognized by most people, all of us, including most engineers, have difficulty in understanding that these stem from the user's relationship to his equipment.

Most of the attitudinal and communications difficulties between engineers and H specialists stem from basic differences in the way in which engineers and behavioral scientists view equipment design and equipment use. The engineer who recognizes some difficulty in operating an equipment views that difficulty largely in equipment terms and does not, as does the behavioral scientist, visualize the possibility that the difficulty is related to the operator and his "human factors."

This is because the engineer is trained to think in terms of physical laws. Even if these laws deal with intangibles (e.g., Ohm's law), they have readily discernable physical effects. He considers that it is sufficient to worry only about equipment functions in designing equipment. To incorporate human factors in design requires a shift in the engineer's characteristic orientation toward equipment. Instead of being concerned solely with how to make his equipment function, he must go one step beyond and anticipate the effect of that functioning on the equipment operator.

Engineering Attitudes Toward Human Error

Some engineers feel that the case for human error and the importance of H is oversold. This feeling is based on a number of engineering attitudes toward human error.

One attitude frequently encountered is that an individual human error may be catastrophic but that such errors are relatively infrequent and, therefore, can be ignored with reasonable safety. It is true that catastrophic errors—errors which result in accidents and death—are comparatively rare although much more frequent than we wish to suppose. One reason for this is that many errors are reversible (i.e., the erroneous action will not damage the system or stop it from functioning). Hence it can be repeated and performed correctly the next time. Much equipment is designed with interlocks or fail-safe devices so that the effect of an error will not be catastrophic. From this standpoint it may appear as if the total effect of error on equipment operation is not significant.

The statistics of human errors and their effects on equipment functioning give the lie to this impression. It is a reasonable hypothesis—validation data are unfortunately lacking—that the cumulative effect of many minor errors will add up to significant decrement in system performance—even if the system does not fail catastrophically. An example is the study cited previously by Meister et al. (1970) in which the effects of human error on equipment reliability were studied at two SAC bases.

The effect of errors leading to reports of equipment malfunctions was determined by analyzing the amount of maintenance down time caused by the errors. It was found that 38 percent of the total maintenance man-hours expended during the five months' study was caused by errors.

Even if an error is infrequent, the potential consequences of that error to the safety of the individual making the error may make it necessary to introduce special features which guard against the probability that the error will be made. This is particularly true of equipment to be used in unusually dangerous or stressful environments, such as outer space, underwater, high speed surface transportation, etc. Even in less obviously stressful situations, the potential consequences of a human error may warrant special provisions.

In one case, for example, a technician in a missile silo had to perform a semiannual inspection of a cable which could be reached only by stopping the silo elevator between levels, lying on the elevator floor, and stretching out halfway over the open pit. The fact that this inspection was only semiannual suggested to the project manager that it could be safely ignored. The potential effect of an error in performing this inspection could, however, have been loss of life.

Another attitude which may influence the engineer's design is the feeling that the potential for human error is so great that it is impossible to anticipate every equipment characteristic which may lead to user error. This is Murphy's law applied to design: it assumes that however well one designs, someone will make an error in operation of the equipment. This attitude rejects the concept that one can protect against operator error. Engineers with this attitude may overlook those design characteristics which might guard against error.

Some engineers assume that the major cause of error in equipment operation is not the equipment configuration but inherent human variability. Under these circumstances they may feel that it is not the engineer's responsibility to compensate in his design for this variability.

The assumption here is that sheer variability produces certain performance fluctuations of a random nature (like noise) which are sometimes great enough to produce error. This may be true in continuous motor tasks, such as tracking, but is unlikely to be true of discrete, procedural tasks. Even if it were true that variability gives rise to error, it would not be enough to explain the major degradation in system performance produced by error. Variable errors do not significantly degrade operator performance because they do not occur repeatedly in exactly the same form in each operating cycle.

Another point of view is that since variable error has apparently no direct causal factor, it cannot be successfully attacked. This is a bit like thinking of error as the result of original sin.

When variable errors do exist, however, they must be differentiated from systematic ones. Systematic errors repeat themselves in much the same way over repeated operating cycles because they are caused by an inadequate system characteristic which, of course, persists throughout operational life unless corrected. Systematic errors can be avoided since the equipment design can be examined and improved. Design analysis to uncover potential systematic errors is an essential part of H practice.

Another common attitude is that even when there are design inadequacies, the human will adapt, will "muddle through." The ability to overcome inadequate design characteristics is a most fortunate result of the same human flexibility which presumably also produces variable errors. Of course, the ability to adapt occurs only when operating conditions do not stress the operator excessively. Stress occurs when undesirable operational conditions such as dim illumination, noisy communications, or the requirement for extreme accuracy (all of these are situational factors) force the operator to respond at or near the limit of his abilities. Under these conditions the operator's ability to adapt to an equipment inadequacy is quite limited, and his accuracy and response time are degraded. A meter which can be read accurately under normal lighting conditions may become impossible to read under reduced lighting. Either the length of time needed to read the meter or the number of errors in reading it, or both, will increase.

One way of noting whether there are human factors design inadequacies in operational equipment is to look for "klooges." A klooge is a makeshift modification of design created by test or operational personnel who are dissatisfied enough with equipment characteristics to redesign the equipment themselves; for example, when a feedback indication is missing and considered vital, someone may hook up a wire to a terminal and attach a simple indicator to it. The klooge is the last resort of the exasperated user. It is precisely this kind of behavior which H seeks to anticipate and avoid.

The H specialist's attitude toward human error is that it is impossible to divorce the man from the machine we build for him. A case can be built for the proposition that the engineer does in fact have a moral, if not legal, responsibility to compensate for human inadequacies because his design may contribute to or worsen these inadequacies. If he builds something for people, he automatically assumes some responsibility for their use of his equipment. This obligation may only occasionally be legal, as when the manufacturer would be liable for damages if he built an electrical motor which, because it was not grounded, electrocuted its users. In every case, however, including those in which there is no legal burden, there is a moral obligation which the engineer cannot escape.

REFERENCES

Altman, J. W., 1967. *Classification of Human Error,* in W. B. Askren, *Symposium on Reliability of Human Performance in Work,* Report AMRL-TR-67-88, Aerospace Medical Research Laboratories, Wright-Patterson AFB, Ohio, May.

Barton, H. R., et al., 1964. *A Queuing Model for Determining System Manning and Related Support Requirements,* Report AMRL–TDR–46–21, Aerospace Medical Research Laboratories, Wright-Patterson AFB, Ohio, January.

Chase, W. P., 1969. *Implementing Human Factors Test Results,* in M. T. Snyder et al. (Eds.), *Proceedings of the Human Factors Testing Conference,* 1–2 October 1968, Report AFHRL-TR-69-6, Air Force Human Resources Laboratory, Wright-Patterson AFB, Ohio, October.

Christian, J. F., 1962. Ergonomics—Palliative or Definitive, *Ergonomics,* **5** (1), 279–284.

Cornell, C. E., 1968. Minimizing Human Errors, *Space/Aeronautics,* 71–81, March.

Ehrlich, J., and D. R. Horner, 1961. *Human Engineering Discrepancy Reports of OSTF–1 Aerospace Ground Equipment.* General Dynamics/Astronautics, San Diego, California, April 28.

Ehrlich, J., and D. R. Horner, 1962. *Personnel Subsystem Reliability During Category II Test Series I OSTF–1 Test Operations.* Report REL–R–146–7–022, General Dynamics/Astronautics, San Diego, California, 13 February.

Fitts, P. M., et al. (Eds.), 1951. *Human Engineering for an Effective Air Navigation and Traffic Control System.* Washington, D.C.: National Research Council.

Gordon, R. B., 1964. *Engineering Safety into Missile-Space Systems,* Presented at the SAE-ASME-AIAA Aerospace Reliability and Maintainability Conference, Washington, D.C., 29 June–1 July.

Harris, D., and R. Chaney, 1969. *Human Factors in Quality Assurance.* New York: Wiley.

Koskela, A., 1962. Ergonomics Applied to Office Work. *Ergonomics,* **5** (1), 263–264.

Lacy, B. A., 1967. The Design of the Operator's Tasks in a Tea Blending Plant, *Ergonomics,* **10** (2), 266–270.

LeVan, W. I., 1960. *Analysis of the Human Error Problem in the Field,* Report 7–60–932004, Bell Aerosystems Company, Buffalo, New York, June.

Majesty, M. S., 1962. *Personnel Subsystem Reliability for Aerospace Systems,* Presented at the National Aerospace Systems Reliability Symposium, Salt Lake City, Utah, April 16–18.

McAbee, W. H., 1969. *Category I Personnel Subsystem Test and Evaluation on the C-5,* Minutes of the Second Tri–Service/NASA Personnel Subsystems Human Factors Test and Evaluation Conference (TESCON), Lockheed Missiles & Space Co., Sunnyvale, California, 3–5 December.

McCornack, R. L., 1961. *Inspector Accuracy: A Study of the Literature,* Report SCTM 53–61 (14), Sandia Corporation, Albuquerque, New Mexico.

Meister, D., 1956. *The Effect of Human Errors on Missile Test Performance,* Report ZX–7–015–T.N. Convair, San Diego, California, 9 April.

Meister, D., 1961. *Analysis of Human-Initiated Equipment Failures During Category I Testing, OSTF–1,* Report REL R–054, General Dynamics/Astronautics, San Diego, California, 21 November.

Meister, D., 1962. *Individual and System Error in Complex Systems,* paper presented at the American Psychological Association meetings, St. Louis, Missouri.

Meister, D., 1967. *Applications of Human Reliability to the Production Process,* in W. B. Askren, *Symposium on Reliability of Human Performance in Work,* Report AMRI–TR–67–88, Aerospace Medical Research Laboratories, Wright-Patterson AFB, Ohio, May.

Meister, D. Personal data.

Meister, D., 1965. Human Factors in Reliability, Section 12, in W. G. Ireson, *Reliability Handbook.* New York: McGraw–Hill.

Meister, D., and G. F. Rabideau, 1965. *Human Factors Evaluation in System Development.* New York: Wiley.

Peters, G. A., and F. S. Hall, 1963. *Missile System Safety: An Evaluation of System Test Data,* Report ROM 3181–1001, Rocketdyne, Canoga Park, California, 1 March.

Rabideau, G. F., et al., 1961. *A Guide to the Use of Function and Task Analysis as a Weapon System Development Tool,* Report NB–62–161, Northrop Corp., Hawthorne, California.

Raffle, A., and R. G. Sell, 1969. The Victoria Line—Passenger Considerations, *Appl. Ergonomics* **1** (1), 4–11.

Rigby, L. V., 1967. *The Sandia Human Engineering Rate Bank (SHERB),* Paper presented at the Man–Machine Effectiveness Analysis Symposium, Los Angeles Human Factors Society, University of California at Los Angeles.

Rigby, L. V., and J. I. Cooper, 1961. *Problems and Procedures in Maintainability,* Report ASD–TNN–61–126, Aerospace Medical Laboratory, Wright-Patterson AFB, Ohio, October.

Robinson, J. E., W. E. Deutsch, and J. G. Rogers, 1970. The Field Maintenance Interface between Human Engineering and Maintainability Engineering, *Human Factors,* **12** (3), 253–259.

Rook, L. W., 1962. *Reduction of Human Error in Industrial Production,* Report SCTM 93-62 (14), Sandia Corporation, Albuquerque, New Mexico, June.

Rook, L. W., 1965. *Motivation and Human Error,* Report SCTM–65–135, Sandia Corporation, Albuquerque, New Mexico, September.

Shackel, B., 1962. Ergonomics in the Design of a Large Digital Computer Console, *Ergonomics,* **5** (1), 229–241.

Shackel, B., 1969. Work Station Analysis—Turning Cartons by Hand, *Appl. Ergonomics,* **1** (1), 45–51.

Shapero, A., J. I. Cooper, M. Rappaport, K. H. Shaeffer, and C. J. Bates, 1960. *Human Engineering Testing and Malfunction Data Collection in Weapon System Programs,* WADD Technical Report, 60-36, February.

Teeple, J. B., 1961. *System Design and Man–Computer Function Allocation,* Presented at ORSA–TIMS Meeting, April 19–21.

Teel, K. S., et al., 1968. Assembly and Inspection of MicroElectronic Systems, *Human Factors,* **10** (3), 217–224.

United States, MIL–STD 1472, 1969. *Human Engineering Design Criteria for Military Systems, Equipment and Facilities.* Washington, D.C.: Department of Defense.

Van Buskirk, R. C., and W. J. Huebner, 1962. *Human-Initiated Malfunctions and System Performance Evaluation,* Report AMRL-TDR-62-105, Aerospace Medical Research Laboratories, Wright-Patterson AFB, Ohio, September.

Urmston, R. E., and C. M. Cutchshaw, 1960. *Human Engineering Principles Applied to the Design of Factory Test Equipment. I: TET–704,* Report AE60–0290, Convair/Astronautics, San Diego, California, 11 April.

Willis, H. R., 1962. *The Human Error Problem,* Report M–62–76, Martin/Denver Company, Denver, Colorado, June.

Willis, H. R., 1962. *Human Error—Cause and Reduction,* Presented to the Joint Meeting of the Midwest Human Factors Society and National Safety Council, Chicago, Ill., November.

Wilson, R. B., and J. L. Gaffney, *Man's Reliability in the X–15 Aerospace System* undated.

III

METHODS OF PERFORMING HUMAN FACTORS ANALYSES

The reader will recall from Chapter One that the designer and the H specialist must, in the process of developing the man-machine system, answer questions dealing with the allocation of functions, the analysis of tasks, and the identification of man-machine interfaces. As a consequence, the following analyses must be performed:

A. *Function Allocation and Verification*
   1. Determine system requirements.
   2. Determine system functions.
   3. Allocate functions between men and machines:
      (a) specify alternative man-machine configurations and functions;
      (b) verify that the human can perform the functions tentatively assigned to him and that his performance will satisfy system requirements;
      (c) select the most effective man-machine concepts by comparing alternative designs.

B. *Task Description and Analysis*
   1. Describe and analyze tasks.
   2. Determine equipment requirements implied by tasks.
   3. Analyze tasks in terms of their demands.

C. *Identify Man-Machine Interfaces*
   1. Select appropriate interface components.
   2. Arrange control-display components in the form of a control panel.[1]

[1] It should not be assumed that the control panel is the only man–machine interface with which the specialist and designer are concerned. However, it is the one most frequently analyzed; hence it is used here as an example of the analytic process.

The purpose of this chapter is to describe these analyses. Although these analyses should be performed both by the design engineer and the H specialist, working in parnership, these analyses will be discussed as if the specialist has primary responsibility for them, at least insofar as they deal with human factors.

The analyses are described in step-by-step procedural form, using a decimal system for denoting steps. The ordering of these steps is as follows:

1-0 Analysis goal (e.g., determine system functions).
  1-1 Major step.
    1-1.1 Minor step.

Before the reader examines these methods, a number of cautions are necessary. First of all, the attempt to describe a logical thought process in the form of discrete steps may suggest that the thought process itself is in reality also discrete and proceeds in step-by-step fashion. This is quite untrue. In reality these analyses are iterative and progressive; the analyst continues to refine previously completed steps, applies new data to factors considered earlier, and continues to test the adequacy of past conclusions.

To complicate the matter, although this chapter can only describe each analytical procedure individually and sequentially, the specialist may actually perform several analyses simultaneously. As a consequence, his analyses are performed much more quickly than might appear from the painstaking, step-by-step manner in which they are described here. However, since they are refined at increasingly molecular levels of system development, they tend to be repeated in detail.

FUNCTION ALLOCATION AND VERIFICATION

1-0 *Determine system requirements.*

System requirements are almost always specified by the customer, although the contractor may suggest additional ones. Hence, to determine system requirements is to analyze them in terms of the parameters they imply.

The starting point for the determination of system functions is the analysis of system requirements. The former are inherent in, and derived logically from, the latter. Thus, in order to implement system requirements, certain functions are necessary. For example, if an aircraft is to deliver a bomb load, it must fly; to fly, it must take off; to take off, engines must be started, etc. The examination of system requirements, therefore, implies progressively more molecular, implementing functions.

At the beginning of the analysis these functions are not differentiated in terms of how they are to be accomplished (i.e., performed solely by equipment or by some man-machine combination). It is enough to say that a function must be performed. However, as was pointed out in Chapter Two, certain system requirements logically demand particular modes of implementation; thus the determination of system functions in such cases is essentially only a matter of deduction.

In many cases, also, the procurement document supplied by the customer (statement of development work to be accomplished by the contractor) specifies the functions to be performed and may even indicate how they are to be implemented. It may appear in these cases that analysis of system functions would be relatively unprofitable for the specialist. Even under these circumstances, however, the analyst will find it useful to start his work by examining system requirements; the reason is that sometimes these requirements have not been clearly expressed and consequently the specification of required human functions and equipment may be inadequate. Required functions may be only implied, when they are not wholly ignored. Without a systematic analysis it is difficult to draw proper conclusions about how such functions should be implemented.

1-1 Determine what information is available concerning the system.

1-1.1 Secure and examine available documents describing the system.

The analyst does not usually work in a vacuum. Almost always there have been previous analyses which can supply useful information. The major documents available to the contractor analyst include the request for proposal (RFP), the statement of work (SOW), and any preliminary feasibility study or test reports. If the analysis is being performed by the customer before the time a contract for the system is being let, planning documents such as the General (GOR) or Specific Operational Requirement (SOR) should be available.

1-2 From the relevant documentation extract and list the following in detail:

(a) the system's mission or goal (e.g., to detect a submarine, to intercept enemy fighters);
(b) required system outputs (e.g., 40 messages must be transmitted per unit time);
(c) required system inputs (e.g., reception of an electronic signal on a designated frequency);

(d) system capabilities and performance requirements demanded by the mission(s) (e.g., store fuel sufficient for 30 days, submerge to depth of 400 feet);
(e) environmental factors which may affect system performance (e.g., anticipated temperature of -64F);
(f) constraints (e.g., maintenance must on the average not take more than 15 minutes).

Comments on Procedure 1-0

In any system analysis one begins with the mission and required inputs and outputs. These in turn may demand certain functions; they may also require certain functional capabilities and impose performance requirements. If special environmental or operational factors apply, they may also demand that the system perform special functions; for example, if the vehicle must perform at very low temperatures, it may be necessary to heat the system, and this in turn may require the temperature to be monitored (with the possibility that the human will be the monitoring agent). The same reasoning applies to constraints, which are in a sense a negative performance requirement.

2-0 *Determine system functions.*

2-1 For each system mission (see Step 1-2) list sequentially the individual major operations that must be performed to implement the mission.

By listing these operations in terms of sequential dependencies (e.g., to fly one must first take off, to take off one must first start the engine) and correlating them with the over-all time frame, they become stages in the accomplishment of the mission. In effect, what one must do to accomplish the mission becomes the individual system functions.

2-2 Describe the resultant system functions in the form of a functional flow block diagram (FFD).

2-3 Determine the effect on system functions of the environmental factors, performance requirements and constraints noted in 1-2.

2-4 When additional functions are required by step 2-3, add the new functions and insert them in the diagram developed in 2-2.

2-5 Specify the inputs to, and the outputs from, each system function.

These are the input actions required to initiate the function and the output actions resulting from performance of the function; for example, to perform the function of submarine detection, the operator must first *scan* the sonar scope. The latter becomes a separate function. One of the required outputs of the detection process is a verbal report by the sonarman that detection has been accomplished so that one has now added a communications function. Step 2-5 is a refinement (more detailed analysis) of step 2-1.

2-5.1 Include each new input/output function in the FFD.

2-6 Indicate alternative functional pathways and feedback loops.

Examine each function in terms of alternative ways in which inputs and outputs may be supplied. These may eventually become design alternatives, one of which will ultimately be selected. Diagram each alternative; these become the pathways. When a function provides an input that modifies a previous function, a feedback loop should be specified. (Feedback may require a display.) When a choice must be made between alternative functional inputs or outputs or means of performing the mission, indicate the choice as a decision point.

2-7 Assign appropriate symbols for decisions and operations to the resultant function flow diagram. (This is purely a technical matter.)

Comments on Procedure 2-0

For each of the system's missions (assuming more than one) the question must be asked: What must the system do to accomplish these misions? The analyst works backward from the performance goal (i.e., system mission) to determine what activities mst be performed in order to achieve that goal. These activities become system functions. Subordinate activities that implement higher order functions are in turn analyzed in the same manner to achieve a finer level of detail (subfunctions).

The analysis so far performed can be illustrated by the manner in which FFD's were developed for a recent study by the author (Meister et al., 1968). The system being developed was the propellant transfer and pressurization system for a space launch vehicle. It had the following primary missions: to store propellants and then to transfer them to a space vehicle. Diagrammatically these two functions could be illustrated as follows:

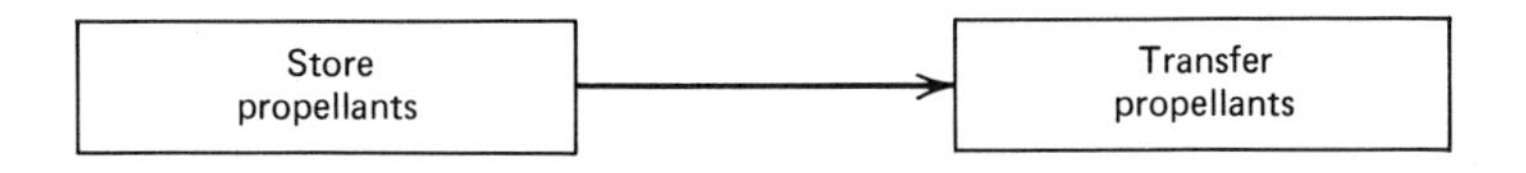

However, in order to store propellants, they must first be received. This prior input activity is added to the previous diagram, thus:

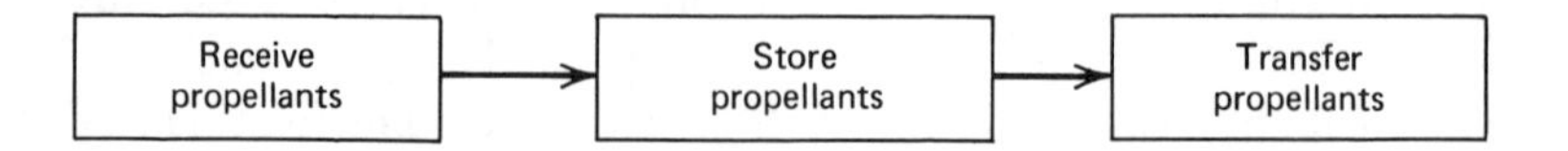

One of the performance requirements is that while the propellants are being stored they must be maintained at a specified temperature. This immediately introduces another function, *monitor propellant temperature,* so that the diagram is now built up to the following:

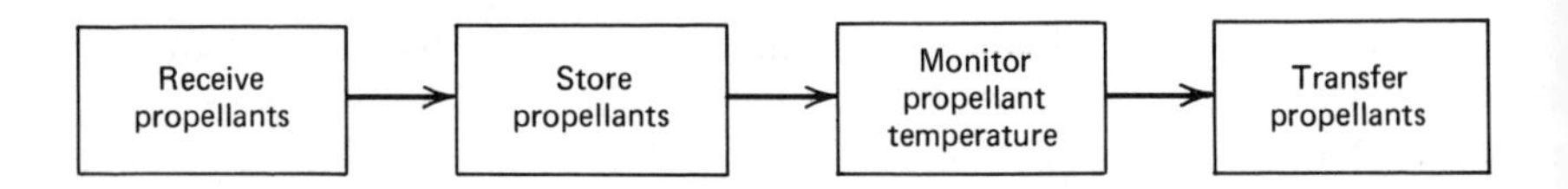

If we then ask how the propellants are to be received and stored, it is apparent that, regardless of the type of vehicle in which they are transported to the storage area, they must be transferred from the conveying vehicle to storage facilities, since another performance requirement is that they may have to be stored for as long as 30 days. Consequently the FFD can be expanded as follows:

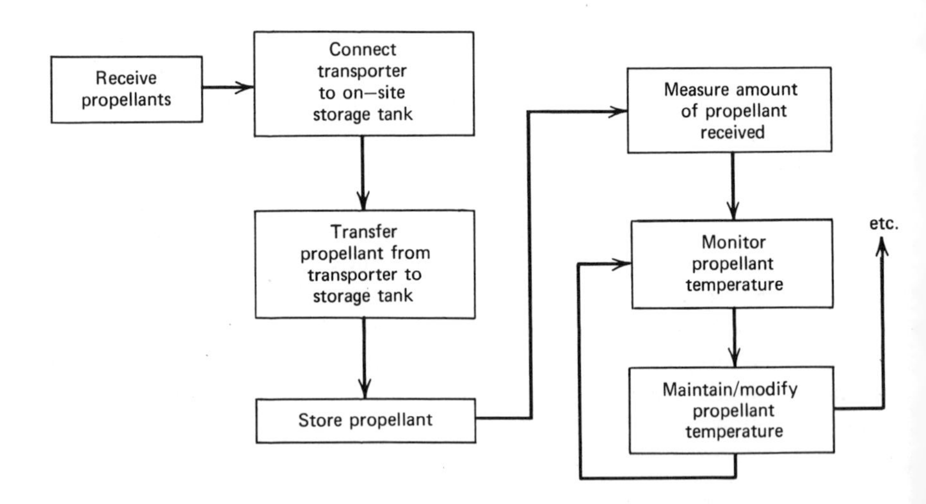

After diagramming the functions required to implement the mission(s) (step 2-2), we must then modify the resultant diagram in terms of factors (step 2-3) that influence these functions. Significant changes in environmental factors, performance requirements, and constraints inevitably result in some change in the system's performance that must be reflected in the diagrams.

In general the procedure for building up an FFD is to analyze each system function in terms of its required inputs (initiating stimuli) and outputs (response consequences). Each input and output is in turn analyzed in exactly the same fashion to elaborate the functional diagram. These inputs and outputs do not, however, imply the mechanism by means of which the inputs and outputs are produced; at this stage in the analysis of functions and development of the FFD, we cannot say whether a function should be performed by personnel or by equipment or specifically in what manner it should be performed.

If one determines the logical inputs/outputs for each function, this effectively defines the sequence in which these functions should be arranged. If, in addition, we know the performance duration required for each function, we can plot the FFD along a time continuum. This is useful later in determining whether the human can perform the function.

3-0 *Allocate functions between men and machines.*

- 3-1 Determine how system functions should be implemented.
- 3-2 Examine all presently available engineering documentation, for example, the SOW, feasibility study reports, etc. to determine if equipment (as distinct from system) functions have already been decided upon by the customer or the contractor management on the basis of previous analyses—as they often are. Even if they have, however, the equipment functions may imply some operator relationship that should be analyzed since, as was pointed out in Chapter One, most equipments require some operation or maintenance.
- 3-3 For those system functions that have not been allocated as yet, differentiate between operator and equipment functions on the basis of $\underline{H}$ criteria.

The $\underline{H}$ criteria referred to are those developed by Fitts (1951) and cited by almost every human factors text, presumably for lack of anything better to recommend. These criteria compare the capabilities of men with those of machines in terms such as, "men are better at inductive reasoning, machines are better at deductive reasoning . . ." As was pointed out in Chapter Two, such criteria are practically useless in making any meaningful, practical function allocation decisions because (a) the criteria are overly general and (b) they assume that functions will be performed either by machines alone or by men alone. However, the Fitts list is a useful starting point (but only that).

3-4 Specify alternative man-machine configurations and functions. A more realistic way of performing the function allocation is to concentrate first on listing and describing all the possible ways that the mission objective(s) can be implemented.

3-4.1 Consider how the set of required system functions could be performed if they were implemented largely by equipment devices (automatic configuration).

This does not mean that there are no operator actions performed in this configuration but that these operator actions are relatively few and simple.

3-4.2 Consider how the same set of required system functions could be performed if they were implemented largely by operator personnel (manual configuration).

For 3-4.1 and 3-4.2 list sequentially and describe in as much detail as possible all the machine plus human operations which would be required to exercise each configuration. Describe as many different configurations as are reasonable. Here the design engineer and the H specialist should work in partnership. Unless the specialist has some engineering background and familiarity with the system requirements, he will have to secure help from the system designer to describe the potential configurations. Actually the procedure being described is what the system designer does on his own in the process of deciding upon a configuration, but, as we shall see in Chapter Seven, he very often performs such an analysis quite unsystematically and may ignore the potential human factors implied by the configuration.

3-4.3 Describe the alternative configurations in the form of a FFD as performed in procedure 2-0. (The new FFDs will, of course, be much more detailed than that of 2-0. The reason for describing each configuration in terms of a FFD is that it permits the analyst to see concretely what each function and mode of implementation consists of.)

3-4.4 For each function in the alternative configurations, list the means of implementing that function, that is, the machine device and/or the operator action that is required to perform that function. An example of alternative configurations at a gross level of detail characteristic of early development stages is shown in Figure 3-1.

3-5 Verify that the human functions can be performed to system requirements.

Alternative 1 (operator primarily)	Alternative 2 (man-machine mix)	Alternative 3 (machine primarily)
Sonarman detects target signal on scope, examines brightness, shape, recurrence, movement, etc., and reports "probable submarine" or "non-submarine target."	Sonarman detects target signal on scope. Associated computer also detects signal, records it, and searches library of standard signals. Computer displays to sonarman original signal and comparison signal on sonar gear, together with the probability of its being a submarine. Sonarman decides on basis of his own analysis and computer information whether target signal is submarine or nonsubmarine and reports accordingly.	When a signal having a strength above a specified threshold is received by the sonar array, a computer associated with the detection apparatus automatically records the signal, analyzes its strength, brightness, recurrence, etc. according to preprogrammed algorithms, compares it with a library of standard sonar signals, and displays an indicator reading "probable submarine."
Operator functions	Operator functions	Operator functions
1. Detection of signal 2. Analysis of signal 3. Decision–making 4. Reporting of decision	1. Detection of signal 2. Analysis of signal 3. Decision–making 4. Reporting of decision	1. Take action on receipt of "probable submarine" signal.
Machine functions	Machine functions	Machine functions
1. Display of signal	1. Detection of signal 2. Recording of signal 3. Searching of comparison signals 4. Analysis of signal 5. Display of information	1. Detection of signal 2. Analysis of signal 3. Decision–making 4. Display of conclusion

Figure 3-1 Illustration of Alternative Man–Machine Configurations

At this stage the analyst has extracted the equipment and operator functions by means of which each system function is to be performed. Eventually it will be necessary to select among these alternative equipment/operator functions, but first the question must be answered: can the equipment device and the operator perform their respective functions to satisfy system requirements?

To answer this question, the anticipated performance of the human function must be compared with that performance required by the system. (We shall not consider here how the engineer determines whether the particular equipment device postulated can in fact meet system requirements. That is the engineer's job. It is the H specialist's job to verify that the operator can perform his functions, and it is this process we will describe.) It does no good to say that an operator can perform a precise motor action if the precision he is capable of is less than that demanded by system requirements. To perform this verification requires answers to two questions: (a) what does the system demand of the human; (b) can the human satisfy the system demand?

> 3-5.1 Determine by reference to system requirements documents and/or conferences with system designers whether a quantitative operator performance requirement exists.

The determination of operator system requirements presents severe difficulties to the H analyst, more so even than determining whether the human can meet this demand. Although some data—admittedly not enough—are available concerning what humans are capable of, almost no procurement document specifies any operator requirements. Moreover, it is sometimes extremely difficult to infer what these operator requirements might be.

It is possible that operator requirements are not included in procurement documents because the precise role of the operator has not yet been decided. However, many procurement documents do contain explicit *equipment* requirements based on no more substantive information about equipment devices. If we do not find a similar set of requirements for human functions, it is often because the customer (e.g., governmental procurement agencies), lacking interest in or capability for human factors analyses, has not done its homework properly.

> 3-5.2 If the set of system requirements does not include an appropriate operator performance requirement, it will be necessary for the specialist to infer what that requirement is, based on his analysis of the over-all system requirements and of the configuration from which the function was extracted. (As was pointed out earlier, the absence of an operator performance requirement is very likely to be the usual case, except when human functions are critical to the system.)

For example, if the projected configuration requires the operator to transmit teletype messages, the specialist will have to determine what the maximum frequency of message transmission is likely to be (this is the inferred requirement). He must then compare this requirement with the maximum frequency of message transmission via teletype which operators on the average can provide (operator capability information). Operator capability information can be found, it is assumed, in the general behavioral literature (Chapter Four will examine the adequacy of the human factors research literature to support development applications).

The standard of required performance must, moreover, be formulated in terms of relevant operator performance criteria; for example, if the operator function to be performed involves lifting weights, the requirement must be formulated in terms of the maximum weight to be lifted. Hence it is necessary to examine the potential criteria that may apply to the human function being evaluated and to select those that are appropriate for the particular function.

General operator performance criteria can be categorized in terms of

(a) frequency of required outputs;
(b) speed of required outputs;
(c) physical requirements (e.g., strength, sensory discrimination capability, decision–making capability) for implementing the function;
(d) accuracy of required outputs.

There may well be other criteria, and certainly the ones listed above are quite general. The requirement the specialist develops (or infers) should, however, be as detailed (and quantitative) as he can make it. When several criteria apply to a particular human function, the requirement inferred must be formulated in terms of all applicable criteria; for example, the teletype transmission function involves speed, frequency, and accuracy.

3-5.3 Compare the operator performance requirement (e.g., speed) with average operator capability.

If operator capability is less than that required by the potential system configuration, the function cannot be performed by the operator, and an equipment solution to the design problem must be accepted.

3-5.4 The comparison process is the same for the other three criteria.

When more than one criterion apply to a particular function (e.g., both frequency and speed or both speed and accuracy), if any *one* applicable criterion fails to satisfy system requirements, the operator function must be discarded even if the other criteria do satisfy system requirements.

Please note that even if we have been talking about *the* human function, we are referring to a configuration which probably includes several functions (see Figure 3-1). The verification comparison must be performed for each human function required by the configuration. Of course, the number of functions extracted will depend on the level of detail to which the specialist wishes to analyze.

3-6 Select the most cost-effective configuration.

Up to this point the specialist has merely been verifying that the operator functions can be performed satisfactorily. If the operator functions required by the various alternatives meet this test and are in fact capable of being performed by operators, it will then be necessary to select the most effective configuration by comparing the alternatives in terms of over-all system criteria.

The criteria employed previously were operator capability criteria. When alternative configurations involving equipment as well as operator functions must be compared, this can only be done on the basis of system criteria (e.g., cost-effectiveness criteria). Because the two types of functions are quite different, they require system criteria.

3-6.1 Weight selection criteria.

To apply system criteria requires the specialist to determine the weight or importance each criterion should have. Cost may be a crucial factor for one system development, reliability for another. In general, the significant criteria to be applied in evaluating alternative means of implementing functions are: cost, performance reliability, maintainability, ability to satisfy performance requirements. Others may also be relevant: producibility or the ease of fabrication, the number of personnel required by a system, power requirements, safety. In some cases one criterion, for example, cost, may be the overriding factor in dictating a design decision; in others, several criteria may each have some influence on that decision.

The determination of the weight or value each criterion should have for a particular system is an entirely subjective judgment on the part of the developer. The following procedure for assigning mathematical weights to these criteria, taken from Hagen (1967) and illustrated in Table 3-1, merely formalizes and quantizes that judgment. It has value in forcing the specialist—but more importantly, the design engineer—to make his decision biases visible. In actual development, few designers quantize their judgments, which makes these easy prey to irrational persuasions. However, the studies performed by the author and his colleagues (discussed in Chapter Seven) indicate that when required to do so design engineers can make formal criterion judgments.

Table 3-1 Assignment of Weights to Functional Design Criteria

Criteria	Choice Tally	Total	Weighting Coefficient
1. Performance requirements	1 1 1 1 1 1 1	7	.25
2. Cost	0 1 1 1 1 0 1	5	.178
3. Reliability	0 0 1 1 1 0 1	4	.143
4. Maintainability	0 0 0 1 0 0 1	2	.0715
5. Producibility	0 0 0 0 0 0 0	0	.0
6. Safety	0 0 0 1 1 0 0	2	.0715
7. Number of personnel required	0 1 1 1 1 1 1	6	.214
8. Power requirements	0 0 0 0 1 1 0	2	.0715
		28	.9995

Weights are assigned by comparing each potential criterion with every other and assigning a value of one (1) to whichever is judged to be the more important of any two criteria considered; zero (0) is assigned to the less important of the two criteria. For example, if performance requirements are more important than cost, it receives a value of 1; cost, 0. The performance requirements criterion is then compared with each of the remaining criteria and values assigned as above. The next criterion, cost, is compared with the remaining criteria, and this process is continued until all criteria have been compared with each other.

The 1's for each criterion are then summed across Table 3-1 and divided by the total number of 1's assigned (28). This gives the specialist a weighting coefficient for each criterion. The weighting coefficient is a relative value (indicating importance of one criterion relative to all others considered), not an absolute value. The sum of the weighting coefficients should approximate 1.00.

3-6.2 Compare alternative man-machine configurations.

Now that the criteria on which the alternative configurations are to be compared have been weighted, the specialist can proceed to compare these configurations. Before doing so, however, it is desirable to specify a quantitative basis for as many criteria as possible. To have an objective, quantitative means of comparing functions reduces the subjectivity of the comparisons. To illustrate the process we shall describe the quantification of cost and reliability, which, along with performance requirements, are probably the most significant decision criteria.

3-6.2.1 Determine the cost to the customer of designing, producing, and maintaining the number of hardware items required.

3-6.2.2 Determine the total cost to the system (including selection, training, salaries, and logistical support) of the number of personnel required by the alternative man-machine configurations being compared.

It should be noted that the derivation of appropriate cost data, both for equipment and human components, is extremely difficult. Appropriate data may not be immediately available. This is particularly the case for determining the life cycle cost to the system of the number of personnel required. The parameters which enter into such a human cost estimate are very many, and their values require a good deal of guess work.

3-6.2.3 Compare hardware cost with total personnel cost. (All other things being equal, the configuration with the lowest cost will be preferred.)

3-6.2.4 Determine the reliability of all equipment devices to be employed in all configurations by referring to the Reliability Engineering group for this information.

The steps involved in computing equipment reliability (see Zorger, 1966) are not described here because they are not human engineering methods. The H̲ specialist would expect to secure this information from the Reliability group.

3-6.2.5 Compute operator reliability for the specified human functions by determining the average predicted operator error rate for each function and subtracting from 1.0 (optimal reliability).

The detailed procedure for performing this analysis is too lengthy to include in this section. However, additional information is given in Chapter Five, and the interested reader can refer to Meister, 1966.

3-6.2.6 Compare alternative functions/configurations.

Assuming that all quantitative data have been gathered, it is now possible to compare each alternative mode of implementing a given function or configuration with every other on the basis of each criterion, giving the favored alternative a value of 1 and the less favored a value of 0. The procedure, illustrated in Table 3-2, is essentially the same as that employed in weighting criteria. Note that alternative configurations A, B, and C are each compared with each other on the basis of the performance requirements criterion first, then cost, reliability, etc. For example, when A is compared with B on the basis of functional performance, B is favored, and given a value of 1 (see choice tally column); when A is compared with C, C is favored and given a value of 1; when B is compared with C, B is favored and given a value of 1. Since there are three alternatives, the total number of 1's in each tally row is divided by three. The resultant quotient for each alternative, called a choice coefficient, can be either .00 (no 1 choices), .33 (one choice out of three), or .67 (two choices out of three).

3-6.2.7 Choose the alternative with the highest criterion weighting.

Table 3-2 Comparison of Alternative Configurations

Alternative Configurations	Criteria	Choice Tally			Total 1's	Choice Coefficient
A	Performance requirements	0	0		0	0
B		1		1	2	.67
C			1	0	1	.33
A		1	1		2	.67
B	Cost	0		1	1	.33
C			0	0	0	0
A		0	0		0	0
B	Reliability	1		0	1	.33
C			1	1	2	.67
A		1	1		2	.67
B	Maintainability	0		0	0	0
C			0	1	1	.33
A		1	0		1	.33
B	Producibility	0		0	0	0
C			1	1	2	.67
A		1	1		2	.67
B	Safety	0		0	0	0
C			0	1	1	.33
A	Number of personnel	0	0		0	0
B	required	1		0	1	.33
C			1	1	2	.67
A		1	1		2	.67
B	Power requirements	0		0	0	0
C			0	1	1	.33

The final step is to construct a matrix with the alternative functions or configurations identifying the columns and the criteria identifying the rows (Table 3-3). The values in the columns under the alternatives represent the multiplication of the criterion weighting coefficients (Table

3-1) times the choice coefficients derived in Table 3-2. Alternative A, for example, had a choice coefficient of .00 (see Table 3-2); this, when multiplied by the weighting coefficient for performance requirements (.25) equals zero (0). The same multiplicative process is performed for every other alternative times weighted criterion cell. The values achieved by each alternative when evaluated by each criterion are summed. The first choice for design implementation is the alternative with the largest value, the second choice is the one with the next largest value, etc.

Table 3-3 Matrix for Selecting Alternative Configurations

Criteria	Alternatives		
	A	B	C
1. Performance requirements × 0.25	0.0	0.168	0.0825
2. Cost × 0.178	0.119	0.0588	0.0
3. Reliability × 0.143	0.0	0.0472	0.0958
4. Maintainability × 0.0715	0.048	0.0	0.0836
5. Producibility × 0.0	0.0	0.0	0.0
6. Safety × 0.0715	0.048	0.0	0.0236
7. Number of personnel required × 0.214	0.0	0.0707	0.1430
8. Power requirement × 0.0715	0.048	0.0	0.0236
	0.263	0.3447	0.3921
	3rd Choice	2nd Choice	1st Choice

Comments on Function Allocation/Verification

The function allocation/verification procedures described are quite crude, and since Swain and Wohl indicated that "there is no adequate systematic methodology in existence . . ." (1961), there has been no substantial improvement. As Whitfield (1967) points out, the various techniques available are simply more sophisticated variations on the "Fitts list" referred to previously.

It is, of course, quite possible that sophisticated techniques are really unnecessary; that if sufficient data on human performance capability were available, function allocation, which is really only a comparison between manual, semiautomatic and automatic means of implementing functions, would become child's play. The subtlety would come in defining what a function is at various levels of detail. In any event, any function allocation procedure, however sophisticated, mathematicized, and computerized it were, would collapse if it did not possess a store of operator capability data. If the method is the engine of system development, then the data is the fuel.

Needless to say, the function allocation/verification procedures we have recommended are rarely carried out as described; in other words, they are largely an ideal. The process as actually performed in system development is unfortunately far less logical and systematic than the specialist would wish. Although specification and analysis of equipment functions may receive careful attention from the system developer, the same cannot be said of human functions, which are too often merely ancillary afterthoughts to equipment-centered activities. To those who may object that this denigrates the role played by the H specialist in system development, we can say only that it represents what *is,* not what should be. There is no question that systematic attention to the analysis of human functions would produce more effective systems. If human functions do not receive this treatment, however, it is necessary to ask why. Knowing why may be the first step to rectifying the problem.

There are several reasons why the procedures described are only incompletely performed:

1. As pointed out in greater detail in Chapter Seven, the initial development of the configuration accepted by the engineer is often so rapid that the time required for the specialist to gather necessary information to analyze that configuration often is not available. It takes time to sketch out alternative configurations in sufficient detail to extract human functions; it takes time to determine what the human functions imply in the way of operator performance requirements and capability; it takes time to gather cost and reliability data. And so on for the other procedural requirements. If the design engineer rushes a single configuration into detailed hardware drawings, the time available for the specialist to analyze alternatives will be severely curtailed.
2. Another problem is that the engineer often does not construct and systematically compare alternative configurations but settles on the one he prefers without adequately considering other potential candidates; evidence for this is provided in Chapter Seven. This may be doing the engineer an injustice, however. Most H specialists, not being engineers themselves, are dependent on the engineer (at least in part) for the delineation of these alternative configurations. Among themselves engineers may, in fact, consider a number of alternatives before deciding on one; but this consideration may be covert, and in any event they may not present these alternatives to others. Hence, lacking the ability to create them on his own, the H specialist may not be aware of the various alternatives. The development and analysis of human functions may be

left to the specialist because the engineer is either too busy or uninterested in analyzing these functions. To do his analytic job properly the specialist may find himself forced to ask the engineer to help him develop configurations that may appear to the engineer to be irrelevant.

3. It has already been pointed out that many procurement documents at the start of contractor design explicitly or implicitly have allocated functions to equipment or personnel. When this allocation has been performed on the basis of systematic analyses of the type described previously, this is to be lauded, for it is the author's position that the customer *should* do his own analyses and make his requirements and desires quite explicit to the contractor.

More often, however, when such allocations have been made before a system development contract is let, they have been done only half-systematically, that is to say, considering only equipment functions. When allocations have not been made by the customer, they are often made by the contractor in the very hurried period while responding to a request for proposal; and again sufficient attention is not usually paid to the analysis of human functions. The central position of the design engineer in developing a proposed system often leads (unwittingly or deliberately) to the exclusion of other specialty disciplines, among which we find H.

4. The essence of function allocation is, as we have seen, the comparison of operator capability with a requirement for operator performance. It has already been pointed out that, except in systems in which human functions are especially critical, most system requirements do not include quantitative operator performance requirements. In part this is because, lacking a sufficient amount of capability data, it may be difficult for the customer to deduce from system requirements what those operator requirements should be. In part also it may be because many system requirements are neutral with regard to the means of implementing these requirements. Hence, if an operator requirement is to be specified, some one must conceptualize alternative ways of implementing functions; unfortunately, the average system developer usually does not think in terms of human functions.

5. Even if an operator performance requirement has been specified or inferred, the specialist may have some difficulty in securing from the behavioral literature the appropriate operator performance data with which to verify the proposed human function. Much of this literature is quite general in terms of the human performance it describes so that one must extrapolate from it to the specific human functions (and tasks) to be performed. If, for example, we wished to predict the maximum number of teletype messages an operator can transmit in an hour, we

would have great difficulty in finding relevant data in the literature. It would require knowing the speed with which an operator can type messages of a specified length and format, such data are generally not available. We may, it is true, do what many H specialists refer to as a "quick and dirty" experiment to secure the data; but even quick and dirty experiments take time.

The lack of such data also contributes to the difficulty of specifying an operator performance requirement for the system because the system developer would have difficulty knowing what can reasonably be required (e.g., 20, 40, 60 messages per hour?) of system personnel.

Does all this mean that the function allocation/verification procedures described are useless? Not at all. There are things that the specialist should, and indeed must, do to ensure adequate consideration of human functions.

First and foremost, to look at the problem from a long-range viewpoint, the need for operator capability data must be satisfied by gearing human factors research to the task of providing these data. The author considers that research to provide these data should have first priority (see Chapter Four for additional research suggestions). In addition, the literature should be examined to determine whatever capability data are available; these should be compiled so that they are readily available to the specialist. It is unreasonable to ask the specialist working on system development projects to research the literature in depth every time he has an operator capability question he needs to answer.

Because it is long range, this research recommendation will not completely solve the immediate problem of applying function allocation/verification procedures to development. However, it is something which the H discipline can do largely on its own and is, therefore, not dependent on the vagaries of engineering or system development.

As to what the specialist in system development can do himself:

1. Even when the design engineer has apparently fixed on a particular man-machine configuration, the specialist can suggest alternatives that can be compared with the one selected by the engineer. Even if these are not accepted by the system developer, these alternatives will cause the latter to analyze his own configuration more closely.

2. The specialist can analyze the engineer's configuration to extract those human functions that the configuration requires but that have not been made explicit. He can point out the human factors implications of the engineer's configuration; he can test the adequacy with which these human functions can be performed by comparing them with explicit or inferred requirements for operator performance.

3. He can analyze system requirements on his own to point out potential human functions that could satisfy those requirements. This would provide the design engineer with a broader choice of alternatives and make implicit human functions more visible.

4. The specialist can question the engineer about his configuration. What are the system requirements which this configuration presumably satisfies? Have all system functions been completely described? Will the functions implied by the configuration satisfy system requirements? What are the criteria for accepting the engineer's configuration? What formal tests have been imposed in accepting the configuration? Such questions may give the specialist the reputation of being a gadfly, but they will undoubtedly stimulate the developer's thinking about the problem.

The formal procedures described in the previous sections should be used at the very least as a *model* to guide the specialist's work during the functional allocation/verification stage of system development. He may not be able to apply the entire procedure, but parts of it may well prove useful.

TASK DESCRIPTION AND ANALYSIS METHODS

1-0 *Describe and analyze tasks.*

1-1 List the system functions to be performed.
(These have been specified as a result of previous analyses.)

1-2 For each function list all the actions that must be performed by operators to implement that function.

In essence, what this requires of the specialist is familiarity with system requirements, with the equipment configuration (insofar as it exists at the time the analysis is performed), and a controlled imagination.

1-2.1 Determine whether any operator tasks have already been specified or implied by customer procurement documents or by previous engineering studies.

In the event that this has been done, the analysis of those tasks has been partially performed for the specialist. We say "partially" because, in any case, the specialist should review the tasks so specified to determine their appropriateness and completeness. The determination of tasks by nonspecialist personnel often lacks detail and fails to draw correct conclusions.

1-3 Arrange all the actions in 1-2 in the sequence in which they should be performed.

This sequence can be developed by referring to already derived FFD's (see preceding function allocation/verification section) with which the task sequence should correspond.

1-4 Apply the following task criteria (Rabideau et al., 1961) to each action listed in 1-2:

(a) The task as defined must include an immediate purpose for the action, a machine output or consequence of the action, and the human inputs, decisions, and outputs necessary for initiating the action and accomplishing its purpose;

(b) The task must involve the output of only one man in some combination with machine components.

Actually the identification of a task poses very few problems. What presents difficulty is the decision on the particular level of detail that task should describe. The analyst must decide whether he wishes to describe the action (task) at the level of the individual control or display manipulation with which that action is involved (e.g., reads thermometer) or at a somewhat more molar level (e.g., determines that temperature is within acceptable range). This is not merely a matter of semantics. In the first case the task has been reduced to its behavioral essentials. In the second case the task description subsumes, but does not make explicit, a number of more molecular subtasks, and it will probably be necessary eventually to break the task down into those subtasks. The particular level of task description is quite arbitrary, hence judgmental; precise rules for this cannot be specified. In the author's opinion it is best to be as detailed as possible because the ultimate goal of task description and analysis, in system development at any rate, is to recommend appropriate hardware (e.g., control-display components) to implement the man-machine interface; such recommendations demand detail. The consequences of describing the task at a more abstract level is that the analyst may overlook the control-display hardware needed to implement the task. In any event the methodology is largely a trial and error process in which previous experience plays an essential part.

1-5 Revise the list of operator actions to fit the criteria of 1-4.

1-5.1 Determine subtasks which must be performed to implement higher order tasks.

When an individual task has several sequential inputs and outputs, these inputs and outputs are subtasks of the task; for example, Rabideau et al. (1961) provides the following illustration: The task "lay portable missile pad" may have as subtasks "remove pad section storage retainers," "obtain pad sections from transporter-erector storage bays," "carry sections to pad area," etc.

1-6 Describe each task in terms of a verb that indicates the nature of the action being performed by the task: for example, monitor, check, read, or throw. Follow each verb with the name of the equipment component that is being acted on by the task; for example, meter, switch, or indicator. The verb-noun arrangement corresponds to the concept of the operator acting upon an equipment interface to cause an effect.

Comment on Procedure 1-0. There is little difficulty in determining which operator tasks are to be performed as soon as we know (a) which system functions have been definitely assigned to the operator, and (b) with what equipment objects he performs the function. The reason why there is little difficulty in specifying tasks is that the process of task derivation is basically quite logical despite any questions which may arise concerning the detail with which the task is described. That logic asks, "Given that an operator function must be performed, what are the things (i.e., tasks) he must do to implement the function?" No specialist training is required for this determination although obviously experience in performing the determination helps the process considerably. On the other hand, the specialist may have difficulty defining tasks if he is unfamiliar with the equipment configuration.

A distinction must be drawn between identifying the task (i.e., to specify that a task of a given class exists), which most personnel can do, and analyzing it in detail, which is the specialist's role. We can draw equipment and behavioral implications from a task only when it is described in terms of meaningful parameters.

The manner in which tasks are described depends on the task taxonomy (i.e., word listing) selected. A great deal has been written about the need for a generally accepted task taxonomy (e.g., Melton and Briggs, 1960; Miller, 1967), and a great deal of effort has gone into the development of such a taxonomy. Controversies among H̲ researchers have usually centered about the level of detail to be provided in the task description. If one decides to describe a set of tasks in highly molecular terms (e.g., "discriminates the difference between two radar blips"), the resultant task descriptions will be different than if one uses more molar terms (e.g., "monitors scope").

This is not to say that task descriptions are unequivocally "good" or "bad"; the suitability of a task description depends on the purpose for which it is being developed. If we wish to compare tasks performed over a variety of systems, then a molar descriptive level is preferable; otherwise we become confused by the system–peculiar details of each task. On the other hand, if we wish to explore task implications for system design, then the more detail the better.

From the standpoint of system development, controversies over which word to use to describe a task and the definition of that word are largely irrelevant as long as the categories applied are reasonably descriptive of system operations. That is because the behavioral analysis that we can apply in system development is somewhat restricted. In point of fact, the task categories should perhaps change as system development proceeds, to make them more detailed. The important thing is not the word or phrase applied to a task but the implications we draw from the task description. Because of the need to derive equipment implications from the task description, the most effective taxonomy in system development is one which is quite detailed. A sample classification of task behaviors, developed originally by Berliner et al. (1964) and slightly revised by Christensen and Mills (1967), is shown in Table 3-4.

Indeed, it is conceivable that no universal taxonomy can be applied because task descriptive needs vary from system to system. However, the availability of a generally accepted taxonomy would largely eliminate the need to develop one anew at the start of each project. The comparison of task structures among different systems would also be facilitated.

How useful is the identification and description of operator tasks? Obviously, until tasks are identified, we can proceed no further in system development. However, even without the H specialist, tasks are identified —poorly perhaps—but identified as part of normal engineering development.

If, however, we ask whether the task description produced by the H specialist (as a distinctive output) is crucial for that development, the answer is more ambiguous. For the specialist merely to *list* and *define* tasks without analyzing them in terms of equipment requirements is not a significant advance over what the engineer provides when he develops an operating procedure. When, however, we come to the matter of extracting equipment and training implications from the task, the specialist's methods become important because the engineer almost never performs a comparable analysis.

Table 3-4 Sample Classification of Task Behaviors

Processes	Activities	Specific Behaviors
Perceptual processes	Searching for and receiving information	Detects Inspects Observes Reads Receives Scans Surveys
	Identifying objects, actions, events	Discriminates Identifies Locates
Mediational processes	Information processing	Categorizes Calculates Codes Computes Interpolates Itemizes Tabulates Translates
	Problem solving and decision-making	Analyzes Calculates Chooses Compares Computes Estimates Plans
Communication processes		Advises Answers Communicates Directs Indicates Informs Instructs Requests Transmits
Motor processes	Simple/Discrete	Activates Closes Connects Disconnects Joins Moves Presses Sets
	Complex/Continuous	Adjusts Aligns Regulates Synchronizes Tracks

2-0 *Determine equipment requirements implied by tasks.*

2-1 For each task list the equipment components which can permit the task to be performed.

Although major equipment items have been specified as part of the development of alternative functions/configurations (see Figure 3-1), this specification is not detailed enough for the demands of detailed design. Hence the need to derive specific equipment requirements from tasks; for example, if temperature must be monitored, the equipment components which could provide this information include a meter, or, if only an overtemperature condition must be determined, an indicator light. If power must be applied to energize an equipment, appropriate equipment components would include a toggle switch or pushbutton or, in other circumstances, a lever or rotary control. The number of types of control-display components available to implement a task is quite limited; hence, the task of selecting an appropriate type of component is not difficult.

Selection of the control-display *part* is more complex because of the large number of parts that can be used with apparently equal effectiveness.

2-2 Select the most appropriate type of control-display component to implement the task.

2-2.1 Apply human engineering criteria from military specifications such as MIL-STD 1472 (DOD, 1969) or handbooks, and refer to standard parts lists (or other catalogues) for recommended controls and displays. Apply human engineering criteria and equipment constraints to recommended components.

2-3 Examine the tentative list of control-display components in terms of the following additional factors:

(a) Communications requirements. Any communication requirement obviously implies some means of communicating; hence a device must be provided.

(b) Contingency requirements. These are usually potential emergencies. The problem with contingency requirements is that they do not always occur; hence, it requires some analysis to decide whether a control-display device should be provided for them.

(c) Maintenance requirements. These are of two types: preventive maintenance controls and malfunction indications, the latter falling also under the heading of contingency requirements. It may be necessary to provide special displays for the latter.

(d) Feedback requirements. Analysis is required to determine what information the operator must have concerning the consequences of his control activations, resultant equipment processes, and effects on other systems and the environment in order to perform his job effectively. Display components will be required to provide this feedback. Only task analysis will provide this information.

(e) Input-output characteristics. These may determine both the selection and the arrangement of controls and displays. Some of the characteristics to be considered are the following:

1. input intensity (which may be low, as in weak radar or sonar pips in which case gain controls may have to be provided to allow the operator to adjust CRT beam or screen brightness).
2. input duration (which may be short so that in selecting a CRT the engineer may wish to select a long persistence phosphor to retain the input longer).
3. output force requirements (i.e., energy which must be expended to activate a control).
4. input/output relationships (e.g., which control activates which display); such relationships guide the arrangement of controls and displays.

2-4 Reduce the number of control-display components required by determining which tasks can be implemented by the same control or display. This can be done by

(a) combining an input with an output (e.g., use of switch-lights or illuminated pushbuttons),

(b) combining multiple inputs (e.g., multilegend indicators capable of presenting several messages),

(c) combining multiple outputs (e.g., use of a rotary switch to replace several toggle switches performing related discrete functions).

2-4.1 Determine which tasks are functionally related (have common inputs and outputs).

Comments on Procedure 2-0. The determination of appropriate control-display component *types* does not require great skill or involve great difficulty as long as we are considering relatively common types (e.g., pushbuttons, toggle switches, indicators). The human engineering criteria for selection of control-display types (see pages 265–280 in Morgan et al., 1963) are relatively few and fairly precise. The problem becomes more complex with less common control-display components like CRT's or joysticks or with uncommon application requirements.

It is only when, having decided on the type of control-display component, we must select a *particular* control or display that familiarity with the varieties of control-display hardware and knowledge of engineering requirements become important. Which manufacturers offer which components; what are the characteristics of that hardware (e.g., the degree of resistance offered by a pushbutton); how well is the component manufactured? Experienced H specialists often maintain catalogues of the latest available control-display hardware and are, in fact, more knowledgeable than many engineers about that special hardware.

It also becomes necessary to consider the nature of the internal components behind the control panel that the controls and displays will activate and of the information flow supplied by the equipment. The latter is particularly important in the selection of displays because the displays must reflect that information provided by the equipment that the operator must have to perform his job. Much of this information flow can ordinarily be secured only from the design engineer and/or equipment descriptions.

The role of the H specialist in the selection of the control-display components is, therefore, ambiguous. He has no difficulty specifying common control-display types, but, because that task is relatively simple, he may find that the engineer has anticipated him. The specialist's greatest utility in component selection lies with relatively complex control-display hardware; here, however, he may find that the engineer, despite the latter's difficulties, fails to consult him.

3-0 *Analyze tasks in terms of their demands.*

3-1 Describe tasks graphically (diagrammatically).

The illustrative diagram we shall use is the Operational Sequence Diagram (OSD). The tasks described by the OSD are those which are performed by two or more operators.

The OSD (Kurke, 1961) is simply a means of displaying graphically task interrelationships. This has obvious advantages as a way of making these interrelationships more visible. However, the specialist should not view the drawing of the OSD as an end unto itself; the OSD should be developed solely to implement further task analysis (see below). A complex OSD is extremely difficult and time-consuming to develop although attempts are being made to automate its development. Its use, therefore, presents difficulties if system development proceeds very rapidly. A diversity of opinion exists concerning its usefulness: many specialists dislike it, while others find it quite useful. The technique is sponsored largely by the Navy and is usually required on Navy system development projects in which human factors programs play a significant role. However, similar graphic tools, for example, time-line drawings, are used by other governmental services.

3-2 Categorize the tasks in terms of their behavioral elements.

3-2.1 Determine which tasks fall into the following behavioral categories:
(a) transmission of information,
(b) receipt of information,
(c) storage of information,
(d) delay,
(e) decision,
(f) control operation,
(g) display monitoring.

To perform this categorization it is necessary to apply criteria of output (what is accomplished) and methodology (how it is accomplished) to the individual task; for example, if the output of the task is a message, then the behavioral element involved in the task is transmission of information. Define the task output in terms of the equipment that is operated or viewed by the man performing the task.

It should be noted that the behavioral categories specified by the OSD represent a task taxonomy that is imposed on the user by the OSD methodology. This is not necessarily desirable. The categories emphasize information utilization; although they are descriptive, the description is relatively gross, which may reduce their usefulness.

At this level of task description the determination of the behaviors involved in the various tasks presents relatively few problems. The behaviors specified (e.g., information transmission) are logically deduced from the nature of the task output and the way in which the task is implemented.

3-3 Develop the OSD.

3-3.1 Apply appropriate symbols to tasks. These are shown in Figure 3-2 which depicts a representative OSD (taken from MIL-H-46855, 1968).

3-3.2 List tasks in required order of performance. (This determination has already been made as a consequence of performing previous analyses.)

3-3.3 Determine task interrelationships.

This is a deduction from analysis of inputs to and from each task. If completion of one task initiates another or if they are highly correlated in time, they are interrelated. This should have already been largely performed as a result of completing previous steps.

3-3.4 Determine man-man (e.g., verbal communication) and man-machine links. (This should be a fallout from accomplishing 3-3.3.)

3-3.5 Assign tasks to the individual operators who perform them.

3-3.6 Arrange tasks horizontally and vertically in the order in which they are performed, as shown in Figure 3-2.

3-3.7 Draw task input/output links with lines connecting the individual task elements; for example, a message from operator 1 to operator 2 would be represented as a line from 1 to 2.

3-3.8 Add explanatory verbal material (where required) to the diagram.

3-3.9 Analyze tasks in terms of a comparison of operator capabilities with task requirements.

This last process is the heart of the specialist's task analysis. Note that it is essentially the same as analyses performed for function allocation/verification; however, the task analyses are performed at a much more detailed level than were the earlier analyses.

3-3.9.1 Examine each task in terms of the questions/categories listed in Table 3-5.

See Chapter Five for a further discussion of the analyses involved in these categories. This analysis should be performed by reviewing system operating requirements to determine the demands imposed on the indi-

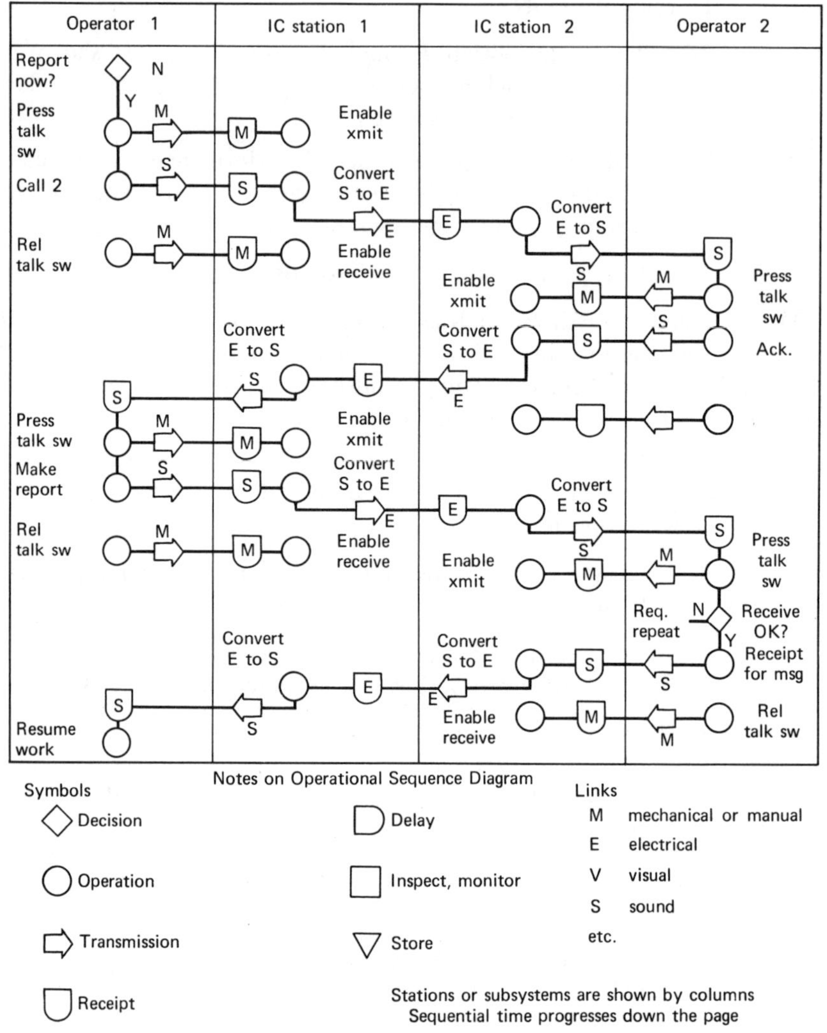

Figure 3-2 Sample operational sequence diagram. Two-station intercom, with station 1 acting as originator.

vidual task and by securing appropriate data on operator capabilities from the behavioral literature.

3-3.9.2 Indicate any task which appears from a comparison of task requirements with operator capabilities to imply potential operator inability to accomplish the task.

Table 3-5 Questions To Be Asked in Performing Design Task Analysis

---

1. Functions/Tasks

   A. Are functions/tasks to be performed within operator capability? Consider requirements in the following functions: (1) sensory/perceptual, (2) motor, (3) decision-making, (4) communication.

   B. Do task characteristics impose excessive demands on the operator?
      1. task duration (possible fatigue effects?)
      2. frequency of task performance (possible fatigue effects?)
      3. information feedback (insufficient operator guidance?)
      4. accuracy (too demanding?)
      5. error probability (supportable?)
      6. error criticality (effect on task performance?)
      7. concurrent multitask requirements (effect of one task on another?)

2. Environment

   A. Events requiring operator response
      1. speed of occurrence (too fast?)
      2. number (too many?)
      3. persistence (too short-lived?)
      4. movement (excessive?)
      5. intensity (too weak to perceive?)
      6. patterning (unpredictable?)

   B. Physical effects
      1. temperature, humidity, noise, vibration (excessive?)
      2. lighting (substandard or special effects?)
      3. safety (problems?)

   C. Mission conditions
      1. potential emergencies (can operator recognize and overcome rapidly?)
      2. mission response characteristics
         (a) accuracy requirements (excessive?)
         (b) speed requirements (excessive?)
      3. event criticality (effect on error probability?)

3. Equipment

   A. Display information requirements
      1. too much to assimilate?
      2. difficult to perceive/discriminate/track?
      3. require excessively fast operator response?
      4. too much memory required?

   B. Control requirements
      1. excessively fine manipulations required?
      2. too much force required?
      3. must be responded to too rapidly?
      4. too many to perform in sequence?

3-3.9.3 When the operator appears on the basis of the comparisons in 3-3.9.2 to be incapable of performing a task, consider alternative ways (man–machine configurations) of implementing the task; it may, for example, be possible to relax demands on the operator by lengthening his required response time or by giving him an additional or a different display, etc.

IDENTIFICATION OF MAN-MACHINE INTERFACES

This section does not refer to development of the over-all man-machine configuration, which was analyzed as part of the function allocation/verification process described earlier. Here we deal only with the immediate man-machine interface which in many cases turns out to be a control panel or a console in which one or more control panels are to be located.

This is not to say that there are no other man-machine interfaces. For example, internal components, test points and test equipment are all interfaces for the maintenance man. Whenever the operator must perform a task involving direct contact with an equipment, that part of the equipment he manipulates or observes becomes for him at that moment the man-machine interface. However, in the majority of cases the most important and frequently encountered interface is the control panel or the console.

In addition, the H specialist's ability to make meaningful recommendations for the design of internal equipment interfaces is extremely limited because he knows comparatively little about the relationship between internal design characteristics and maintenance behaviors.

1-0 *Select appropriate interface components.*

This task has been performed as part of task description and analysis. This is because it is impossible to describe the task without defining, at least grossly, the equipment object of the task activity. As a consequence, the activity described in this section deals with the arrangement of components to form the interface.

2-0 *Arrange control-display components in the form of a control panel.*

2-1 Determine the size of the panel on which control-display components are to be arranged.

2-1.1 Refer to SOW requirements and/or to equipment designer for required panel dimensions.

Ordinarily the specialist is not called upon to evaluate the adequacy of panel dimensions except where these are so restricted that they present problems in incorporating required controls and displays.

2-2 Determine whether spacing between controls and displays may present a problem.

This may not be a problem if the number of controls and displays is small relative to over-all panel size. In many cases, however, spacing does present difficulties. If spacing presents no problem, go immediately to step 2-3.

2-2.1 If spacing is a potential problem,

(a) review the list of control-display components,
(b) count the number of required controls and displays, and determine their dimensions,
(c) determine other constraints on arrangement of components (e.g., internal component dimensions).

2-2.2 Determine the minimum distances required to separate various types of control and display (see MIL–STD 1472 for relevant tables).

2-2.3 Calculate spacing as a ratio between the size and number of control-display components and available panel space.

2-2.4 If spacing is a problem, consider combining related controls and/or displays.

2-3 Examine alternative principles of control panel arrangement.

There are four generally accepted principles of control-display arrangement. These are

(a) arrangement by sequence of operating the equipment (always the preferred method because it is the easiest for the operator),
(b) arrangement on the basis of common functions (next most preferred),
(c) arrangement on the basis of the frequency with which the control-display must be activated or viewed,
(d) arrangement on the basis of the relative importance of the individual control or display.

When there is a definite flow of equipment events, controls and displays may also be arranged on the basis of a symbolic (e.g., schematic) representation of that flow, as in process control panels. In its simplest form control panel layout is a matter of selecting the most appropriate one of these four or five principles. This is, of course, a tremendous oversimplification since very few control panels can be ordered unequivocably by one or the other principle. (See Meister and Farr, 1966 for a more detailed discussion of control panel design.) However, it serves as a good starting point for the layout analysis.

2-3.1 Apply the criterion of arrangement by sequence of activation.

Examine the list of control-display components in terms of the sequence in which they must be activated or viewed. The sequence in which controls and displays must be operated can be determined by referring to the OSD (if one exists); if this does not provide the sequencing information, refer to the design engineer to determine if the equipment forces an operating sequence on controls and displays. (If a forced sequence exists, arrange the components on the basis of that sequence. If not, consider other principles.)

2-3.2 Apply the principle of arrangement in terms of similar or related functions.

Examine sources of information (e.g., FFD's, Task Descriptions, OSD's, etc.) to determine which human functions are similar or related; list the control-display components which perform similar or related functions. Arrange all controls and displays on basis of their functional similarity, taking into account other constraints.

2-3.2.1 Compare the functional layout with the layout based on sequence of activation.

Which one requires more manual activity? Apply link analysis procedures to compare relative manual operations. Which layout appears more likely to be subject to error? Obviously there is a certain amount of intuitive judgment involved in making these decisions. Any attempt to describe the judgmental process in hard and fast rules appears to be fruitless at this time. If it appears that the two layouts should be modified to bring them into consonance, do so.

2-3.3 Apply the principle of frequency of use.

Perform the procedures described in steps 2-3–ff as they apply to the principle of frequency of use. The essential element is to ascertain the frequency with which each control-display must be activated. This may be determined by referring to the OSD, operating procedure, or design engineer. Arrange the controls-displays in terms of frequency of use (e.g., those most frequently employed in the center, those less frequently employed at the periphery of the control panel).

2-3.3.1 Compare the arrangement by frequency with the drawings organized by sequence and functional relationship. If it appears that the sequence and functional relationship drawings should take frequency of usage into account, modify the former in accordance with the frequency principle.

2-3.4 Apply the criterion of importance.

Define importance in terms of error–activation consequences. Examine sources of information for the importance principle (e.g., FFD's, Task Descriptions, OSD's, and other documentation) or consult with the design engineer to determine error consequences. If alternative man-machine configurations have been analyzed as described previously, some data relating to importance of controls-displays should be readily available. Rank the control-display components in terms of importance. Arrange the components in terms of importance (e.g., locating those of greatest importance in the midquadrant of the control panel, those of less importance in the periphery), taking into account other constraints.

2-3.4.1 Compare the drawing based on the importance criterion with all other layout drawings.

2-4 Modify all drawings to develop a compromise layout which includes all arrangement principles as relevant with emphasis given to sequencing and functional relationships.

2-5 Ensure that the control panel has been arranged in accordance with anthropometric requirements.

Select anthropometric principles which apply to control panel layout (e.g., display faces shall not be less than 45 degrees from operator's line of sight). Modify the selected layout where required to ensure conformance to anthropometric requirements.

Comments on Procedure 2-0. When we examine the four human engineering principles of control panel arrangement, we find that they are rather general in nature; consequently, they cannot be used as the sole criteria for layout. They must be modified by other general principles, such as controls-displays that have a common function should be separated from controls-displays that differ in function. In addition, these principles are qualified by a number of relatively specific rules which cannot be utilized for arrangement but which affect that arrangement. The rule that display faces shall not be less than 45 degrees from the operator's line of sight is an example of such a rule.

In most cases panel arrangement does not lend itself unequivocably to one principle or another. In fact, most layouts are based on some compromise among all applicable design criteria. Consequently, all human engineering design principles must be subordinated to an over-all design logic for which precise rules cannot be specified. In fact, human engineering principles cannot be correctly interpreted except in the context of the design logic applied to the development of the individual equipment.

The basic input datum to the determination of the panel arrangement is a formal or informal operating procedure or, if available, an OSD. The operating procedure by itself, however, will not permit the selection of the appropriate arrangement principle because answers to certain questions must first be derived from analysis of the procedure: (a) what is the sequence in which controls and displays must be operated; (b) which controls and displays are functionally related; (c) how frequently is each control-display activated; (d) which controls and displays are most important? When an operating procedure is available, frequency information is usually quite explicit, but a discernable sequence may be difficult to establish when many controls and displays are variably operated. It is even more difficult to deduce functional relationships and component importance because this information is not inherent in the procedure; the specialist may have to refer to his task analysis information or go to an equipment description or the engineer for information bearing on these points.

Although the analysis described in this section suggests that each arrangement principle is sequentially selected and applied, this is an artifact of the description. In actuality all arrangement principles are considered by the human engineer more or less simultaneously. However, based on the H specialist's biases, certain arrangement principles may take precedence over others.

The process of determining which of several layouts is most acceptable is as subjective as the entire arrangement process itself; perhaps even more so. How do we know when we have achieved the most desirable layout? Several alternative arrangements may be equally satisfactory or at least the differences among them may not be significant. Criteria for evaluating the adequacy of a layout are geared to uncovering some design feature which *violates* H principles. As is common in human engineering as a whole, it is easier to know when we have designed improperly than when we have designed correctly. In addition, since human engineering criteria deal with individual design characteristics rather than the layout as an organized whole, they are not very efficient in dealing with the control panel as a unit. Moreover, keeping in mind the inevitable subjectivity of the whole process, what is acceptable to one H specialist may be less desirable to another. The adequacy of our evaluational methods (e.g., checklists) is discussed in Chapter Six.

It should not be assumed from what has been said that the author is downgrading human engineering design activity. All engineering design is essentially a subjective process qualified to a certain extent by the application of logic and data. Principles, rules, and checklists help to formalize and structure this subjectivity; however, to the extent that the design process is a creative process, it may be impossible to structure it as much as we wish.

REFERENCES

Berliner, C, D. Angell, and J. W. Shearer, 1964. *Behaviors, Measures and Instruments for Performance Evaluation in Simulated Environments,* Presented at Symposium on Quantification of Human Performance, Albuquerque, New Mexico, 17–19 August.

Christensen, J. M., and R. G. Mills, 1967. What Does the Operator Do in Complex Systems, *Human Factors,* **9** (4), 329–340.

Department of Defense, 1968. MIL–H–46855, *Military Specification: Human Engineering Requirements for Military Systems, Equipment and Facilities,* 16 February.

Department of Defense, 1969. MIL–STD 1472, *Human Engineering Design Criteria for Military Systems, Equipment and Facilities,* 1 December.

Fitts, P. M., et al. (Eds.), 1951. *Human Engineering for an Effective Air Transportation and Traffic Control System.* Washington, D.C.: National Research Council.

Hagen, W. C., 1967. *Techniques for the Allocation and Evaluation of Human Resources,* Report OR 8735, Martin Marietta Corp., Orlando, Florida, March.

Kurke, M. I., 1961. Operational Sequence Diagrams in System Design, *Human Factors,* **3**, 66–73.

Meister, D., 1966. Human Factors in Reliability, Section 12, in W. G. Ireson (Ed.), *Reliability Handbook.* New York: McGraw–Hill.

Meister, D., and D. E. Farr, 1966. *The Methodology of Control Panel Design,* Report AMRL–TR–66–28, Aerospace Medical Research Laboratories, Wright-Patterson AFB, Ohio, September.

Miller, R. B., 1967. Task Taxonomy: Science or Technology? *Ergonomics,* **10** (2), 167–176.

Melton, A. W., and G. E. Briggs, 1960. Engineering Psychology, *Ann. Rev. Psychol.,* **11**, 71–98.

Rabideau, G. F., J. I. Cooper, and C. Bates, 1961. *A Guide to the Use of Function and Task Analysis as a Weapon System Development Tool,* Report NB–60–161, Northrup Corporation, Hawthorne, California, January.

Morgan, C. T., J. S. Cook, A. Chapanis, and M. W. Lund, 1963. *Human Engineering Guide to Equipment Design.* New York: McGraw–Hill.

Swain, A., and J. G. Wohl, 1961. *Factors Affecting Degree of Automation in Test and Checkout Equipment.* Stamford, Connecticut: Dunlap and Associates.

Whitfield, D., 1967. Human Skill as a Determinate of Allocation of Function, *Ergonomics,* **10** (2), 154–160.

Zorger, P. H., 1965. Reliabiliy Estimation, Section 5, in W. G. Ireson (Ed.), *Reliability Handbook.* New York: McGraw–Hill.

IV
HUMAN FACTORS RESEARCH

This chapter describes the research needed to supply the basic data for performing the analyses of the preceding chapter. Whether or not these analyses are accomplished and how they are accomplished, depends on the availability of data relating system variables to operator performance. If this research is not performed, or is performed inadequately, the H specialist's contributions to development will dry up.

THE RELATIONSHIP BETWEEN PSYCHOLOGY AND H

Before discussing the specifics of a human factors research program, however, it is necessary to indicate those elements that differentiate human factors research from other behavioral research, particularly *applied* or *experimental psychology*.[1] For, if H is simply a special application of psychology, then its methods should be those of psychology in general, and it needs no special research goals or priorities. If, on the other hand, significant differences exist between H and psychology, it must develop a distinctive approach to its research.

To say that human factors research has goals, priorities, and methods of its own does not imply a rejection of psychology as the parent discipline. Like any other behavioral discipline, H builds upon principles describing how individuals perceive, learn, and think. Psychology provides H with information on basic human capabilities, for example, sensory/perceptual, cognitive, and physiological, and this information is, as we have pointed out elsewhere, invaluable for human factors analyses. It

[1] For those who complain that we emphasize the behavioral aspect of H as against its engineering aspects, it should be fairly obvious that, whatever H is, it is not an engineering or physical science although it contains elements of both. We also do not overlook contributions from anthropology, physiology, or medicine, but all these inputs pale beside the importance of psychology.

is inevitable, therefore, that psychological and human factors research overlap at many points.

Psychological research does not, however, answer a set of new questions posed by the man-machine interaction. The primary focus of psychology is the *man,* and its search is for principles which govern the man's behavior. Physiological psychology, for example, seeks to understand the man's internal functioning. Other special interest areas of psychology are concerned with man in relation to objects being learned, responded to, or communicated with. In all of these the goal is to understand how the man functions. The objects with which the man is involved are thought of primarily as stimuli that evoke his behavior, and *it is the behavior which is of interest rather than the objects.* Among these objects are tools and machines; this is a point of overlap with H. However, outside of its role as a stimulus that evokes behavior, *the tool or machine in the psychological framework is relatively unimportant.*

If, for example, a psychologist performs a study in which a new tool or machine is presented to a child or adult, the focus of the study will be in terms of how long the subject takes to learn to use the tool or machine or how he conceptualizes the characteristics of the tool. The characteristics of the tool or machine, except as these provide an effective means of learning about the subject's responses, will be essentially irrelevant in this type of study. The human is the primary focus of interest; the machine is merely another stimulus, which can be replaced, if we wish, by non-machine stimuli like puzzle boxes, visual displays, or cryptograms.

Herein lies a basic difference between psychology and human factors. Human factors research conceives of an ineluctable relationship between the man and his machine. Together they form a system which does not exist in psychology, where the man alone is the system. The test of the proposition is the answer to the question: would there be a distinctive discipline—H—concerned with human performance if the machine did not exist? Clearly not; this would be the province of psychology.

The system as consisting of two major elements—the man and the machine—therefore forms an entity which distinguishes human factors from psychological research. The human factors researcher is interested in the man *only* in his relation to the machine (defined very broadly, of course, as including the environment and culture in which the man operates his machine, the operating procedures, operator training, etc.), and he is interested in this relationship on a reciprocal basis. The psychologist may be interested in the effect of the machine on the man, but the human factors researcher will be concerned also about the effect of the operator on the total system. (This latter effect is quite significant, as evidenced by the many systems in which the human input determines the total output of the system, for example, postal letter sorting machines.)

Psychology is, of course, concerned about systems, but these are what can be termed "man-man" systems; the focus of its system-interest is the role of the individual in, and his effect on, groups of men. Only H̲ pursues the interaction of the human with equipment and equipment operations as a primary element in the system.

There are other differences between H̲ and psychology as related to the system concept; for example, for H̲ the system is always an artificial construction, that is, created by and for men. We can therefore establish specific performance requirements (goals) for man-machine systems; a primary focus of H̲ interest is how well the man can accomplish these goals. We cannot usually establish requirements for man-man systems (except where the psychologist creates his own artificial system, e.g., special languages, learning and testing materials, therapy and play groups). Because the man-machine system is an artificial creation, it can be more readily modified than the natural systems of the psychologist. In psychology only the modification of the human (primarily by training) is of interest; in the man-machine system the modification of the equipment and its operations is a primary goal in preference to modification of the operator. Another way of saying what we have been striving to express so far is that psychology is *man-centered* regardless of the context in which the man functions.

Because the system that the H̲ specialist deals with is an artificial one, with predetermined goals, human behavior is of interest to him largely in terms of the means provided to the operator to accomplish system goals and how well he does so. For the psychologist, performance in relation to a system goal outside the human (particularly one in which the requirements are man-made) is relatively unimportant except in learning.

Since man and machine form coequal elements of the system, the development of the machine by adapting its characteristics to user limitations has as great interest to H̲ as the modification of the man through training. Psychology lacks such interest.

The differences we have identified between psychology and H̲ are, of course, differences in degree, not in absolute kind; under certain circumstances it may be difficult to distinguish these differences.

What are the implications of this difference in viewpoint for the performance of human factors research?

1. Obviously psychological research to determine human capabilities and limitations will be required but cannot solve all or even most

human factors problems because it does not take into account the man-machine interface. Hence research of a specific human factors nature must be performed.

2. This research must deal with

(a) the ability of personnel to perform in the machine environment (e.g., how well operators can perform specified tasks);

(b) the characteristics of the machine and the machine environment as these influence the capability of personnel to perform (i.e., the effect of machine characteristics on the operator performance, how machines should be developed to maximize operator performance);

(c) the relationship between operator performance and system output (system output conceived of as more than or different from the individual contributions of the human and the machine).

3. Because the machine environment is essential to operator performance, it is necessary in testing that performance to reproduce the essential characteristics of that environment as faithfully as one can. This problem does not have the same significance in psychological research because to the psychologist the machine is only a stimulus and can be replaced by a different—presumably equivalent—stimulus. To the extent that the machine environment cannot be easily simulated in the laboratory, it becomes necessary to perform human factors research in the "normal" machine environment, that is, in the operational situation. Unfortunately, the controls possible within the laboratory are much less available in the operational situation; this poses a dilemma for the human factors researcher.

THE HUMAN FACTORS RESEARCH PROGRAM

The single fundamental assumption which underlies the research program proposed is that this program must be oriented toward answering questions posed by system development. System development is not, of course, the only possible framework around which the research program can be stretched. As we saw in Chapter One, one of the goals of the H discipline is to explain how men function in a technological culture; from that standpoint any investigation of the relationship between men and their machine environment (e.g., how much satisfaction machines give their users) is important.

Most critical, however, is the need to support system development. Any research program must, as a minimum, answer the major questions raised by system development (including use of that system). Described in general terms only, of course, these are listed in Table 4-1.

Chapter Three indicated that the major tasks of system development are to allocate responsibility for implementation of system functions to the most efficient man-machine combination and to identify the equipment which will permit the most efficient implementation of these functions. It is assumed that no meaningful allocation can be performed without knowing what the operator is or is not capable of (items 1, 4, and 5 in Table 4-1). Similarly, no man-machine interface can be correctly specified without knowing the effect of equipment parameters on operator performance (items 2, 3, and 6 in Table 4-1). The remaining questions in the table deal with the manner in which operators (singly or as a group) function over time.

Table 4-1 System Development Questions for Which Human Factors Research Is Required

1. What is the operator's capability:
   (a) to perform various functions;
   (b) under various task and environmental conditions?
2. What is the effect of various types of equipment and equipment characteristics on the operator's performance?
3. What equipment characteristics are most critical for operator performance?
4. What physical and physiological limitations does the operator impose on equipment and system design?
5. What is the operator's probability of error: *
   (a) in performing specific tasks;
   (b) under various environmental conditions?
6. How is that probability of error affected by various equipment characteristics?
7. What is the effect of number of personnel on performance of specific tasks?
8. How does the performance of one task affect the performance of concurrent or subsequent tasks?
9. How does the operator's performance of particular functions/tasks vary over successive trials
   (a) in learning;
   (b) in fatigue?

* Error defined as a deviation from required performance.

The major elements of this program are the elements of the man-machine system as described in Table 4-1: *equipment, personnel,* the *functions* and *tasks* they perform (or alternatively, the procedures they employ), and the *environment.* To these have been added as essential elements in examining research, *measures* and *methodology.* These elements have been categorized in a series of tables (4-2 through 4-7), one for each element. These categories permit the classification of any human factors study and will, therefore, serve as a guide to our analysis of the suggested research program.

It is apparent after reviewing these tables that human factors research can be highly varied particularly in the interactions among the six parameters discussed. An almost infinite variety of studies can be performed which can be described as "human factors research." The author considered the desirability of showing these interactions in the form of a matrix table but gave up, appalled at the resultant complexity. Consequently, in line with the criterion of relevance to human factors in sys-

Table 4-2 Classification Scheme for Human Factors Research Studies: The Equipment Parameter

1. Descriptive Dimensions
   (a) design concept [1]
   (b) design characteristic [2]

2. Components
   (a) external [3]
   (b) internal [4]

3. Type [5]
   (a) information-sensing (e.g., radar, sonar)
   (b) information-retrieval/display
   (c) command/control
   (d) weapon
   (e) navigation
   (f) transportation
   (g) communication
   (h) simulator/trainer

4. Size/number [6]
   (a) system
   (b) subsystem
   (c) equipment
   (d) component

[1] The design concept is an attribute of the equipment as a whole or of assemblies of equipment (a system); it is also a way of designing the equipment or system. Examples of design concepts are: level of automation (e.g., automatic vs. semiautomatic), modularity, unitized packaging, self-check capability.

[2] The design characteristic is an attribute of the single component (e.g., toggle switch length, scale markings, color coding of displays).

[3] Largely controls and displays.

[4] For example, amplifiers, circuit boards, power supplies.

[5] Also can be termed function or use. Characteristics in this category tend to overlap; for example, a naval Combat Information Center (CIC) is a combination of (a), (b), and (c). The size of the system described by any of these categories may also vary from very large to very small; a weapon, for example, is both a rifle and an ICBM.

[6] Only those major categories which will be of interest to the H researcher are noted; for example, piece parts are not included in the list because the specialist does little with these.

tem development, only those research areas that can benefit that development have been examined.

The following sections are divided according to the major parameters of equipment, personnel, functions/tasks, etc. Each section is preceded by a table, which lists the variables included in the relevant parameter, and is followed by a discussion of the parameter. This discussion emphasizes the research that has been performed relative to the parameter and the research that needs to be performed. Special attention is paid to the relevance of the recommended research to the problems faced by the industrial H specialist. In other words, we attempt to answer the question, why is this research important?

One disclaimer before we begin. This chapter does not attempt to provide a comprehensive review of the human factors research literature despite the generalizations presented. This would require a separate book larger than the present one. In consequence the references supplied in the discussion are illustrative only.

In fact, the author is not aware that a completely comprehensive review of human factors research has ever been attempted. For partial reviews the reader should refer to articles in the *Annual Review of Psychology* (Poulton, 1966; Melton and Briggs, 1960; Chapanis, 1963; and Fitts, 1958).

EQUIPMENT

A fundamental assumption underlying the H discipline is that equipment characteristics are a major determinant of human performance in systems and that design concepts and characteristics should be selected in accordance with the operator performance to be anticipated from these. We would, therefore, expect the discipline to have readily available a substantial body of research describing equipment-performance relationships. A body of research does exist, but it has several deficiencies: there are gaping holes in that research; much of it is inappropriate, and that which is appropriate describes primarily design characteristics as differentiated from design concepts. Design concepts refer to the logic with which the entire system or equipment has been designed; design characteristics describe the attributes of individual equipment components such as controls and displays.

Perhaps because it is easier to deal with molecular rather than molar attributes of equipment, human factors research has dealt largely with design characteristics rather than design concepts. We can therefore find in the literature such data as the effect of scale markings on performance (Grether, 1947) and the optimal separation between knobs (Bradley, 1969). However, the literature deals sparsely, if at all, with major organizing principles (design concepts) which guide the design of the total equipment or system.

There are, for example, relatively few data concerning the organization of controls and displays (although see relevant research on control panel layout principles by Chapanis and Lockhead, 1965, and Fowler et al., 1968). There is practically none about design concepts. Some work has been performed on level of automaticity (Swain and Wohl, 1961) and troubleshooting processes (Highland et al., 1956, and Rigney, 1964).

It is, of course, more difficult both conceptually and experimentally to handle design concepts, yet it is precisely data relating design concepts to operator performance that the specialist must have if he is to do his job completely and adequately. At the start of design the parameters the engineer manipulates are relatively gross ones (e.g., number of assemblies, level of automaticity); consequently, the H inputs which must be supplied for this design stage must describe these more molar parameters. To do this requires appropriate research.

Before the needed research can be performed, however, the parameters themselves must be specified and classified. In effect, this requires the development of a taxonomy, a way of descriptively classifying from a behavioral orientation the attributes that characterize equipment. This has not yet been done in any satisfactory manner.

The first stages in the development of a science is the classification of events, phenomena, or stimuli. The way the scientist classifies reveals the way he conceptualizes the operations of his discipline. In the case of H it is difficult to discern a distinctly behavioral cast to the way in which system elements (and particularly the equipment parameter) are ordered.

A number of taxonomies of equipment characteristics exist (see Chambers, 1969, for a representative sample of taxonomies not only for equipment but for human performance in general). However, they stem from engineering usage and have relatively few behavioral implications. We need only look at the parameters in the Data Store (Munger et al., 1962) to see what human factors researchers have been studying: joystick length, amount of toggle switch throw, number of indicators, etc. Although these are important for detail design, they are too molecular for the initial stages of development in which major design decisions are made.

To talk of equipment in terms of subsystems, assemblies, components or amplifiers, circuit boards, and controls does not imply operator performance. The taxonomic categories needed to describe equipment from a behavioral standpoint are attributes of equipment which presumably influence operator/technician performance, in other words, something other than a listing of components and component elements such as meters and scales. (We do not suggest that the original terms be discarded; they are needed to communicate with engineers. Something more behavioral is needed, however, for research guidelines.)

What equipment attributes are important then? The following list contains only some of the dimensions which *may* be important; we do not know what is important because relevant data are largely nonexistent. It is presented only to indicate what an equipment/behavioral taxonomy *might* contain. Some of the possible behavioral implications are shown in parentheses.

1. *Number* of controls-displays and internal components. (The larger the number, the greater the likelihood of error.)
2. *Arrangement* of equipment:

   (a) layout of controls and displays;
   (b) location of test points;
   (c) accessibility of internal components.

   (When arrangement is nonoptimal, error probability increases.)
3. The *amount of information provided* by equipment (including feedback displays). (The more information provided—within limits, of course—the more efficient the operator's responses.)
4. *Demands* imposed by the equipment on the operator/technician (behavioral requirements):

   (a) behavioral functions required of personnel;

      (1) perceptual requirements;
      (2) motor requirements;
      (3) cognitive requirements;

   (b) required accuracy of response;
   (c) required frequency of response;
   (d) required speed of response.

   (The greater the demand, the poorer the performance.)
5. *Type of stimuli* presented by displays:

   (a) structured (e.g., alphanumerics or scale values) as contrasted with
   (b) unstructured (e.g., raw sonar/radar signals).
   (Unstructured stimuli impose a greater perceptual demand on operators; hence they lead to greater error-likelihood.)

6. *Packaging* of equipment:

    (a) individual assemblies;
    (b) unitized equipment.

    (Individual assemblies may be associated with greater error-probability.)
7. Equipment *complexity,* which is difficult to define but which may represent the interaction of the number and arrangement dimensions. (Complexity is inversely related to performance, for obvious reasons—see Harris, 1966.)
8. *Level of automaticity* (amount of operator interaction required with equipment).
9. *Level of troubleshooting* (component level to which troubleshooting is required). (The more molecular the level, the greater the behavioral demand and the greater the error-probability.)

These are, of course, not the only equipment-behavioral dimensions that can be conceptualized, nor are the performance implications of these dimensions anything more than illustrative suggestions. It is possible also that each of the dimensions listed are themselves complexes of lower-order dimensions and upon further experimental study would be broken down further.

Note that each parameter listed above has potential implications for operator performance. The primary criterion for development of an appropriate taxonomy of equipment dimensions should be that it is possible to deduce behavioral implications of the dimensions included in the taxonomy. Obviously, if the researcher cannot conceptualize a behavioral consequence from the effects of an equipment attribute, it is probably pointless to include it as a dimension.

If it appears as if the author has unduly emphasized the need for conceptualizing a behaviorally-oriented equipment taxonomy, it is because such a classification scheme must be developed before the researcher can attack his problems meaningfully. In the past researchers have made do with relatively obvious equipment characteristics, and the results have not been overly productive. Most of the specific quantitative data we have on the performance correlates of equipment characteristics, as represented in the Data Store (Munger et al., 1962), for example, is a fallout of experimental studies in which it just happened (fortuitously) that some equipment variable was included in the study.

The goal of human factors research on equipment parameters should be to develop equations which permit us to translate these parameters into anticipated operator performance. Implicitly and explicitly, the specialist is asked to make such transformations every day; for example, when he critiques a design, it is assumed that he does so because he knows what characteristic *X* means for operator performance; not merely because he knows that characteristic *X* is important or that, as it occurs in the design, it is inadequate but because he knows *how* it is important and what the effect of that inadequacy—large, small—will be on performance. This is a far cry from what we are able to do today. Presently our design evaluative principles are almost all subjectively and nonquantitatively based.

Moreover, since he is rarely faced with just a single equipment characteristic when he looks at a design, but rather with a configuration of such characteristics, the specialist must also know how the individual characteristics relate to each other and how they combine to form higher order design concepts. A design concept like complexity probably cannot be tied unequivocally to any single equipment characteristic, yet it impacts upon a variety of characteristics. As a consequence, the specialist must know, for example, how the number of controls interacts with their arrangement in terms of operator performance in operating those controls. Unless this is known, it is almost impossible to evaluate a design configuration and make meaningful recommendations. Again, unfortunately, the data we have seem to relate solely to the individual characteristic.

The need to supply quantitative performance equivalents for equipment characteristics means that the researcher cannot be satisfied simply with determining that a parameter is or is not important to performance. The system developer will ask, "Important how? In terms of what performance effects?" To determine that display size, for example, has an effect on operator performance, it may be sufficient for the researcher—but not for the engineer—simply to compare performance with two widely different display sizes. If the performance differences are significant, the question is apparently settled; if they are not, the question is also settled in reverse; it is assumed that the parameter is unimportant and further inquiry is suspended.

To say, however, that display size *A* is more desirable than display size *B* in a particular system a series of displays representing roughly equivalent intervals on a scale of display size must be tested. It is important to be able to plot a graph of display sizes against operator performance because the shape of the curve may well be nonlinear. Generally, however, studies of the effect of equipment parameters on operator performance have presented only two or three points on the dimensional continuum, which makes them difficult to use for predictive purposes; for example, if the specialist wishes to advise an engineer on joysticks, he should have available data on the entire range of joystick lengths rather than a selected few. (This is not an endorsement of research on joysticks.) This has rarely been done.

The fact that it is necessary to know the relationships among equipment characteristics forces our research to assume a multivariate character. To use display size again as an illustrative parameter, performance as a function of differences in display size is also influenced by symbol resolution, by ambient lighting, by the size of the displayed symbol, etc. Hence display size can be studied meaningfully only in interaction with these variables. If the experimenter includes all these variables in the same study, that study may be overly complex, so the experimental situation is artificially simplified by controlling these variables, that is, by not allowing them to vary. However, this merely produces an inefficient answer since the engineer cannot hold these variables constant when he makes tradeoffs in the actual design situation.

Necessarily this suggested approach to data collection is much more lengthy than the one commonly employed, but human factors information requirements demands such an approach. Note that it does not require a change in the basic strategy of experimentation, merely in the thoroughness with which that strategy is implemented.

It is characteristic too of human factors research that it has dealt much more with external (e.g., controls, displays), than with internal, components (e.g., circuits, amplifiers). This is understandable since the most obvious man-machine interface in operating equipment involves controls and displays. The basic concepts with which initial design is concerned, however, often involve equipment characteristics other than those in controls-displays (e.g., how does packaging influence performance time and errors?). Moreover, with increasing automation of operating functions the focus of human involvement, and hence the system development questions raised by the specialist, often pass to maintenance operations (internal components). A question that must be answered is whether there are equipment dimensions that are significant for maintenance as distinguished from operator behavior; and if so, what are they?

Although equipments and systems differ in terms of the functions they perform, research to contrast performance on the different types of systems (e.g., communications vs. command/control) does not seem overly profitable if only because many systems are functional hybrids.

Another variable is, however, quite important; this is the size variable. This can be phrased in terms of a question: do the factors influencing operator performance change systematically as the size of the equipment unit (i.e., as we proceed from component to equipment to subsystem to system) increases; and if so, in what way? Obviously, different behaviors are required as one ascends the scale from component to equipment (e.g., troubleshooting vs. operation), but do we find similar major differences in moving from equipment to subsystem to system? Are there equipment attributes which are significant for system operation which are not for equipment operation? The question is important because the basic design concept which the specialist wishes to influence often deals with a subsystem or indeed an entire system for which behavioral concepts and performance data are quite vague.

Table 4-3 Classification Scheme for Human Factors Research Studies: The Personnel Parameter

1. Physical characteristics	3. Number
(a) age	(a) individual
(b) sex	(b) crew
(c) race	
(d) physiology/anthropometry	4. Background
	(a) training
2. Psychological characteristics	(1) amount
(a) aptitude	(2) type
(b) skill	(b) experience
(c) personality	

PERSONNEL

Since the following discussion deals with personnel variables listed in Table 4-3, the reader would do well to refer from time to time to this table.

It would be quite untrue to say that their *physical characteristics* exert only an unimportant effect on personnel performance, but, with the exception of *anthropometry*, the need for research on these characteristics is minor. There are, of course, age variations among operators, and it is well known that faculties like vision and hearing change with age (see Edholm, 1967, Chapter 16, for a representative discussion), but these changes do not bulk very large in most man-machine operations except to serve as selection criteria in very skilled and hazardous occupations like piloting.

The same is true of *sex* and *race:* the differences between men and women exist largely in physique and strength and in any event do not limit design significantly especially since most equipment operators and maintainers are men. There may be a few studies to determine differences between the races in terms of their ability to handle machines; the author simply knows of none and, in any event, except in considering the relatively undeveloped countries of Africa or Asia, race and nationality appear not to be important factors in man-machine operations (although see Lippert, 1968).

Research on the physiology of work output has been comparatively lacking in this country although it plays an extremely important role in Europe (from an ergonomics standpoint) (see Murrell, 1969). One reason for this lack of interest is that American research has aimed at the development of highly advanced systems in which physical exertion has been minimal.

In contrast, ergonomic research has to a much greater extent focused on industries in which physical labor plays a much more important role, such as steel factories (Laner, 1961). It is unfortunate that more ergonomic research is not performed in this country because there are still many jobs involving strenuous physical activity (e.g., longshoring, the post office). In any event, much more should be known about the degradation of performance resulting from prolonged physical activities.

However, if one were to assign priorities to research, ergonomic research, particularly that involving physiological measures of output, would occupy a rather secondary position on the author's list. It is difficult to see how changes in blood pressure and breathing, for example, can be very enlightening about the way in which computer or keypunch operators function (except possibly as indicators of fatigue, but performance indicators signal the onset of fatigue even earlier than physiological indicators do). Of course, physiological changes do occur in parallel with more overt performance indices, but the information the former provide relative to the *design* of equipment is rather obscure.

Anthropometry occupies a rather special situation in human factors research. The fund of information available is fairly extensive (see Chapter 11, Morgan et al., 1963) but must be periodically updated because of continuing changes in body structure. The influence of anthropometry on design is quite obvious; physical characteristics are the most obvious limitations to which equipment is designed (e.g., for two-handed rather than three-handed use). However, anthropometry acts largely as a constraint on that design so that once certain basic requirements (e.g., arm reach) are complied with, the importance of this variable diminishes.

In any event the design questions which anthropometry can answer are relatively simple compared to those described in the Equipment section. Hence the need for extensive anthropometric research is not considered of overriding importance.

If research on the physical characteristics of the operator occupies a rather secondary status in our view, the same cannot be said for *psychological* characteristics, principally aptitude and skill, which play an important role in system design and operations. The primary distinction between the two is in phasing. Aptitude is capacity before it is trained; skill is capacity after training. Alternately, we can think of aptitude as a potential for performance whereas skill is actualized performance.

Very little human factors research has been performed in exploring aptititude for machine operation as a factor affecting performance, although personnel psychologists have created tests that fall into that category. The H specialist is not much concerned with aptitude since systems are not currently designed for personnel of varying aptitudes. He is, however, vitally interested in the equipment-skill relationship.

The specialist utilizes the equipment-skill relationship in either or both of the following ways:

1. He wishes to minimize the skill level and type of skill required by including equipment characteristics in his design which require reduced skill and noncognitive functions.
2. Once equipment has been designed, its characteristics may require a particular level and type of skill, which must then be described in terms of selection and training requirements.

In either case it is necessary for the specialist to be able to pinpoint certain equipment features that are related to skill characteristics. In view of the generally accepted proposition that high skill demands impose a severe burden on the upkeep of systems (e.g., human errors leading to equipment malfunction), equipment design that minimizes skill requirements has obvious cost-effectiveness. Research on the relationship between skill parameters and design should therefore have high priority.

The research requirement exists for other than purely theoretical reasons; almost all military customers demand of their contractors (explicitly or implicitly) that equipment be designed to minimize the need for highly qualified, highly trained personnel. They may even go so far, as does the Air Force (but only occasionally), to specify that the personnel who will operate and maintain the system must be of such and such a skill level.

The only problem is that, as Meister et al. (1969) have shown, engineers do not know how to design to meet specified skill levels or even how to reduce skill level requirements generally. When they consider the matter at all, they operate on the basis of certain stereotypes: that, for example, highly skilled maintenance men should be able to troubleshoot to the circuit level or receive information in the form of quantitative displays; or that unskilled men should be given go-no/go (qualitative) displays and asked to troubleshoot only to modules. They conceptualize maintenance skill in terms of knowledge, troubleshooting, and flexibility, with troubleshooting representing a common thread running through the other factors; but these are *logical* rather than empirically derived relationships, and beyond everything else they are qualitative rather than quantitative relationships.

In truth, we have almost no useful research that describes these relationships. There are, of course, studies analyzing the skills required to perform certain complex man-machine tasks (e.g., Locke et al., 1965), but none of these relates the skill dimensions investigated to specific equipment variables or tells the engineer how to include skill variables in design.

As in the case of the equipment parameter, it is first necessary to conceptualize the behavioral dimensions of skill before one can proceed to tie these to design features. Finley et al., 1970, point out that this is an analytic task of great difficulty. The general psychological literature on skill (see Guilford, 1967, as a representative example) has classified literally dozens of highly specific skill dimensions, including dynamic strength, numerical adequacy, finger dexterity, etc.; but it is difficult to see their relationship to equipment variables that are far more complex. Not only is there something nonparsimonious, and hence distasteful, about the concept of 75 or 100 skill dimensions, but the more specific a skill dimension is, the more difficult it becomes to relate it to a design variable. For example, what are the implications of such skill dimensions as concept fluency or gross body equilibrium for equipment design and operation?

The *personality* of the man-machine operator has attracted little research attention although obviously personality has some effect on system operations, particularly in the maintenance area, where it may form an essential component of the technician's approach to troubleshooting. The more unstructured the job, and the more idiosyncratic factors are allowed free play, the more we would expect this parameter to influence performance. In the scale of relative priorities, however, research on personality relationships to man-machine operation should take a secondary place if only because there are many more pressing basic problems.

Many researchers have felt that additional (or at least different) behavior occurs when men work in groups as opposed to when they work as individuals. Certainly there are additional functions to be performed, most obviously, communication and task coordination, but other peculiarly group dimensions, such as dependency, cueing, and saturation (Glaser, 1955), have also been conceptualized.

Hence the importance of the *number* variable and the reason why a fairly respectable amount of research has been performed on workgroups or crews. Much of this research is again, unfortunately, not very relevant to requirements for the design of equipment although it may well be for the development of system procedures.

For the H specialist working in system development, the *personnel-number* questions for which he needs answers are directly related to system variables:

1. Given an equipment configuration, how many men are required to perform a given task?
2. Given a specific number of men as operators of an equipment, how should that equipment be configured to produce more efficient performance?
3. How is the probability of error-occurrence influenced (increased or decreased) by the number of men working as a group?
4. How do we *arrange* the job (in terms of developing operating procedures) to increase group performance efficiency?
5. How do we *train* the group (as distinct from the individual) in terms of required personnel interactions?
6. Assuming we need to test group efficiency, how do we measure the interacting processes involved in crew performance (as opposed to measuring the individual)?
7. Is it possible to trade off number against skill so that a larger number of less skilled operators will perform as effectively as a smaller number of more highly skilled operators?

It may well be that these questions are extensions of more basic ones and that the starting point for research to answer these questions is to conceptualize (and then verify by testing) the underlying dimensions involved in crew activity. The fact of the matter is that no research presently available suffices to answer these questions adequately. Certainly, if we were to ask the question, how should a system be designed to take account of significant group factors, the specialist could provide no meaningful answers. It is of course conceivable that the group variable is not important in system design, but implicitly we feel that it should be.

Although the author would not assign the highest priority to research on crew variables, the problem is important enough to warrant more sustained examination than it has received in the past.

The *training/experience* variables function much as does the *skill* variable. From a productivity (cost-effectiveness) standpoint, the H specialist would like to design equipment to minimize the amount of training and on-the-job experience required to operate and maintain the system effectively. As in the case of skill, the specialist has a two-fold task

(a) to influence design during development in such a way that the duration of required training is reduced;
(b) to predict the appropriate training content and duration for a system configuration that has just been designed.

The first problem has never been empirically solved because the relationship between equipment and procedural characteristics and training is only hypothetical at the moment, and in fact few, if any, training specialists have been able to influence the design engineer to modify his design concepts.

The second problem is solved largely on the basis of subjective judgments that constantly require trial and error revision. It is characteristic that, after the training curriculum for a system is developed, it is repeatedly modified to satisfy operational demands for revision. The essential element which is missing is a set of concrete, empirical relationships between equipment, procedures, and training requirements (content and duration).

Despite the large amount of what can only be called "mythologizing" literature, which assumes that specialists know how to influence design by training considerations and how to derive training requirements from that design, as far as the author knows there has been no useful research relating equipment characteristics to the training required although the fact that the relationship exists appears to be obvious. Articles have been written describing the methodology involved in deriving training requirements from equipment (e.g., Crawford, 1962), but these are largely theoretical. Perhaps the relationship between equipment/procedures and training requirements is assumed to be so completely logical that, given the equipment and its procedures, the required training automatically falls out. We suspect that in actual practice the process is highly intuitive. Mackie and Christensen (1967) suggest on the basis of their evaluation that the very extensive research on training methods has had little impact upon actual training processes.

As in the case of the Equipment parameter, the starting point for establishing any valid relationships between training and equipment characteristics seem to be the definition of the dimensions (both for equipment and training) to be associated. Some preliminary work has been done in attempting to develop a taxonomy of tasks based on learning functions (Cotterman, 1959), but no effort has been made to develop a comparable list of equipment characteristics which may have an effect on training.

The area of training is important for human factors research because it strikes at the heart of the assumption that equipment design must be tailored to the operator. Part of that "tailoring" involves designing the equipment in terms of its training implications. It is obvious that, even though one simplifies equipment as much as possible, the operator will require training in its use. Consequently, if we cannot establish a valid relationship between equipment design and the training needed to operate that equipment, then the "tailoring" that the specialist and the engineer perform on equipment is quite restricted.

In contrast to training, experience is a much less feasible parameter for research since it varies with the individual (if only because it is viewed through and transformed by the operator's attitudes) and cannot readily be controlled. Very little is known about the effect of experience on performance except that its effect is assumed to be positive since so many people swear to its efficiency and refer to it as the source of their accomplishments. Recent research by Meister et al. (1970b) on the performance of aircraft autopilot maintenance suggests, however, that experience as a function determining that performance is important only in the relatively short period of time between the technician's graduation from training school and his achievement of some minimal level of proficiency—an on-the-job apprenticeship as it were. That time period varies with the complexity of the tasks to be learned, of course, but for moderately difficult work we can think of a period of about six months. Once the technician has become reasonably proficient in his tasks, further experience does not seem to influence his performance significantly. This does not mean that there are no significant differences between students and experienced personnel because there are; but it does indicate that experience as a factor to be considered is important primarily in the early stages of a technician's working life. Similar conclusions can be drawn from a study of operator error leading to equipment failures (Meister et al., 1970a) and from a study of maintenance by Crumrine (1968).

The fact is that we simply do not have enough information to deal with experience objectively; perhaps we should not even try until more basic questions are answered. It is possible also that, if training could be improved substantially, the experience variable would become less important. Presumably we rely on experience only because training is less effective than it should be.

From a purely pragmatic standpoint (i.e., the cost of training and its relative lack of effectiveness) we must assign a relatively high priority to any research that attacks the relationship between training and system design in concrete terms.

Table 4-4 Classification Scheme for Human Factors Research Studies: Functions/Tasks *

1. Type
   (a) operate equipment
   (b) maintain equipment
   (c) perform manual tasks in support of (a) and (b) [1]
   (d) communicate information
   (e) instruct personnel
   (f) two or more of the above in combination.
2. Behaviors required
   (a) sensory/perceptual
   (b) motor
   (c) sensory/motor (psychomotor) [2]
   (d) cognitive
   (e) communication
   (f) two or more of the above in combination.
3. Task Characteristics
   (a) task composition [3]
      (1) individual
      (2) group
   (b) task duration [4]
   (c) task criticality (effect on mission performance, system availability, etc.)
   (d) task frequency of occurrence
   (e) task interrelationships [5]
      (1) independent
      (2) dependent
      (3) concurrent
      (4) sequential
   (f) task length (e.g., number of procedural steps)
   (g) task pacing [6]
      (1) system/machine paced
      (2) individual/self-paced

* It is assumed that the reader is aware of the definition of these terms; they are defined in Chapters Two and Three.

[1] For example, recording failure information on forms preparatory to or following troubleshooting or drawing replacement parts from supply. In other words, activities which, in the strict sense of the term, are not actually equipment operation or maintenance but which are required to perform them.

[2] This category involves (2a) and (2b) in combination as in tracking a target using a cathode-ray-tube display or typing a letter; (2a) and (2b) individually refer to sensory and motor activities performed alone.

[3] Is the task one which an individual or a group performs?

[4] How long the task takes either in absolute time (e.g., 5 seconds) or in descriptive comparative terms (e.g., prolonged, momentary).

[5] The performance of an individual task may depend upon the performance of one or more prior tasks (the task then is dependent); or it may not depend upon prior tasks (it is then independent). One task may be performed at the same time a second task is performed (it is then concurrent). A task may be either dependent or independent as well as concurrent or sequential.

[6] Refers to speed and accuracy criteria imposed on task performance. If the task is self-paced, the operator performs with the speed and accuracy he feels are most comfortable; if it is system/machine paced, the operator works to speed and accuracy standards established by someone other than himself.

FUNCTIONS / TASKS

System/equipment design and the functions/tasks performed by operators are reciprocally related. In very early system development the major operator functions to be performed often determine the equipment to be developed (e.g., to detect sonar signals requires that, in the present state of the art, the operator be provided with a CRT among other devices). Later, when equipment has been developed, its configuration often determines the detailed tasks to be performed in operating and maintaining the equipment (e.g., a control panel layout in part determines its operating procedure). On the other hand, the system cannot (without risk of mission failure) require that a function beyond operator capabilities be performed, nor can the particular function/task assigned to an operator unilaterally determine the nature of an equipment configuration.

The H specialist's need to know about the relationship between system/equipment characteristics and functions/tasks (hence his need for research data describing that relationship) is dictated by the problems he must solve during the development process. Each of the system development questions cited in Table 4-1 involves the performance of functions/tasks.

Function Allocation

Theoretically (and we say theoretically only because the goal is rarely achieved) the specialist assists in deciding the mix of functions performed by the operator and his equipment. Traditionally this is called "function allocation" and has been described in some detail in Chapter Three.

In function allocation theory the decision to use a man in a certain role presumably depends on how efficiently that man will be able to perform certain functions and tasks. This efficiency is compared explicitly with that of an equipment which might perform the same function. The theory says that, if human capability for performing a system function is high, the chances of the man being assigned that function is high (all other things being equal); and vice versa. This means that a body of research is required which permits the specialist to say: under specified conditions the average operator can perform such and such functions/tasks to the ______ level of performance.

In the real world of the H specialist the comparison between human and equipment capability rarely occurs for two reasons:

1. The data needed to assess the efficiency of human performance often do not exist. True, comparable quantitative data for equipment are also scarce but are not quite as improbable as human performance data.
2. Moreover the engineer is rarely unbiased enough to make a conscious comparison of human and equipment capabilities.

The two reasons interact: although the engineer's bias toward equipment solutions of man-machine problems is quite evident, it is heightened by the specialist's lack of usable human performance data. We may ask therefore whether, assuming the specialist had the requisite human performance data, the engineer would pay any more attention to these data than he does at the present time to the general principles (Fitts, 1951) advanced by the specialist? The author believes that the engineer is rational enough that, confronted with adequate quantitative evidence, he will consider the data in selecting a man-machine configuration.

We have emphasized the specialist's role in function allocation since it is in this area that he can most effectively influence basic design (assuming he has the tools). Elaborate procedures (e.g., Teeple, 1961) have been and are being developed to enable the specialist to perform function allocations, and a certain mystique has grown up about this activity; this is, however, hardly justified. The myth exists that function allocation requires special methods and decisions when in fact the process is rather simple, provided the specialist has the requisite data.

We see the function allocation methodology as being largely a problem of having operator performance data. If we can specify how effectively a man can perform different types of tasks under varying conditions, the methodology of function allocation is reduced to comparing quantitatively alternative ways of mixing man-machine elements (as in Chapter Three) and giving the design engineer the relevant performance data. In short, the requirement for function allocation as a highly sophisticated mathematical technique or model is essentially a false problem which may disappear if the specialist has appropriate quantitative data on task performance. Given the requirement to indicate how the man should be used in the system, he can say to the engineer: "Here are the different tasks which would be required of the man under the various system configurations which you are considering. Here are their expected levels of performance (e.g., in frequency of expected error or error rate). From the human factors standpoint, all other things being equal, configuration Y is better than X and Z because the error rate is least and because fewer catastrophic failures can be anticipated."

Equipment Design

Once the human has been allocated a particular functional role in the system, it is necessary to determine how equipment should be designed to implement this role. In its commonest form we see this in the requirement to lay out a control panel to perform functions like navigate, track, or initiate.

In order to make meaningful recommendations for design, the specialist must know the effect of the recommended design characteristic on the operator's task performance. The choice of a control-display component, for example, should be based on the operator performance to be anticipated with that component. Similarly the location of that component (e.g., in a control panel) should be determined by data on the effects of different locations on operator performance.

Although it may be permissible to be intuitively creative in one's design, the choice among alternative design configurations can be made meaningfully only on the basis of the operator performance expected from the configuration. This requires data—of which we have too few.

Procedure Development

Later the specialist will assist the design engineer with the development and (much more so) the evaluation of operating and maintenance procedures. This too is more a goal than an accomplished practice, again because the specialist lacks the tools to do the job. Procedure development and evaluation are also matters of knowing how well the operator is likely to perform a given task.

To evaluate the adequacy of a procedure requires the specialist to determine whether the demands imposed by the procedure are excessive. Beyond the intuitive use of his experience in making such judgments, the specialist must rely on data which inform him how well the operator can perform tasks under specified conditions (including those which are excessive). When the equipment configuration permits alternative ways of operating/maintaining the equipment, the specialist's assistance in procedure development consists of evaluating alternative procedures by contrasting operator performance with these alternatives; for example, performance on procedural task A is expected to be .9950; on procedural task A′, .8743. All other things being equal, the choice of task procedure to be selected (A) is obvious. Just as in function allocation and equipment design, procedure evaluation requires information on operator performance to determine whether the procedure is acceptable.

Training

Finally, from the nature of the task description the specialist would like to draw certain conclusions concerning the training which should be provided the personnel who will operate and maintain the equipment.

Here the question is not a matter of performance capability but of determining those characteristics of the task which have implications for training. It was pointed out in the previous section that the most crucial need with regard to deriving training requirements from equipment was the conceptualization of certain attributes or dimensions which were meaningful for training. Exactly the same thing must be said for deriving training requirements from task data. If we consider the various characteristics of the task (e.g., frequency of task performance, task criticality, and basic operator functions), which of these imply what and how training should be conducted?

The bland assumption in governmental specifications like AFSCM 80-3 (USAF, 1967), that training content and methodology *will* be derived, regardless of how much or little we know about the area, seems to imply that the methodology already exists; if so, it is superficial and highly intuitive. Characteristically we derive a training requirement in the following way: assuming the task is described as "Switches fuel from in-board to out-board fuel tank (in aircraft)," the training requirement is usually phrased as, "Requires knowledge of fuel management procedures." In other words, the task description is changed slightly to reflect the need for knowledge to perform the task. There is usually little consideration of the possibility that the task may make multiple and concurrent demands on the perceptual, motor, or cognitive abilities of the operator or that the task is not necessarily, or even usually, the same thing as that which must be trained to perform the task.[1] An adequate method of deriving training requirements from tasks simply does not exist at the moment.

It is possible now to identify certain research requirements in relation to functions and tasks. The first and certainly the most critical is *to develop a body of human performance capability data describing how well the average man can perform tasks under specified conditions.*

The requirement for specific research merely to record performance may strike the reader as being somewhat naive. Does not every human factors study provide performance data? Unfortunately, no. One reason is that the tasks utilized in many human factors studies are not meaning-

[1] To give the devil his due, though, this lack of consideration derives more from an absence of a way to conceptualize these factors than from lack of interest on the part of the training specialist.

ful in man-machine terms. If the task whose performance is being measured is an artificial one (e.g., tossing quoits), it supplies no usable information for the specialist who is concerned with functions of operation, detection, etc. Obviously, underlying all tasks is the same behavioral substrate. Attempts are being made to equate man-machine tasks with behavioral ones, but this process, even if successful, will require a prolonged effort.

Another reason is that in many studies task performance data are gathered for only a relatively few trials. A third reason is that many human factors studies do not report raw data but rather the statistical transformations of these data. Sometimes we can work backwards to the raw data, but many times we cannot. Fourth, data are gathered largely under the artificial conditions of the laboratory rather than in the real world environment.

Unfortunately, as was pointed out earlier, research which *specifically* addresses itself to the question of how well operators perform tasks and hence deals with meaningful tasks measured over a reasonable number of trials is almost completely lacking. H̲ research, like most psychological research, has usually been performed merely to determine whether a parameter is significant or not. Consequently the research not only deals with relatively few data points, but seems curiously unrelated to real system development questions.

The only significant movement to secure performance data has been emphasized by workers in what has been termed "human reliability" (Meister, 1967); and that effort has been oriented toward the performance correlates of equipment variables rather than tasks. When applicable task/performance data are secured, it is primarily as a fortuitous fall-out from studies performed for other purposes. Thus, in their compilation of available data up to 1962 Munger et al. (1962) could find only 164 studies which supplied usable data.

There is also one additional problem. We pointed out in Chapter Three that the way in which we describe the task is largely a matter of our purpose in describing it. However, it makes a difference to any performance inventory we develop whether we describe the task at a relatively molecular level (e.g., adjust torque with wrench) or at a grosser level (adjusts equipment). The major difference between these two task descriptions are the parameters implied. The former assumes that the wrench manipulation as a factor influencing operator performance is important; the latter, that it is not.

If we developed research specifically to gather task performance data, it would first be necessary to specify the behavioral parameters that are assumed to be important for the task performance. Just as there is a need for a set of equipment dimensions, so there is a need for a set of dimensions which characterize tasks. Although there have been attempts to define underlying task dimensions (e.g., Miller, 1967), much of this research has been directed toward underlying aptitudes or skills. The behavioral dimensions that have been generated as a consequence do not seem to be particularly useful as a starting point for the collection of performance data perhaps because the purpose for which they were developed was not that of data collection.

Table 4-5 Classification Scheme for Human Factors Research Studies: The Environment

1. Physical location
   - (a) space
   - (b) air
   - (c) sea surface
   - (d) sea subsurface
   - (e) ground
2. Parameters involved
   - (a) temperature
   - (b) noise
   - (c) lighting
   - (d) acceleration
   - (e) vibration
   - (f) atmosphere
3. Work context [1]
   - (a) civilian
   - (b) military
   - (c) both
4. Measurement context [2]
   - (a) operational
   - (b) quasi-operational (e.g., field test)
   - (c) design/development
     - (1) mockups
     - (2) simulators
     - (3) prototype equipment
   - (d) laboratory

[1] The system may be a military or a civilian one or a military system operated by civilians (often found in prototype testing of advanced equipment).
[2] Refers to the environment in which measurements are made. Operational refers to measures taken of a system performing in its designated functioning environment (e.g., on a war mission). Quasi-operational refers to a field test or exercise situation which resembles, or may even be, the operational environment (e.g., field trials for a new submarine), but which is experimentally controlled in its operations or the manner in which they are carried out. Design/development is the contractor's engineering environment, in which tests may be performed using mockups or simulators and on prototype equipment. Laboratory tests are self-explanatory.

A second major research task is to organize and compile whatever quantitative data on task performance are to be found in the literature and to make these available to the specialist working in system development. Whatever quantitative information on this parameter is to be found is scattered through many journals and reports. It is unreasonable to expect the specialist to search this literature every time he is requested to provide an answer to a specific question. Moreover, since the available data have not been systematically organized in the form of a "bank" suitable for information-retrieval, it is difficult to determine just what usable information these data provide. Data in masses or unorganized studies do not translate into usable information.

Fortunately this research problem does not require the performance of new experiments and hence can be solved, although not without difficulty. These comments about compilation apply to all the other research parameters discussed in this chapter.

ENVIRONMENT

Not a great deal of time need be spent in discussing environmental research because to a large extent it is of a special character, that is, designed to answer questions about performance in severe environments, such as space, underwater, and high altitudes; questions which, although important in themselves, have limited usefulness for anyone except those who ask the question.

Most human factors research implicitly assumes an environment that does not differ significantly from that in which most operators can function; this in fact means either a ground or surface sea environment. When we read reports of laboratory experiments and there is no mention of special conditions such as temperature and atmospheric extremes, it is because the researcher consciously (or more likely unconsciously) assumed a surface-earth environment. Most research is therefore not directed at any particular environment; this means that it can be generalized to the range of earth environments in which most operators function.

The kinds of questions raised by special environments have limited applicability because the effect of the environment is the focus of interest, not the behavioral principles involved in the performance being measured. Such studies describe essentially deviations from the "normal" environment. In that sense deviation studies that test the effect of special environmental conditions depend on prior investigations that utilized a normal environment. In another sense such special environmental studies are, except for problems of instrumenting the research situation and the subjects to reproduce the special environment, easier to perform than other research because they build upon research techniques tested earlier. Special environmental research is essentially comparison research: results found in performing tasks in the "normal" (earth) environment are compared with results found in performing the same tasks in the special environment; for example, one studies reaction time under extremes of cold (Teichner, 1959); another, the effects of anoxia on handwriting (Kennedy, 1952). Such studies assume that data gained from experiments performed under normal conditions will be first available.

Because special environmental studies are largely comparison research, their adequacy depends on the adequacy of human factors methodology in general. For example, if we wish to determine the influence of anoxia on the ability to perform reasoning tasks, we must first be sure that the ability test actually measures reasoning. If the performance measure used is, as is common, a physiological one like heart beat or brain wave patterns, it is necessary to ensure that the means of recording the measure is effective. If this is not done, it is possible that any variance in performance from the normal environment will be uninterpretable.

Because of the nature of unusual environments, their study is almost always experimental and in many cases involves extensive simulation. It is impossible at the present time to utilize these environments without first developing a highly controlled experimental situation; for example, we would hardly consider performing a questionnaire-type survey of behavior in outer space or 50 feet below the sea surface.

In giving environmental research a lower priority than other areas, we do not mean to imply that environmental questions are unimportant. It is entirely reasonable to ask (a) what is the effect of special environmental variables (e.g., temperature, acceleration) on operator performance, (b) what are the limits within which the operator can function in the environment, and (c) how must design be modified to take into account variations in human performance resulting from the environment.

When, however, we consider the research questions which the H̲ discipline as a whole must answer, it is difficult to see an overriding urgency for environmental research however important that research may be to the particular agencies that need the answers. The usefulness of the answers provided will continue to be limited at least for the forseeable future in which man will remain essentially a surface planetary animal.

One special environment needs to be noted. We have considered (Table 4-5) the measurement context as a form of environment. A discussion of this context is delayed until we deal with the *Methodology* parameter.

MEASURES

The Choice Of Criteria

Measures, like Methods discussed in the next section, have relevance to the H̲ specialist in two ways. First, the specialist depends for his research data on the results of studies whose efficiency is influenced by the researcher's choice of criteria and measures. Second, during the development process he may find it necessary to conduct mockup, simulator, or operational feasibility tests. Since these tests are in fact experiments, he is faced with the necessity of selecting appropriate measures. In this selection the specialist encounters an embarrassment of riches. He has a large choice of measures, many of which appear at first glance to be similar, whereas others are vastly different. On what basis can he make a choice?

Obviously he does not choose his measures at random; he will select a measure only after analyzing the information he needs and the information he can expect to receive from that measure. Ultimately the choice he makes will be—or should be—determined by analysis of the system. This is because the system can (as we see in the notes to Table 4-6) be viewed in a number of different ways, each of which may require different measures. Because of this the specialist may make use of very dissimilar measures.

There may be some benefits from all this. It is possible to advance the proposition that the researcher *should* select as many different measures of the same system as he can. For one thing, since it is difficult to anticipate the sensitivity of the selected measures to the performance to be measured, some measures may be more effective than others, and perhaps the specialist, by using as many as possible, can maximize his chances of a successful investigation.

It is also possible that data secured with certain of the measures may be artifactual, that is, the act of measuring in a particular way may elicit data which are more a function of the characteristics of the measure than of the phenomenon being investigated. Under these circumstances the specialist can *check* the validity of his data by examining the consistency of study results gathered with alternative measures. On the other hand, different measures of the same phenomenon may provide different results; this may create problems for the researcher if varying answers make results ambiguous.

The criteria among which the researcher must choose as the basis of his selection of measures are primarily influenced by one overriding fact: that man-machine behavior in the system context occurs at *different system levels* simultaneously.

Man-machine research is—or should be (always that disclaimer!)—profoundly influenced by the fact that behavior is produced by the system as a whole and also, and at the same time, by an individual (or crew-group of individuals) within the system. Hence the varying system goals. Further each type of behavior can occur at different stages during system performance. Of these stages two are most important:

Table 4-6 Classification Scheme for Human Factors Research Studies: Measures

1. Criteria employed [1]
   - (a) system
     - (1) terminal [2]
     - (2) intermediate [3]
   - (b) individual
     - (1) terminal
     - (2) intermediate
   - (c) behavioral [4]
   - (d) psychological [5]
   - (e) physiological [6]
2. Types of measures
   - (a) objective
     - (1) individual measures
       - (a) performance accuracy
       - (b) errors (magnitude/frequency/rate)
       - (c) event occurrence
         - (1) frequency
         - (2) percentage
         - (3) mean
       - (d) response time (duration, reaction time)
       - (e) accidents
       - (f) critical incidents [7]
       - (g) physiological (e.g., heart rate)
     - (2) system measures [8]
       - (a) performance accuracy (e.g., "miss distance")
       - (b) performance reliability
         - (1) probability of error occurrence
         - (2) probability of task completion
         - (3) percentage/frequency of human-initiated malfunctions
       - (c) Event occurrence [9]
         - (1) frequency
         - (2) percentage
         - (3) mean
       - (d) Performance duration (e.g., time required for system to complete mission)
   - (b) subjective
     - (1) ratings/rankings
     - (2) opinions (e.g., survey/interview responses)
3. Descriptive characteristics [10]
   - (a) quantitative
   - (b) qualitative

[1] System criteria describe system (equipment plus crew) as distinct from individual (operator alone) performance measures.

[2] Terminal criteria are those that describe functions and tasks representing mission completion. For example, in the Air Antisubmarine Warfare (AASW) mission the end product of a series of complex tactical operations by the crew is the dropping of depthcharges against the submarine. The terminal criterion is whether or not a kill is achieved; this is in turn a function of other variables such as distance from the submarine when the charge is dropped.
[3] Intermediate criteria describe functions/tasks that lead up to or implement the completion of the mission but do not themselves describe mission completion. For example, in the AASW mission referred to, the mission segments preceding the kill are search, detection, localization of the target, and dropping of barrier sonobuoys.
[4] A behavioral criterion is one which relates to or describes human performance in a system context (e.g., the speed with which the AASW operator detects the target and the accuracy with which he classifies its "signature").
[5] The psychological criterion refers to individual human performance which is not necessarily mission-related, such as motivational or attitudinal reactions (e.g., boredom, satisfaction, or even physical reactions like squirming).
[6] A physiological criterion relates to the human's body functions (e.g., blood pressure changes in astronauts during lift-off) which are correlated with and in part describe his performance in accomplishing system requirements. In considering a system event like the AASW mission referred to above, all or some of these criteria may apply simultaneously, depending on which aspect of the event we are considering (e.g., the mission as a whole, a segment of the mission, a crew task, or the behavior of a single crewman).
[7] A critical incident is some distinctive, representative, or frequently occurring behavior which illustrates in summary fashion a particular facet of the individual's behavior almost in the way the few penstrokes of an artist will catch the essence of his subject's face.
[8] In differentiating between performance accuracy and performance reliability, we define accuracy as referring to the single event (e.g., the accomplishment of a task or mission goal); reliability refers to the frequency or probability of accomplishing that event (e.g., the task over repeated occurrences).
[9] Event occurrence describes phenomena occurring during the mission which may or may not implement the system mission. An individual event occurrence may be the number of updating range reports made by an AASW operator during a mission. System event occurrence may describe the frequency of AASW crew communications as a whole during that mission.
[10] Measures may be either quantitative or non-quantitative (i.e., qualitative). Qualitative measures may be either objective or subjective just as quantitative measures may be. A running verbal description of events occurring during a test is an example of a qualitative objective measure. The temperature of the room in which work is being performed is an example of a quantitative objective measure. The ranking of a group of maintenance technicians in terms of skill level would be an example of a quantitative subjective measure. What makes a measure quantitative is the metric in which it is expressed rather than the means by which its data are gathered. Interview responses are qualitative (verbal) and subjective (because they are expressions of subjective feeling) but become quantitative if, for example, the researcher were to make a frequency count of the number of times the pronoun "I" was used by interviewees.

those occurring at the end of system operations (hence *terminal criteria*) and those occurring at some point in time prior to the end of the mission (*intermediate criteria*).

Since distinctions in level of performance and mission stage exist, they require that behavior occurring at one level or stage relate to behavior occurring at other levels and stages. Unless this is done, behavior at the more molecular (individual) level or intermediate stage is uninterpretable. It is necessary, for example, to assess the effect of an error made by one operator in an intermediate mission segment on terminal system performance of the total crew because the error is unimportant except as it influences the crew's accomplishment of the terminal system goal. (The reverse is not true; system behavior is comprehensible without data on individual performance because the questions asked of the system—was the mission accomplished or not—are less detailed—rather simplistic, as a matter of fact—than those asked about individual performance.)

Why is the interrelationship of performance at different system levels and mission stages important? Not only is the question theoretically interesting, but it is a significant—perhaps the overriding—question to be answered in human factors measurement. The answer in part determines the measures the researcher will select.

Behavioral measures are measures of the efficiency of the operator in system performance. Unless operator performance has an influence on over-all system functioning, it need not be considered. Such a statement assumes that the goal of H activity is the improvement of system efficiency; whatever does not contribute to that efficiency can, according to that principle, be ignored. (However, there are other H goals: to make the operator's job easier or more satisfying to him. From that standpoint, even if improving operator performance will not improve system performance, it may be justifiable to make such improvements.)

It is unacceptable, however, to modify the individual operator's performance without examining the effect of that modification on the total system output. It is conceivable, for example, that although the individual's performance varies markedly during a mission, as a function of certain parameters, the crew's performance does not or vice versa. To determine these interrelationships it is necessary for the researcher to measure at system, crew, and individual levels simultaneously and to integrate performance at more molecular levels (operator performance) into that of more molar levels.

It is, of course, possible to measure individual operator performance without relating that performance to system outputs or, conversely, to study system performance without considering individual performance.

However, both these solutions are inadequate because individual measures are largely uninterpretable unless they are related to system outputs, and system measures alone provide only restricted information.

One of the stated goals of H is to improve the efficiency of the total system output by adapting design to maximize the efficiency of the individual. From that standpoint, it is necessary to understand the relationship between the individual's and the system's performance; unless we understand this, we cannot make meaningful design recommendations.

The human factors researcher has a choice of other criteria as well although they are not as important as the system-individual dimension. Measures may also be *behavioral, psychological,* or *physiological.* We need pay little attention here to physiological criteria. As was pointed out earlier, in this country most measures of individual performance are likely to be behavioral (e.g., errors, response time) rather than physiological as is the case in European ergonomic research. The reasons for this have been explained elsewhere.

It is important to differentiate psychological from behavioral criteria. Behavioral criteria (e.g., errors, time) describe *efficiency* in accomplishing system goals. Psychological criteria relate to operator *satisfaction* with the equipment he uses and the ease of using it. Both should be considered in system design and testing. Performance efficiency in system operation may be high but at a cost to the individual user. In general, psychological criteria have been overlooked in human factors studies. This exclusion ignores a critical factor, one which will become increasingly important as technology expands and impacts increasingly on the average user.

In a sense psychological criteria are implied when the researcher interviews the operator to discover any difficulties he may have in using an equipment. We shall see also that such self-report measures are needed to interpret performance efficiency measures. Moreover operator dissatisfaction with equipment may lead to nonuse of the equipment or attempts to develop rough and ready substitute equipment (i.e., klooges).

We consider that one of the aims of H (parallel to increasing system efficiency) is the reduction of the discomfort experienced by equipment users. This aim has been largely overlooked because H has been geared to assisting engineering development, which is directed to increasing the efficiency of the total system, no matter what it does to reduce the satisfaction of the equipment user.

In our increasingly technological society people are forced to use equipment whether they like that equipment or not. They have no choice

—witness the ubiquitous automobile. Many people who drive out of necessity dislike the automobile; the difficulty of driving on modern freeways undoubtedly contributes to much physiological strain and discomfort. In time, as technology becomes increasingly demanding of human capabilities, the goal of operator satisfaction may become more important than that of improving system efficiency.

If, as we believe, H has the goal of increasing operator satisfaction as well as increasing system efficiency, then psychological criteria should be emphasized more than they have been. Whether this goal is as, or more, important than that of increasing system performance efficiency is a question which the reader himself must answer.

Choice of Measures

Because system performance must be viewed simultaneously on both the individual and system levels, it is necessary for the human factors researcher to utilize measures on both levels. However, system complexity may make it difficult to define precisely what an individual measure consists of. Notice in Table 4-6 that certain types of measures (i.e., event occurrence) are the same in both the individual and system categories. The context in which the measure is recorded may make it appear to be simultaneously an individual and a system measure.

Take error, for example, which researchers tend to use more than any other measure. For all its ubiquity it is difficult to define what it really means. It is something done (errors of commission) or not done (errors of omission). The error may have consequences to system performance, or it may not. When it has consequences, it may affect the operator alone or the operator and the system together. The operator may be aware of the error after it has been made, or he may not. The error may be an act consciously but not deliberately committed, or the operator may be completely unaware of making the error. The operator may correct his error, or he may not. Are all of these the same error?

We have been implicitly talking about error as an individual operator phenomenon, but the system too has its error measure (e.g., miss distance). System performance errors have not only human components (i.e., the individual operator error) but also equipment components (equipment performing out of tolerance), and how is one to differentiate the operator, from the equipment, aspects in this system error? Is the sum of all operator errors equal to the sum of all system errors? It is hard to say. What is obvious is that research is needed to examine all of these questions.

Even in dealing with individual operator error, the significance of the error varies as a function of its system context. In measuring operator performance we can record error (a) in all tasks (in which case the researcher refuses to differentiate important errors from nonimportant ones); (b) in critical tasks only (this says in effect that some errors in noncritical tasks are of no importance whatever); (c) in terminal tasks only (this negates the importance of errors in intermediate mission segments); etc.

Most researchers use the error measure indiscriminately without considering these underlying questions. (Of course, if a study is performed in the laboratory and there is no system performance to which the individual operator error relates, the question does not arise, but in that event we must believe that all errors reported in the study are crucial or that none of them are.) The question of the significance of the measure selected (significance relevant to some system output) does not thereby go away. Ignoring these questions simply provides data whose meaning has been insufficiently examined. The only solution to the problem is to analyze the measures selected in terms of the system interrelationships discussed previously or at least to hypothesize their meaning with reference to some anticipated system usage.

The other dimension of measurement the human factors researcher must consider is the distinction between *objective* and *subjective* measures. We present this distinction with some trepidation and do so only because the distinction is traditional and has certain consequences for measure selection. It may in fact be an artificial distinction. Even objective measures gathered through mechanical means are "tainted" with subjectivity since they must be recorded—or interpreted—through a data collector or analyst. A measure may be objective or subjective depending on how directly the stimulus events to be measured can be observed. This may in turn reduce to the number of dimensions those events possess, and how many one wishes to examine.

Unfortunately, because of the positive valences associated with objective data (and the negative ones, with subjective data), specialists tend to look down their noses at subjective measures. The strange thing is that they never fail to make use of subjective data and would as a matter of fact be at a loss to interpret their results if they did not. In point of fact it is impossible to perform human factors research without using subjective data because they are almost always essential to the interpretation of the objective data. This is true not only of "field" studies but also of laboratory experiments. The researcher who performs an experiment without asking his subjects about their interpretation of the stimuli presented is deliberately wearing a blindfold. (This does not, of course, refer to studies on infra-humans.)

The reason for this is that the subject is a very sensitive instrument for responding to the stimuli presented in the study as well as to system events. It may be difficult to extract that information from him because we lack the tools to do so, but that is our fault as researchers and does not reflect on his unique capabilities as a transducer. Moreover, much of his behavior is determined by his perception of the situation; in unstructured situations like troubleshooting, he is responding primarily to his own stimuli, only secondarily to the researcher's.

In a system test situation objective measures (e.g., accuracy, response time) are useful as a means of comparing operator performance with some required standard of performance, whether this standard be explicit or implicit. Such measures answer the question: can operators perform as required? This is an important question to answer. However, in evaluating the adequacy of man-machine performance, objective measures have restricted usefulness because *this is the only answer they provide.* If difficulties arise during the test of prototype equipment (as they almost always do), the specialist wishes also to determine *why* the system has experienced difficulties if only so he can make some recommendations for elimination of the difficulties. Objective measures do not necessarily supply "why" answers; for these a deduction is required, and deductions from objective measures are often severely limited.

From a research standpoint the specialist also wants to determine the relationships between equipment and task variables and performance. Objective measures are severely limited in this respect unless we have extracted in advance the specific variables to be tested and unless the objective measures (e.g., errors) *point unequivocally to those variables.* The last asks a great deal of the study situation especially if one is dealing with systems in the less than optimally controlled operational environment.

The "why" of performance can often be determined by the operator's description of the difficulties he has in performing an operation and in perceiving system events. This is not to say that it is easy to extract such subjective data because it is not. Subjects have difficulty expressing themselves and may in fact not be aware of what they really "know."

Unfortunately there may be considerable inconsistency in the subjective responses received. However, we should not necessarily expect consistency between operators in subjective responses when they reflect the individual's idiosyncratic perception of events. Researchers become disturbed by this inconsistency because they consciously or unconsciously expect the subjective data to reflect stimulus events outside of the individual. Subjective data may be highly consistent but only when the individuals' perception of external events coincide with each other. Consistency is most often found in relatively simple, structured situations, such as the estimation of performance times. If we accept subjective data for what they are—the representation of the individual's perception of the event, not the event itself—a good deal can be done with such data.

However, our subjective-report measures (with the possible exception of rating scales) are inconceivably crude. We do not know effective ways of eliciting subjective information from operators, and the amount of research on such tools as interviews and questionnaires is quite small (although see Kahn and Cannell, 1961). In particular the interview, which is the most common subjective method used, is not based on any research that points out the most effective ways of securing man-machine information. Specialists commonly make up their own.

Many of the measures described in Table 4-6 (e.g., errors, the occurrence of events, critical incidents) are reported by means of observation. Although these measures are considered objective, they are derived by means of a technique that is essentially subjective since the observer's perception serves as the medium through which the data are gathered.

This presents relatively little difficulty to the researcher when the behavior being observed is highly structured as, for example, a step-by-step operating procedure. Under these circumstances the definition of an error or an event can be pretty much unequivocal: the operator either performs a particular step, or he does not. The difficulty arises when the behavior is not highly structured as we find, for example, in troubleshooting. Under these circumstances the definition of an error, indeed the determination of what has been performed, is often obscure.

Difficulties also arise because the phenomenon being observed may have multiple dimensions and may therefore be viewed in a number of different ways with varying amounts of detail. Since the task being observed is always performed in relation to an equipment and possibly some dimension of that equipment, we must know in advance of observation what equipment characteristics the subject is likely to respond to. This may be a relatively obvious deduction for a task involving a single control but quite difficult for complex equipment. An observation can be therefore considered objective only when the behavior to be observed has been specifically and unequivocally defined in advance of the observation. By that criterion, most observations are unconscionably subjective.

Because they are not based on any previously conceptualized relationship between what is to be observed and equipment/task parameters, what we routinely see, paradoxically, may not be what should be observed. As we note from task analytic categories (e.g., frequency, criticality, physical action, purpose), any observable event may have a number of different dimensions. Which of these should be the basis for the observation? The research sayeth not.

Since observation is a—perhaps the—significant technique for securing behavioral data, the need for developing observational dimensions should be quite apparent. To do so, however, we must first conceptualize a relationship between one or more characteristics of the equipment and task and the resultant operator performance.

Unfortunately very little attention has been paid to observation as a major data gathering technique. Despite the evidence to the contrary, H specialists have assumed that observation is an entirely "natural" or direct process and therefore does not require artificial guidelines.

Design Significance

Assume that the researcher has carefully considered the choice of measures and has collected all the requisite data; the question still arises, what do the data mean? We refer here not to the interpretation of data as describing system events or what occurred during the system mission but rather to what the researcher can do with the data, what recommendations for action the data suggest.

Granted, for example, that operators consistently make certain errors in some phases of the system mission, what does this really mean? Here the specialist faces a most difficult problem. Most behavioral measures are measures of discrepancy such as errors or accidents and merely alert him that something is wrong. They do not indicate *what is wrong*. This can best be illustrated by describing an admittedly oversimplified hypothetical example.

Assume that the specialist is attempting to check the adequacy of a proposed control panel layout by means of performance measures. To do so he has created a functional mockup of the panel, taught subjects to operate it, and run them through the operating procedure on the mockup for a statistically appropriate number of trials. He has response time and error measures for each control-display on the panel, and the frequency of error for one such control is significantly greater than that for all other controls. What does this mean for design significance? Should the specialist assume that the type of control or its location is inadequate because of the excessive number of errors associated with it especially when—and here we add something to make the problem harder—overall performance on the panel is not unacceptable?

The fact that performance has failed to satisfy system requirements or that a substantial number of errors has been made does not indicate *why* the failure or the errors occurred; or what the causal factors are. Measures of discrepancy such as errors or accidents merely alert the human factors specialist that something is wrong. Since the error is the *end product* of a condition that led to the error, it does not directly reveal what the error-predisposing condition is. That condition must be deduced from other evidence.

It is assumed that a discrepancy measure such as error or the difference between achieved and required performance is associated with and hence points to a particular feature of system design. This should tell us what went wrong. But is this really true? It is certainly not true of duration measures. The fact that one performance takes longer than another does not necessarily indicate a design deficiency unless some time requirement has been exceeded. The same thing is true of reaction times. How about errors? If a high frequency of errors is associated with a particular component or maintenance action, it would suggest that we should look into that action or component. However, it would not indicate what was wrong. In the case of errors in highly structured, proceduralized operations such as operating a control panel the number of things that might be wrong with a control are limited to the type of control or its position on the panel, but with more complex behavior the number of possibilities is much larger, and the nature of the error does not help to reduce the number of possibilities.

There is no easy solution to this problem. It may be that no general answer can be supplied and that each data situation will have to be examined on its own system-specific merits. The point we wish to make is that data are not an infallible indicator of design action, and that, when all the data have been accumulated and interpreted, there will still be a need for the exercise of the specialist's own judgment.

METHODOLOGY

As was pointed out in the beginning of this chapter, some of the assumptions underlying the H discipline have significant implications for its research methodology. The three assumptions whose implications interest us in this section are the following:

1. Operator performance in man-machine operations has no meaning unless it is measured with reference to equipment and systems.
2. Man-machine performance automatically assumes at least three levels, the individual operator with his equipment, the crew, and the total system.
3. The interactions among the various man-machine levels are inherent in performance at any single level and must be considered if we are to explain performance at that level.

Certain implications follow from these assumptions. It is necessary in measuring man-machine performance to reproduce the equipment/system context with reasonable fidelity. As pointed out earlier, this is one of the differences between psychological and man-machine research. Because of its emphasis on the individual, the necessity for reproducing the situational context in psychology is less than it is in man-machine research, where the emphasis is on both the operator and his machine. The implications of this for generalizing laboratory studies to operational situations has been discussed by Chapanis, 1967.

Table 4-7 Classification Scheme for Human Factors Research Studies: Methodology

1. Empirical methods [1]
   (a) experimental design
      (1) parametric
      (2) nonparametric
   (b) nonexperimental techniques [2]
      (1) observation
      (2) interviews
      (3) questionnaires
      (4) test batteries [3]
      (5) ratings/rankings/checklists
      (6) system performance data [4]

2. Analytical methods [5]
   (a) descriptive
      (1) quantitative
      (2) qualitative
   (b) predictive
      (1) quantitative [6]
         (a) simple statistical relationships, measures of central tendency
         (b) probability estimates
         (c) models
      (2) qualitative [7]

3. Measurement objective: description/prediction of [8]
   (a) individual performance
   (b) crew performance
   (c) system performance
   (d) interrelationships

4. Data collection agency [9]
   (a) human
   (b) instrumentation
   (c) combination

5. Measurement context [10]
   (a) laboratory
   (b) quasi-operational
   (c) operational

[1] Refers to any method used to gather data. The experimental designs referred to obviously imply, even if they do not specifically mention, a great variety of statistical techniques such as Analysis of Variance, multiple regression analysis, etc.

[2] Techniques in which subject behavior is relatively uncontrolled or manipulated. Notice we say, relatively; there is a degree of subject control and manipulation in any situation in which stimuli are presented. Any of the nonexperimental techniques may be used in conjunction with or as an adjunct to the experimental ones; for example, we commonly interview subjects as a follow-up to experimental sessions. Observation is an essential part of any experimental technique.
[3] A group of tests often (but not invariably) of a paper and pencil type.
[4] If human performance tests are conducted as part of an over-all system test, the fall-out from test results will contain certain nonbehavioral data (e.g., failure reports, critical incidents) that not only reflect the performance of the system but can also be used to help understand individual behavior.
[5] The term "analytical" refers to the description or prediction of operator behavior but not to empirical data gathering although the description or prediction may be based on previously acquired empirical data. In descriptive analysis the analyst is describing a present state of affairs (e.g., the task error rate is .9988; this task is ordinarily performed by three men), whereas in predictive analysis the analyst refers to a future state or requirement (e.g., the following skill level will be required to perform this task). When the analysis is qualitative, the two categories often overlap.
[6] The listing under "quantitative" implies a hierarchy of complexity from simple statistical relationships (e.g., correlation) to the most complex (full scale models).
[7] A common type of qualitative predictive analysis is task analysis.
[8] The goal of the human factors measurement is to describe and predict the performance, *individually and collectively,* of the operator, the crew as a collection of individuals, and the total system as an entity including both equipment and personnel.
[9] How data are recorded: by personnel alone (solely in a manual mode, for example, by observation, interview, self-report); by instrumentation (in which, even though there is an operator for the instruments, the actual data sensing is done by the instruments); and by both (which is most often the case).
[10] Measurement context can be ordered on two dimensions: fidelity to the operational situation in which personnel will perform and amount of experimental control available. The laboratory environment is familiar to every one and permits the most control with least fidelity; the quasi-operational context attempts to reproduce the operational environment as much as possible while controlling conditions as much as we can (e.g., testing of missile ground equipment at a specially-constructed launch pad that reproduces a military launching site [see Meister and Rabideau, 1965]). The operational environment is precisely that in which the operational equipment will be deployed and as a consequence combines the greatest fidelity with the least experimental control.

The need for operational fidelity requires us either to reproduce the operational environment by simulation or to make direct use of that environment. Since simulation of major systems (i.e., equipment and personnel) is both difficult and expensive, the operational environment assumes for human factors research a greater importance as a measurement context than it does for other behavioral disciplines. Since the operational situation in which the H researcher must work is poorly controlled, it follows that he must develop techniques that enable him to function under such conditions. It also follows that for data to be maximally useful to the human factors specialist it must be formulated in such a way that it can be referred to some object or situation outside of the individual.

Data that describe phenomena occurring solely within the individual (e.g., physiological measures) or perceptual phenomena like after-images have limited usefulness to the human factors specialist except as general principles or background information. One reason (although not the only one) is that it is hard to translate such data into physical (e.g. design) correlates. For this reason, when psychological research is concerned solely with intra-individual phenomena like physiological concomitants of behavior, it has little application to human factors problems.

H data are—or should be—used to modify man-machine systems because these systems are artificial constructions that can be modified in a way in which natural systems (e.g., social classes) cannot. Because there is (at least hypothetically) a direct application of human factors data to system development and use, their predictive precision is expected to be greater than that of data supplied by other behavioral disciplines.

It is therefore not sufficient for the specialist to say that such and such parameter affects design; in order to influence that design he must provide highly specific quantitative predictions of the nature of the parametric effect. (This characteristic is not unique to H. As other behavioral disciplines are utilized much more to modify actual behavior (e.g., in education or urban planning), they too will require more specific predictions.)

The demand for specificity of prediction imposes on H the necessity for developing *quantitative data storehouses;* this in turn requires a subtle modification in the specialist's approach to research.

The Operational Environment

The importance of investigating major system characteristics (e.g., level of automation), which describe an entire equipment or system, has been pointed out. It is these characteristics which are the system developer's initial and basic concern during the crucial early days of development. If the H specialist is to influence design significantly, he must do so in relation to these characteristics.

Unfortunately, in contrast to minor characteristics (e.g., individual control and display features), which can be divorced to a certain extent from other interactive equipment features, it is difficult to abstract major attributes, to divorce them from their hardware and operational context, and to incorporate them into the usual laboratory setup.

The use of the laboratory (including both physical and mathematical simulations) to study the effect of major equipment and system characteristics seems therefore in many cases to be precluded. Only the largest laboratory could include masses of men and equipment. This forces the specialist to use the operational environment, in which systems can be found *in vivo*, as the source of his data.

In consequence, if the specialist is attacking problems involving major system design characteristics, he will—or should—(a) find most of his hypotheses for investigation stemming from operational situations; (b) find himself conducting most of his research in an operational or quasi-operational environment; and (c) validate his research conclusions by testing them in the use-situation if he is given the opportunity. Even if he performs his research in the laboratory, steps (a) and (c) are still required.

The human factors researcher is fortunate in having available to him the entire complex of military operations as a laboratory in which to test and verify hypotheses (assuming governmental cooperation, of course). If we look at the range of man-machine functions performed in military systems, it seems possible to select from among them those that represent parameters in which the researcher is especially interested. This is particularly so because most of his research is funded by military organizations whose research questions stem from these operations. There is, of course, a limitation—not to interfere with on-going operations—but much can be done even within this limitation.

A much more serious source of frustration to the human factors researcher and one which has limited his work within the operational environment is his lack of experimental control over variables and measurement conditions in the operational environment. It would be pointless to deny that this is a serious handicap and probably the one major reason for the popularity among specialists of simulation devices as research vehicles. The handicap, however, is most distressing to those researchers who adopt an uncompromisingly experimental strategy for their research.

Even with limitations on the amount of control that can be exercised, it is possible to use the operational situation more effectively than it has been. It should, for example, be possible to approximate experimental control by selecting equipment, tasks, and personnel that represent the experimental design conditions we would have created in the laboratory. The operational researcher would exercise control by selecting his test conditions in such a way that they correspond to the treatment conditions he would otherwise be testing in the laboratory.

This is somewhat idealistic, it is true, and we do not suggest that it exactly replicates the degree of control possible in the laboratory, but it is not too far off the mark. We assume that the operational environment is so rich in situations (hundreds of thousands of machines and men performing thousands of tasks) that we could select a subset of the operational conditions available as those that represent the parameters to be investigated. If, for example, we wished to study performance of two types of task with two extremes of equipment complexity, we would first examine the variety of tasks and equipment available operationally and select two that have the required differences. It may not be possible to equate groups of personnel or to control other conditions quite so nicely.

It may, however, be possible to cancel out these contaminating factors by sampling data over a large enough range of conditions in such a way that the contaminating factors (e.g., differences in aptitude, experience, etc.) will be effectively randomized. The experimental design utilized in a recent study by the author and his colleagues (Meister et al., 1970b)[2] is a case in point. The technique involved measurement of performance in a situation in which predictor and performance variables were allowed to vary in an uncontrolled manner. Then correlations among the predictor and performance variables were derived. These were then tested in a multiple regression analysis, and the percent of variance accounted for by each variable was determined. The pattern of effective variables, their *b* weights, and the significance of their effect on the performance index was determined by Analysis of Variance.

The essential element in the use of the operational environment for research purposes is to find that situation that contains the equipment and task parameters to be studied. Of course, this assumes that the various systems, tasks, and operating conditions found in the operational environment have been catalogued in advance in terms of relevant variables, the equipment and task dimensions referred to in earlier sections of this chapter.

Intra-System Interactions

As indicated earlier, the reason for examining man-machine performance at the various system levels—the individual, the crew, the system—simultaneously is that unless we do so the *significance* of the performance is lost.

There are two major implications of this:

1. If, to make sense of his man-machine data, the researcher must function on several levels simultaneously, he can no longer afford to perform studies at the individual level alone. This does not mean that

[2] The design in this study was developed by Mrs. Dorothy L. Finley.

he will no longer be able to conduct studies involving the single operator and the single equipment. It is, for example, perfectly meaningful to study the performance capability of the sonar operator with the single operator at his SQS-26 gear. The specialist will be able to make such determinations as, "SQS-26 operators are capable of performing detection functions with a reliability of .9973," or "The factors affecting the operator's performance are such and such."

The only thing he will not be able to say is what the significance of that performance is, whether .9973 performance as part of an over-all system mission is excellent, satisfactory, marginal, or poor performance; or whether the variables that significantly affect the operator's performance are also significant in affecting system performance. To determine this he must measure the operator's performance as part of the total system context and determine the effect of his performance on his fellow crew members and the total system.

Since operator performance has only limited significance at the individual level, it means that all research, even if conducted at first on the individual operator level, must be carried on into the additional system levels. Although certain parameters by their very nature can be investigated only at the individual operator level (e.g., troubleshooting the individual equipment), the importance of these parameters is obscured unless they are related in further research to higher order system performance. What does it mean, for example, if mean time to repair components in system X is T hours? For this we must know whether T hours is acceptable or not and how it influences system availability. Research at the individual operator level is therefore necessary but not sufficient; necessary *and* sufficient research requires all three levels.

2. If necessary and sufficient research is to be conducted at the system level, it means, as we have indicated earlier, a much greater shift in emphasis to research in the operational environment. The implications of performance at the system level can in many cases be studied only when the system is operationally exercised. This does not suggest the elimination of the laboratory, but it does mean that the laboratory alone is not sufficient.

Man-Machine Prediction

One of the consequences of the fact that he attempts to construct and to modify systems is that the H̲ specialist needs a large mass of quantitative predictive data.

When we are developing an object (an equipment or system) or a phenomenon (the training of personnel to operate or maintain that system), extremely precise data are needed. Operating without such data is like trying to build a house with a tape measure or ruler that is only an approximation, however close that approximation may be. To develop an equipment or system with the kind of H data presently available is like building that "approximate" house.

Because system development demands highly specific statements of fact, the specialist is obligated in his research to provide a range of quantitative values. For example, the question may be asked, what is the probability of an error occurring when one control of the 5 (or 8, 10, 12, etc.) is operated on a specific control panel? It is not enough to make the general statement that error affects performance or that the greater the number of controls on a panel, the higher the probability that operating any single one of them will be performed erroneously; the precise frequency of error anticipated for each operating condition must be provided. Only with precise data can the developer make meaningful tradeoffs between alternatives.

For this reason the traditional experimental study which concentrates largely on demonstrating that parameters are or are not significant in their effects is not sufficient for H purposes. This does not of course eliminate the need for determining whether parameters are significant, but this must be coupled with the systematic exploration of parametric effects.

To determine the range of effects does not require a different type of experimentation or study, but it does require that studies not confine themselves to a limited number of parametric points. The gathering of more data is of course a general goal of science, but, whereas it can be stated as a general goal for other disciplines, it is an immediate, urgent necessity for a discipline that specifically attempts to put its data to practical use in modifying systems, events, and phenomena.

The need for more detailed prediction makes it necessary to develop performance data banks. An example of a rather primitive type of data bank (taken from Munger et al., 1962) is shown in Figure 4-1. Because the number of variables and conditions that influence operator performance is large, such banks must be extensive. It also goes without saying that they must be quantitative and that as a minimum they must be formulated in terms of such measures as anticipated accuracy, expected error-likelihood, and performance response times. (The author has difficulty conceptualizing useful data in terms other than the probability that a task or job will be accomplished because this is the only measure which makes system development sense.)

A very limited number of such data banks have been developed (Munger et al., 1962; Blanchard et al., 1966; Hornyak, 1967), but the number of studies on which they are based (164 for Munger) and the number of parametric conditions which they describe are pitifully few. Manifestly there is a need for the human factors researcher to modify his goals of data collection to give higher priority to the gathering of data for these banks.

Joystick
(May move in many planes)

Base time = 1.93 sec

Time Added (sec)	Reliability	
		1. Stick length
1.50	0.9963	(a) 6–9 in.
0	0.9967	(b) 12–18 in.
1.50	0.9963	(c) 21–27 in.
		2. Extent of stick movement (extent of movement from one extreme to the other in a single plane)
0	0.9981	(a) 5–20 degrees
0.20	0.9975	(b) 30–40 degrees
0.50	0.9960	(c) 40–60 degrees
		3. Control resistance
0	0.9999	(a) 5–10 lb
0.50	0.9992	(b) 10–30 lb
		4. Support of operating member
0	0.9990	(a) present
1.00	0.9950	(b) absent
		5. Time delay (time lapse between movement of control and movement of display)
0	0.9967	(a) 0.3 sec
0.50	0.9963	(b) 0.6–1.5 sec
3.00	0.9957	(c) 3.0 sec

Figure 4-1 Sample Data Store Card

If it seems strange that the many human factors studies performed over the years do not supply the necessary data, it should be remembered that the goals of these studies and the manner in which they are conducted reduce their usefulness; for example, the tasks on which performance is recorded are often contrived (do not replicate real world tasks). This renders their data useless for data bank purposes.

The author is presently engaged in a project to determine the feasibility of utilizing general behavioral research (e.g., laboratory studies) for expanding human performance predictive data banks. Initial results indicate that this is feasible, but only with great difficulty. The primary problem is the diversity of treatment conditions and measures applied to the same parameters. This suggests a need for greater methodological standardization in behavioral research.

SUMMARY

An effective human factors research program if we define "effective" as meaningful answers to the questions posed by the system developer (Table 4-1) and the system user, should center around the following major points. (See Table 4-8 for a detailed list of priority research questions.)

Table 4-8 Listing of Priority Research Questions

1. What behavioral dimensions describe equipment in terms that are meaningful for operation and maintenance of that equipment and that can be reliably observed?
2. How is operator/technician performance influenced by equipment with these dimensions?
3. How are equipment characteristics related to the aptitude, skill, and training needed to operate and maintain that equipment?
4. What behavioral dimensions describe task performance in terms that relate to equipment dimensions and training?
5. What is the distribution of operator/technician performance in tasks described by these dimensions?
6. How can the specialist derive training content, duration, and methodology from task/equipment descriptions?
7. How do tasks interrelate in terms of their sequential effect upon each other and the probability of achieving a system output?
8. How do the various measures of human performance interrelate; how can we integrate the performance of an operator into a single measure (e.g., skill)?
9. What is the design significance of human performance measures like error?
10. What is the relationship between measures of individual, group, and system performance?
11. How can the operational environment be utilized as a laboratory for human factors research?

Equipment-Behavioral Dimensions. The starting point for an effective research program is the development of a set of behaviorally relevant dimensions that describe equipment and tasks both individually and in terms of their interaction. (Not a set of a priori taxonomic categories—

although these would be a starting point—but empirically derived dimensions, as Miller, 1967, suggested. It should be possible to sample a variety of man-machine tasks and to develop an appropriate taxonomy by extracting their common dimensions.)

Operator-Performance Capabilities. The ability of personnel to perform tasks described in terms of these task/equipment/behavioral dimensions must be determined. The starting point for any human factors research program should be the determination of basic capabilities: how well can the operator perform specified tasks? This implies not only an average performance but the upper or lower bounds of capability that limit that performance.

The fact of the matter is that we know very little about how personnel perform in actual operations. In the same way that handbooks of equipment reliability data (e.g., MIL-HDBK 217A, DOD 1965) have been developed, similar handbooks for operator performance data should be compiled.

Inter- and Intrasystem Relationships. Important as it is to develop a body of data describing the effectiveness of individual operator and task performance, it is as important to investigate the interrelationships among tasks, among equipments, and between individuals and systems. Since the individual error and task are not performed in isolation, they have significance only in terms of how they impact upon each other and upon the accomplishment of the system purpose. For this reason it is necessary to determine quantitatively the following factors:

1. The impact of an error (defined as any discrepancy) on the performance of the task.
2. The effect of a task upon the tasks which follow or are concurrent with it.
3. The effect of completing or not completing a job upon the performance of other jobs being performed concurrently or sequentially.
4. The effect of individual operator performance on system performance.
5. The interrelationship of design with task parameters.
6. Interactions among the total number of parameters influencing task performance.

This is not to say that there are no problem areas other than the three listed previously, but rather that with the restricted effort available to solve a large number of problems, attention should be centered on these.

Because of its limited facilities, it is essential that the discipline adopt for itself a set of priorities for the research it will perform. At present human factors research is a melange of separate interests almost completely dictated by the interests of the individual governmental agencies that support the research. It may be foolish to expect anything else as long as H is dependent on these agencies for funding. It would help, however, if the senior members of the discipline (perhaps sponsored by organizations like the Human Factors Society) attempted to develop a set of priorities (by consensus, if no better method exists). System development representatives should be represented among these policy makers.

Far too little attention has been paid to the solution of methodological problems that are basic to the more immediate problems for which government contracts research. Much of this research is highly dubious because answers cannot be secured without the prior solution of methodological problems that are ignored.

In this connection there has been a tendency for human factors research to fly "too far too fast"; for example, the use of highly sophisticated mathematical or computer techniques to attempt to solve behavioral problems whose basic concepts have never been explored in detail is not only inefficient but serves to divert attention from underlying problems.

Human factors research should be oriented to solve problems which arise from the operational environment, that is, research stemming from the need to answer questions raised by the system development and use process. If system development requires that task performance be predicted or that training requirements be derived from equipment characteristics, then research directed to these ends should be performed. This should not conflict with what was said earlier about the need for methodological research, which is usually inherent in system development requirements.

Research solutions achieved through work in nonoperational (e.g., laboratory) environments must be validated by testing their relevance in the operational situation. All too often researchers have avoided the necessity of demonstrating the significance of their research conclusions to the problems that presumably elicited the research.

In this connection the military environment with its thousands of personnel performing a great variety of man-machine tasks is a perfect research laboratory—better in fact than the specialist may have for his present work. Obviously there are difficulties in attempting to make use of such a laboratory, primarily lack of control over personnel and task variables. Researchers may, however, be compensated by richness of equipment, tasks, and subject material.

One last word. All the strictures in this chapter notwithstanding, a great deal of research has been performed on man-machine relationships. In fact, so much has been done that it is difficult to (a) assimilate the conclusions derived and (b) determine what gaps in that research need to be plugged by new research. With all the attention presently being given to conservation (ecology), perhaps a form of human factors research conservation ought to be practiced; that is, additional efforts should be made to compile the research that has been performed on various topics, to analyze that research in terms of the man-machine problems that need solution, and to point out the new research needed to answer questions that so far have been imperfectly answered.

REFERENCES

Blanchard, R. E., M. B. Mitchell, and R. L. Smith, 1966. *Likelihood-of-Accomplishment Scale for a Sample of Man–Machine Activities.* Santa Monica, California: Dunlap and Associates, 30 June, AD 487 174.

Bradley, J. V., 1969. Optimum Knob Crowding, *Human Factors,* **11** (3), 227–238.

Chambers, A. N., 1969. *Development of a Taxonomy of Human Performance: A Heuristic Model for the Development of Classification Systems,* Tech. Report 4A, American Institute for Research, Silver Spring, Maryland, October.

Chapanis, A., 1963. Engineering Psychology. *Ann. Rev. Psychol.* **14,** 285–318.

Chapanis, A., 1967. The Relevance of Laboratory Studies to Practical Situations, *Ergonomics,* **10** (5), 557–577.

Chapanis, A., and G. R. Lockhead, 1965. A Test of the Effectiveness of Sensor Lines Showing Linkages Between Displays and Controls, *Human Factors,* June.

Cotterman, T. E., 1959. *Task Classification: An Approach to Partially Ordering Information on Human Learning,* Report WADC 58-374, Wright Air Development Center, Wright-Patterson AFB, Ohio, January.

Crawford, M. P., 1962. Concepts of Training, in R. M. Gagne (Ed.), *Psychological Principles in System Development.* New York: Holt, Rinehart and Winston.

Crumrine, B. E., et al., 1968. *A Study and Investigation of the Quantification of Personnel in Maintenance.* Willow Grove, Pennsylvania: Philco-Ford Corp., September (AD842 878).

Department of Defense, 1965. Military Standardization Handbook 217A, *Reliability Stress and Failure Rate Data for Electronic Equipment,* Washington, D.C., 1 December.

Edholm, O. G., 1967. *The Biology of Work,* World University Library. New York: McGraw–Hill.

Finley, D. L., et al., 1970. *Human Performance Prediction in Man–Machine Systems: I. A Technical Review,* NASA CR-1614, National Aeronautics and Space Administration, Ames Research Center, August.

Fowler, R. L., et al., 1968. *An Investigation of the Relationship Between Operator Performance and Operator Panel Layout for Continuous Tasks,* Report AMRL-TR-68-170, Aerospace Medical Research Laboratory, Wright-Patterson AFB, Ohio, December.

Fitts, P. M., et al. (Ed.), 1951. *Human Engineering for an Effective Air Navigation and Traffic Control System,* National Research Council, Washington, D.C.

Fitts, P. M., 1958. Engineering Psychology, in *Ann. Rev. Psychol.*, **9**, 267–294.

Glaser, R., et al., 1955. *A Study of Some Dimensions of Team Performance,* American Institute for Research, September.

Grether, W. F., 1947. *The Effect of Variations in Indicator Design Upon Speed and Accuracy of Altitude Readings,* Report TSEAA-694-14, Aero Medical Laboratory, Wright Air Development Center, Wright-Patterson AFB, Ohio.

Guilford, J. P., 1967. *The Nature of Human Intelligence.* New York: McGraw–Hill.

Harris, D. H., 1966. The Effect of Equipment Complexity on Inspection Performance. *J. Appl. Psychol.,* **50,** 236–237.

Highland, R. W., et al., 1956. *A Descriptive Study of Electronic Troubleshooting,* Report AFPTRC-TN-56-26, Air Force Personnel and Training Research Center, Lackland AFB, Texas, January.

Hornyak, S. J., 1967. *Effectiveness of Display Subsystem Measurement and Prediction Techniques,* Report RADC-TR-67-292, Rome Air Development Center, Griffiss AFB, New York, September.

Kahn, R. L., and C. F. Cannell, 1961. *The Dynamics of Interviewing.* New York: Wiley.

Kennedy, J., et al., 1952. *Handbook of Human Engineering Data.* Institute for Applied Experimental Psychology, Tufts University, Medford, Massachusetts.

Laner, S., 1961. Ergonomics in the Steel Industry, in *Proceedings of Conference on Ergonomics in Industry.* H. M. Stationery Office, London, England.

Lippert, S. (Ed.), 1968. Special Issue: Developing Countries, *Human Factors,* **10** (6), 557–662.

Locke, E. A., et al., 1965. Studies of Helicopter Pilot Performance: II. The Analysis of Task Dimensions, *Human Factors,* **7** (3), 285–302.

Mackie, R. R., and P. R. Christensen, 1967. *Translation and Application of Psychological Research,* Report 716-1, Human Factors Research, Goleta, California, January.

Meister, D., 1967. Development of Human Reliability Indices, *Proceedings Symposium on Human Performance Quantification in Systems Effectiveness.* Sponsored by Naval Material Command and National Academy of Engineering, Washington, D.C.: January 17–18.

Meister, D., and G. F. Rabideau, 1965. *Human Factors Evaluation in System Development.* New York: Wiley.

Meister, D., et al., 1969. *The Design Engineer's Concept of the Relationship Between System Design Characteristics and Technician Skill Level,* Report AFHRL-TR-69-23, Air Force Human Resources Laboratory, Wright-Patterson AFB, Ohio, October.

Meister, D., et al., 1970(a). *The Effect of Operator Performance Variables on Airborne Equipment Reliability,* RADC-TR-70-140, Rome Air Development Center, Griffiss AFB, New York, July.

Meister, D., D. L. Finley, and E. A. Thompson, 1970(b). *Relationship Between System Design, Training and Autopilot Maintenance Performance,* AFHRL-TR-70-20, Air Force Human Resources Laboratory, Wright-Patterson AFB, Ohio.

Melton, A. W., and G. E. Briggs, 1960. Engineering Psychology. *Ann. Rev. Psychol.,* **11,** 71–98.

Miller, R. B., 1967. Task Taxonomy: Science or Technology? *Ergonomics,* **10** (2), 167–176.

Morgan, C. T., et al., 1963. *Human Engineering Guide to Equipment Design.* New York: McGraw–Hill.

Munger, S. J., et al., 1962. *An Index of Electronic Equipment Operability: Data Store.* Report AIR-C43-1/62-RP(1), American Institute for Research, Pittsburgh, Pennsylvania, January.

Murrell, K. F. H., 1969. *Ergonomics: Man in His Working Environment.* London: Chapman and Hall.

Poulton, E. C., 1966. Engineering Psychology. *Ann. Rev. Psychol.,* **17,** 177–200.

Rigney, J. W., 1964. *Human Factors in Maintenance,* in A. S. Goldman, and J. B. Slattery, *Maintainability.* New York: Wiley.

Swain, A. D., and J. C. Wohl, 1961. *Factors Affecting Degree of Automation in Test and Checkout Equipment.* Report TR-60-36F, Dunlap and Associates, Stamford, Connecticut, March.

Teeple, J. B., 1961. *System Design and Man–Computer Function Allocation.* ORSA–TIMS Meeting, April 19–21.

Teichner, W. H., 1959. Environmental Factors Affecting Human Performance, in *Human Engineering Concepts and Theory,* P. M. Fitts (Ed.). Ann Arbor: University of Michigan Press.

United States Air Force, 1967. Report AFSCM 80-3, *Handbook of Instructions for Aerospace Personnel Subsystem Designers,* Air Force Systems Command, Washington, D.C., revised.

HUMAN FACTORS: Practice

Overview of Part Two

Chapter Five describes what the H̲ specialist usually does and should do during the proposal and predesign stages of development. Factors influencing H̲ outputs during these stages are indicated. The reader is told how to respond to the human factors requirements of a request for proposal, how to develop a human engineering program plan, what is involved in task analysis and design tradeoff studies, and how to apply predictive techniques to system development. Special attention is given to guidelines for successful human engineering.

Chapter Six carries the story of Chapter Five into detail design. Subjects covered are: the development and evaluation of procedures; the use of criteria, checklists, and mockups for design evaluation; and the evaluation of prototype hardware. The chapter concludes with the contents of the H̲ test plan and rules for verification testing.

Chapter Seven emphasizes the importance of the design engineer's attitudes toward human factors and the H̲ specialist. Experimental studies of how the engineer designs and of how he incorporates human engineering into design are described. The utility of H̲ standards and handbooks is examined.

Chapter Eight describes the procurement of the human factors research supporting system development, the researcher's relationship to the governmental customer, and the factors determining the adequacy of the research.

Chapter Nine concentrates on the administrative details of H̲ functioning: the organization of the H̲ group within engineering, qualification and selection of H̲ personnel, facilities, finances, and indoctrination of engineers. The factors that determine whether or not an H̲ input is likely to be accepted by engineering are examined. Finally, the contents of the controlling H̲ specification and standard are analyzed.

Chapter Ten concludes the book by presenting the views of 12 distinguished H̲ specialists and researchers on the major issues of human factors theory, research, and practice. The outlook for H̲ in nondefense industry is—as far as it can be—anticipated.

V

HUMAN FACTORS IN PREDESIGN

The purpose of this chapter and the one following is to describe what the H̲ specialist (a) *usually* does, (b) what he *should* do, and (c) what he *does not* and *should not* do during system development.

This chapter was written with several goals in mind.

1. Engineering management should be aware of those outputs it can reasonably expect to receive from its H̲ group at different stages of system development. Since the program described is a minimal one, the program is not living up to its potential if fewer outputs are received.
2. Engineering should understand the *reasons for* and *activities required* by H̲ functions. In turn, this will permit engineering management to help the H̲ group perform these functions more effectively. Obviously that group can perform more effectively if management understands what it is attempting to do and why.
3. When the company lacks H̲ specialists, engineering management should know those minimal H̲ activities that are required either contractually or simply to develop an adequate system. This will permit non-specialist engineering personnel to satisfy those requirements.
4. This description may also help the H̲ specialist to understand why he does what he does and the engineering and budgetary factors that often determine just how effectively he can perform.

FACTORS INFLUENCING H̲ OUTPUTS

What the H̲ specialist does and how he does it depends on several things:

1. Stage of system development.
2. Number and type of human functions.
3. System "newness."
4. Cost.

Stage of System Development

Activities are determined by the stage of system development in which H̲ must function. Analysis of personnel functions, for example, is performed in very early developmental stages when the basic design concept is still hazy; the need for this analysis does not ordinarily exist after drawings have been developed (except possibly for training applications). Similarly, an operational H̲ test plan cannot be developed until after operating procedures have been written. Consequently, in the description following particular attention will be paid to the influences and constraints of the individual developmental phases and the information available from them.

Although they overlap and blend into each other, several stages of development are generally recognized. These are the following:

1. *Precontract award.* Since this book is written from the standpoint of the engineering contractor, it does not describe those activities that would be performed before the time a request for proposal (RFP) arrives at the contractor's plant. Activities prior to the receipt of an RFP are ordinarily performed by governmental personnel (or others under contract to them specifically to perform these activities). This phase is often referred to (by the Air Force) as the concept formulation phase.

If the contractor is active on a project during precontract award, his activities center on those analyses needed to respond as effectively as possible to an anticipated RFP. Sometimes the government provides relevant advance information about a new system procurement weeks or even months prior to issuing an RFP. In that case an interested contractor may perform pre-RFP analyses in expectation of receiving the RFP.

2. *Proposal period.* Once the contractor receives the RFP he may have anywhere from two weeks to several months to respond. Since his future work on the new system hinges on a winning bid, extensive design analyses involving H̲ activity (among other disciplines) are performed during this period.

3. *Predesign.* Assuming that a contract is won, the predesign period (also known as the contract definition phase) is devoted to developing the fundamental concepts of the system design concept. This is largely a decision-making period in which alternative design concepts are examined, accepted, or rejected.

4. *Detail design.* The fundamental design decisions made in predesign are elaborated and detailed in the detail design or system acquisition

phase. The gross outlines of the design concept are filled in to the detail level needed to produce a functioning equipment.

5. *Production.* Prototype equipment is fabricated in the company's production facilities. After it is tested, necessary modifications to the equipment are made, and a continuing production run is started.

6. *Testing.* The prototype equipment or the first unit or units of the production run are tested under laboratory or operational conditions to confirm the adequacy of the design concept. Production and testing are usually considered part of detail design or acquisition.

7. *Operation.* After testing has confirmed that the system will function as required, it is turned over for the customer to use. At this stage the contractor's relation to that system is ended except for supplying logistical and technical assistance.

The system development process is far from being a serial one as the stages listed above might suggest. Design as a whole is an *iterative* process; for example, in the course of analyzing alternative design configurations in predesign, detailed drawings may be made to see what the system will look like if fabricated. If a design concept is scrapped, the design team may start again with very gross drawings.

There is also considerable overlap among the various stages; for example, design modifications may be made during testing as a result of information gained from that testing; as a result, design activity (although slight) continues throughout development.

Number and Type of Human Functions

Another factor influencing the amount and nature of H̲ activity is the degree of human involvement required by the new equipment. If the system is one which by its nature demands considerable personnel involvement, the amount of H̲ activity required will probably be greater than in a system which demands few manual operations.

Personnel involvement is defined by two factors:

1. The number of manual functions needed to exercise the system.
2. The criticality of these functions to mission accomplishment. Of these, function criticality is more important in determining the nature of H̲ activity. An equipment may require the operator to perform many motor operations, but, if these are simple and discrete (e.g., throwing switches), H̲ activity may be limited to arranging these switches in proper control panel layout. Indeed the layout itself may be quite simple.

On the other hand, when manual functions are highly critical (even if there are only a few of them), H activity must be increased. An ECM equipment may need only be turned on with a single switch, for example, but requires the operator to analyze complex, shifting wave patterns. The H task of selecting the appropriate display characteristics to implement operator functions may demand extensive research, design, and testing.

It is, of course, very difficult to define criticality in objective terms. Function criticality is obviously influenced by the engineer's preferences for modes of operation. If, for example, he elects to automate equipment functioning, operator functions become less critical. The specification of the functions that will be implemented by personnel (hence their criticality) is, therefore, partially determined by nonbehavioral engineering considerations. Many human functions are, however, implicit in the nature of the equipment being designed. For example, a system like radar aircraft detection and tracking, which cannot be operated without extensive human participation, forces critical human functions upon the designer. To that extent the equipment itself will define the H activity to be performed.

Actually it is easy to overlook the importance of human functions in a system particularly when they are implicit rather than explicit. When human functions are explicitly required by a procedural step that demands a manual action like the activation of a control or the reading of a display, it is difficult to ignore them. Implicit, conditional functions (mentioned in Chapter Two) are often unrecognized. Hence the H activity that would have been concentrated on the implementation of these implicit functions may never be exercised.

Moreover the importance or criticality of a human function is even easier to misunderstand. Criticality is defined in terms of the probability of error (p_i) times the consequence of an error (f_i) to equipment functioning; either of these may be difficult to deal with.

Error probability data for human functions are in many cases unavailable. At early design stages too little may be known about the design concept to do more than guess at the human functions whose error probability we are attempting to estimate. Even if the functions are known, the error probability data that describe them are in many cases unavailable. Even if the functions and their error probability are known, the design engineer and the H specialist may only be guessing at error consequences unless each possible consequence is examined in detail by something like a reliability failure modes and effects analysis. Often such an analysis is not conducted either by the engineer or the specialist.

Quite apart from other considerations, the engineer has a tendency to underrate the significance of human involvement in a system whereas the $\underline{H}$ specialist has a tendency to overvalue that involvement. A reasonable evaluation of the amount of that involvement lies in splitting the difference.

System "Newness"

A system may require many and critical human functions, but, if that system is relatively similar to one that the contractor has developed earlier, then the amount of $\underline{H}$ effort required will probably be less, perhaps even minimal. This is particularly true when the system being developed is only a model variation upon the ones preceding it; for example, when the Atlas missile was converted to a space launch vehicle, only minor changes in valving, control panels, and launch procedures were required. Of course, this assumes that the model preceding the one under development was "worked on" by the $\underline{H}$ specialist or had no special human factors problems. If the earlier system suffered from problems that were not solved, the next model will probably face the same problems.

On the whole, highly "advanced" systems (those that extend the state of the art) are more likely to require intensive $\underline{H}$ effort than ones that do not, simply because they tend to raise questions of operator capability that have not been answered previously. The $\underline{H}$ effort involved in developing the Apollo space module was undoubtedly much more extensive than that required by the development of the Boeing Model 747 aircraft because the former required its personnel to do much more in more severe environments than the latter.

Cost

The amount of money the customer is willing to spend for the $\underline{H}$ effort also determines the amount of that effort. If the customer decides that he cannot afford the necessary cost, regardless of the need for the $\underline{H}$ activity, that activity will not be performed. Since $\underline{H}$ is usually not considered by engineering management as a basic design discipline (however much the specialist and the customer may think it is), its cost is usually not computed as a mandatory development item. Hence the service is not usually provided unless it is specifically called out and paid for as a distinct contractual item.

Even when an extensive $\underline{H}$ effort is initially desired by the customer, it is often trimmed down (sometimes to nothing) in the formal contract negotiations between the customer and the contractor. The cost of the system proposed by the contractor may exceed the amount the customer is prepared to pay even though the latter wants the technical effort proposed. In the ensuing seesaw negotiations, those tasks considered by both parties to be less than essential often go by the board. Moreover, even when the H̲ effort is funded contractually, among the first things reduced or eliminated, if the contractor runs into trouble and exceeds his estimated cost, is the H̲ effort. (The effect on the H̲ specialist's ego can be imagined.) All of this, of course, without prejudice to the need for the H̲ effort.

Cost, however, is not an all or nothing factor. Even if the customer does not buy all the functions described in this chapter or those which specialists consider necessary, he often buys some. The size of the H̲ effort is naturally scaled down to what the customer will pay. Although this is classic capitalistic process (the customer buying what he can afford) it is inefficient from a total system development standpoint when the customer lacks, as he often does, the proper background to determine those H̲ activities he really needs.

SUMMARY

Depending on the factors previously described, all, most, some, or none of the H̲ functions described in this chapter will be performed in the individual system development project. They are discussed here fully for the sake of completeness, but the *complete* H̲ program is usually a "design goal" and is accomplished only in part. Engineering management and the customer select those functions they want implemented, based on the needs they recognize, the cost they are willing to bear, and the advice given them by the H̲ staff. This chapter describes what they can get for their money.

HUMAN FACTORS IN THE PROPOSAL STAGE

Introduction

It may appear from the length of this section that a disproportionate emphasis has been placed on H̲ proposal activities. (For a list of these activities started in the proposal and then carried on in subsequent development, see Table 5-1.) If this is true, it is only because a successful proposal is the one critical element without which the need for H̲ activity does not arise. Moreover many of the basic design decisions which influence H̲ activity are made during the proposal. Tessmer (1967) analyzed a number of actual system development case histories and found that

Table 5-1 H Activities Initiated in the Proposal Stage

Activity	Activity Duration Proposal	Predesign	Detail design
1. Analyze RFP for H implications			
(a) determine human factors implications of system requirements.	———		
(b) determine which human factors specifications apply to system.	———	———	
(c) analyze effect of applicable H specifications on contractor response.	———	———	
2. Develop H section of proposal			
(a) determine and describe personnel functions and tasks.	———	———	
(b) determine major equipment characteristics required by personnel tasks.	———	———	———
(c) indicate personnel capability limits affecting design.	———		
(d) perform tradeoff studies involving human factors.	———	———	
(e) estimate manpower required for new system.	———	———	
(f) assist in development of initial control-display and workplace layouts.	———	———	———
(g) evaluate alternative configurations in terms of human factors implications.	———	———	———
(h) budget contractor manpower and cost for H system development program.	———		

"in practice, most tradeoff areas are identified and tentative decisions made during *preproposal and proposal efforts* (italics supplied by the author). These decisions are solidified or modified within the first few months after contract award. It is remarkable that so many tradeoffs are typically resolved in so short a time." (p. 3-3)

It is important, therefore, for the H specialist to know what he should contribute to the proposal.

Characteristics of the H Requirement

Increasingly, governmental RFP's for development of new systems contain some section, however vaguely phrased, that requires the contractor to consider the human factors aspects of the system in his response. The section may be captioned in various ways: the terms human factors, human engineering, personnel subsystem, personnel requirements, and operability are frequently used, but in most cases it will be phrased somewhat as follows:

"The system will be designed to minimize the need for large numbers of highly skilled personnel."

"Equipment design will be performed in accordance with applicable human engineering principles (or MIL-STD 1472)."

"Equipment design will be such as to minimize the opportunity for error."

The implication of the requirement is that equipment should be designed not only to secure the most effective mechanical, electrical, hydraulic, etc. performance but also the most effective performance of the personnel who will operate and maintain the equipment.

Phrased in this manner, the requirement is largely a design goal. Except when an applicable human engineering standard like MIL-H-46855 (DOD, 1968) is called out, no formal methods of accomplishing the goal are usually described in the RFP section. In the case of large scale advanced system development, the section may require a program with specific inputs and outputs, such as the Air Force's Personnel Subsystem Program (AFSCM 375-5) (USAF, 1964). If a system analysis is required as part of the development work, the RFP may request the contractor to analyze his design in terms of the following:

1. The impact on system design of the functions to be performed by operators and the number and types of personnel required.
2. The impact of the contractor's design concept on the numbers and types of personnel needed and what they are to do.
3. How the system will be designed to ensure the most effective personnel performance.

Occasionally the RFP will specify the maximum number and skill level of personnel for whom the system should be designed. This may be done when the government's analysis of system requirements prior to issuance of the RFP has been extensive enough that specific personnel needs can be described. In that case it will be necessary to determine how these personnel requirements will affect the design proposed.

Upon occasion an RFP for a new system will fail to contain a specific H requirement. Even under these circumstances the contractor would be well advised to develop an H response to the RFP in order to indicate that he is aware of personnel-design relationships. When the RFP contains a section specifically designated for H, the contractor should give additional emphasis to developing the H response because the customer obviously believes that some H activity is necessary. The opportunity for major H participation in the writing of the proposal is, therefore, considerably greater when such a section is included in the

RFP. Otherwise the necessity for H contributions to the proposal effort may be overlooked by engineering management, which is often not oriented to such activity.

Because the H requirement is usually described in general terms such as those noted earlier, it is often interpreted by the contractor in rather vague and simple generalities: make the design simple and easy to operate and maintain; reduce the opportunity for error; cut down on the need for large numbers of personnel and especially skilled personnel; design in accordance with human engineering military standards. The simplistic nature of the contractor's interpretation of the H requirement makes it appear to engineering that no substantive H work need be done to satisfy the requirement.

The generality of most requirements makes them very difficult to enforce since criteria of accomplishment are almost never included and in any case would be difficult to agree on. Particularly lacking in the RFP (or anywhere else, for that matter) are explicit relationships between personnel requirements and the kind of design which will satisfy those requirements. In effect the requirement is a "design goal," an objective that the contractor will work to accomplish, if sufficient money and time are provided, but for which there is little compulsion.

For the same reasons the H requirement is rarely very detailed or lengthy (except in a very few cases where the involvement of the human in the system is so critical that the system literally succeeds or fails on the basis of its personnel performance).

Reasons for Responding to the H Requirement

It may appear from what has been said that the contractor need spend very little time and effort responding to the H section of the RFP. Although many contractors treat this requirement in a perfunctory manner, it is a mistake for the following reasons:

1. If we think of an RFP as a form of test of the contractor's ingenuity and intelligence, it is entirely possible that the evaluator will count the contractor's H response more than the latter imagined. The contractor may get demerits for turning in a poor H response even if he does not receive "brownie" points for turning in a very good one. Although almost no system development competitions are won or lost solely on the basis of the contractor's H response, it is possible for such a response to *influence* the final decision.

2. Increasingly the government is paying additional attention to the manner in which the contractor will handle the problems of design for personnel use. Measured on an absolute scale, the number of system

development competitions in which H plays a significant role is not great, but the percentage increase of RFP's in which H is called out as a significant design factor has increased markedly over the last 10–20 years.

So far we have been talking about impressing the customer. There is an additional reason for writing a competent H response: the RFP proposal period is the one that will largely determine the basic design concepts and how the system will finally appear. From that standpoint the H response ought to be just as much a guide to engineering design of that system as any other type of equipment analysis.

In some cases the customer may display a distressing lack of knowledge about the requirements and problems involved in developing the requested system. When these involve human factors, it may be necessary to "educate" the customer before he is capable of appreciating the contractor's proposal.

In summary, there are at least five reasons for responding as effectively as possible to the H section of RFP.

1. To impress the customer with one's capability to handle the area.
2. To provide an analysis that, in the event the contract is won, will guide engineering in those aspects of design that are affected by human factors.
3. To supply answers to engineering for problems in design that may be caused by human factors requirements.
4. To justify the design being proposed on the basis of its human factors aspects.
5. To teach the customer certain of the human factors information he needs to know to understand the system.

Methods of Responding to the H Requirement

The specific methods (such as function, task, and time-line analyses) by means of which the analyses needed for the H response can be performed have been described in Chapter Three. These methods are formal ones, so called because they can be described by a specific, step-by-step procedure. Often, however, the response time available for many RFP's, sometimes no more than 2–3 weeks for even complex systems, may not permit the use of a formal analytic procedure. Nevertheless the logical processes demanded by these procedures (processes that are the only really important part of these procedures) can be employed informally —in the head, so to speak. The concept that a human factors analysis for an RFP response need necessarily be a prolonged one is incorrect.

Although Table 5-1 suggests that H proposal analyses are a primary responsibility of the H specialist, he cannot and should not perform it isolated from what the design engineer is planning in the way of equipment. In fact, he will often find that major decisions are forced upon him by equipment considerations. For example, if an accuracy requirement is too exacting for human performance, then an automated system becomes necessary. The point is that the H specialist must be aware that there are system requirements and equipment factors that constrain the deductions he can draw from his analyses. Like everyone else participating in a development project, he must work within these constraints (assuming, of course, that the constraints are real and not invented). Sometimes, however, these constraints are not made known to the H specialist soon enough; consequently he may blunder down the wrong pathway.

The proposal effort is a system development project in miniature. The H specialist should therefore perform in miniature functions similar to those he would ordinarily perform during full scale system development; that is, he should determine the functions and tasks to be performed by personnel, note their design implications, and recommend suitable design concepts. He should advise on the best way in which operating procedures should be performed, develop or review control panel layouts, etc.

Apart from his writing of the H section of the proposal, however, the most common functions performed by the H specialist during the proposal are to develop on his own (or assist in the development) and review control-display or workspace layouts required by the system being proposed. He may also be asked to specify the manpower required by the system and to describe the tasks these men will perform. Unfortunately, in too many cases the proposal manager sees the H specialist on the proposal team only as a reviewer of control panel layouts; the proposal section which the H specialist is assigned to write is too often confined to a description of the human engineering characteristics of the control panel.

What The H Response Should Contain

During the development of the RFP response, the H specialist should supply engineering with the following information: [1]

[1] The degree to which we expand on each of these items obviously depends on the need for exploring the questions described by the item. For example, if the equipment being designed is a power regulator consisting of a single on-off switch and meter to indicate the amount of power output, the specialist will probably not describe in very great detail the tasks required of the operator or the problems involved in monitoring the meter. This is merely common sense; the specialist should not attempt to inflate his analysis.

1. If there is a specific personnel constraint on design such as the number and type of personnel for whom the system is to be designed, what is the potential effect on the design concept?
2. If there is a design concept constraint such as a specified or implied level of automation, what is the potential effect on numbers and types of personnel and the jobs they are to do?
3. If a specific human engineering military standard is called out as applicable to the design, are there any requirements in the standard that the engineer cannot live with or that he must seriously consider as a factor affecting his design?
4. If the customer has not advanced his own ideas about manning the system, what kind of personnel structure (i.e., numbers and types of personnel and tasks to be performed) should be recommended to the customer?
5. If system development requires training of new personnel, what should that training consist of?

To answer these questions and to prepare to write a suitable response requires an H analysis of the RFP requirements and the company's design concept. This analysis should produce the following information, which should be explicitly stated in the response, avoiding answers which are only implied.

Specifications. List either the human engineering specifications required by the customer or, if he has not indicated one, a specification which is in general use in the industry or a company human engineering standard. The DOD specification MIL-STD 1472 (DOD, 1969) is one which is likely to be called out in the RFP; if not, it is entirely safe to cite it. A sample response might be: "The X Corporation will generally follow the provisions of MIL-STD 1472 and MIL-H-46855 and particularly sections ___ and ___." This indicates that the company knows which H specifications/standards and which sections of these are applicable to the particular problems of developing the new system.

It is not necessary to repeat these sections verbatim but to note that paragraphs so and so will apply to the problem, why they apply, and what implications they have for design of the proposed system.

It is, of course, a matter of form only, but it is useful to indicate compliance with the specification *in toto*. If the company cannot for any reason comply with particular sections of the specification, list the sections to be waived and describe why the waiver is necessary. (Since most H specifications and standards are rather general, the necessity for waiving overly stringent H requirements will not often arise.)

2.0 *Description of Man-Machine Operations*

As indicated in the section on Personnel Task Analysis (2.3) the shipboard automated data processing system requires in normal, routine operations no complex or difficult personnel skills. Almost all daily operations require little more than the activation of one or two machine switches and/or the monitoring of an equal number of discrete indicators. Unless a malfunction occurs the system operates in the following sequence: (1) initialize the system by setting up the conditions for operation (e.g., loading cards and inserting magnetic tapes); (2) perform a readiness test of the system which consists simply of programming the system to run with the aid of a standard set of cards; (3) commence operations by exercising the machine on selected data cards and tapes with the aid of program cards; (4) where necessitated by the volume of data, change the program run by stopping the system, changing tapes and cards, and resetting the controller; (5) whenever required (which will be infrequent) entering additional information by typewriter; (6) shutting down system operations by in effect reversing step (1). During routine operations the operator is required to monitor data output from the printer and to distribute data through designated Navy distribution channels. Additional minor activities of a routine preventive maintenance nature, such as servicing and cleaning, are also required.

Figure 5-1 Sample analysis of functions and tasks.

H Implications. Describe what the H section in the RFP implies. Rather than respond to the H requirement in general terms, it is important to interpret the requirement back to the customer in terms of specific system development operations; for example, the response might say,

"Operability in the context of this project is understood to mean that (1) controls and displays will be laid out in accordance with operational procedures; (2) only that information required will be presented in the control panel," etc.

This does not tell the customer anything he would not find in the applicable human engineering standard, but it helps to indicate that at least we know the appropriate words.

System Functions and Tasks. Describe the functions and tasks to be performed by system personnel (see Figure 5-1). Sometimes the RFP is quite explicit about these; at other times, it is vague. In the latter case it is useful to define the functions and tasks required and the design reasons why they are required. Particular attention should be paid to functions of the conditional or emergency type that are only implied.

Equipment Design Features. Indicate what equipment design features must be incorporated in design to implement these functions and tasks (e.g., a temperature gauge in order to monitor temperature). Although

Task loading investigation clearly indicates that the right hand should manipulate the cursor ball, light gun, select and change projector slides, and initiate the console power switch and teleprinter readout functions. The left hand will accomplish the function of setting up a program on the program keyboard and actuating control keyboard switch functions as required. These tasks are less stringent and frequent than those to be done by the right hand . . .

Monitoring and interpreting the CRT is one of the most critical interfaces the operator has with the device.

Pertinent human engineering considerations relative to the critical design parameters of the CRT display to be incorporated in the console are the following: the CRT must be flicker free to reduce visual fatigue, have high resolution, brightness, contrast ratio, no hot spots, and proper alphanumeric character size. These considerations will be optimized in order to provide CRT presentations that are quickly and efficiently legible to the operator. Provision will be made for reduction of glare on the surface of the CRT by a filter coating on the implosion shield, tilting the face of the CRT in the console and providing a hood or shield.

An anthropometric study has been made (Figure 5-3) to determine optimum placement of the CRT display in the console. The CRT will be centered on the 50th percentile sitting operator's normal line of sight with the center of the CRT 42.0 inches above the floor surface. . . .

Since the only anticipated interface the operator will have with the teleprinter is to tear off typed messages and do other small routine tasks, the teleprinter will not be made an integral part of the over-all console configuration.

Figure 5-2 Sample analysis of equipment design features for a computer-aided display console.

it is unnecessary to provide a rationale for each common control or display, major equipment characteristics (such as built-in test equipment or go/no-go indicators rather than meters) should be pointed out as having been selected to meet personnel needs. (See Figure 5-2 for a sample analysis of a computer-aided display console.)

***H* Problem Areas.** Describe equipment problem areas which arise as a result of what personnel may be required to do. For example, if personnel must read CRT displays under daylight conditions (as they may if they are controlling catapult aircraft via TV displays from the flight deck of a carrier), special long persistence CRT phosphors or some sort of display shielding may be required. This question arises largely in advanced systems where traditional design solutions are not readily available or in systems which are highly complex even if they are not advanced. Such systems make demands on personnel, the nature of which (demands) may be inadequately appreciated because of the newness of the technology involved.

Personnel-Imposed Limitations. Indicate the limitations on equipment design imposed by personnel capability. Such limitations are to be found primarily in highly advanced systems where the nature of the environment in which the operator must function or the tasks he must perform severely stress him. For example, if we are bidding on the development of an Apollo space suit, the discussion should include statements that sufficient degree of flexibility will be included in design to permit the wearer to operate controls when the suit is worn even under full pressure conditions. Specific factual material (e.g., flexibility defined in terms of number of degrees of arm and finger movement) should be included to support the statements made.

H Rationale. If a specific design is proposed by the company, the H response should contain whatever human factors reasons exist (or can

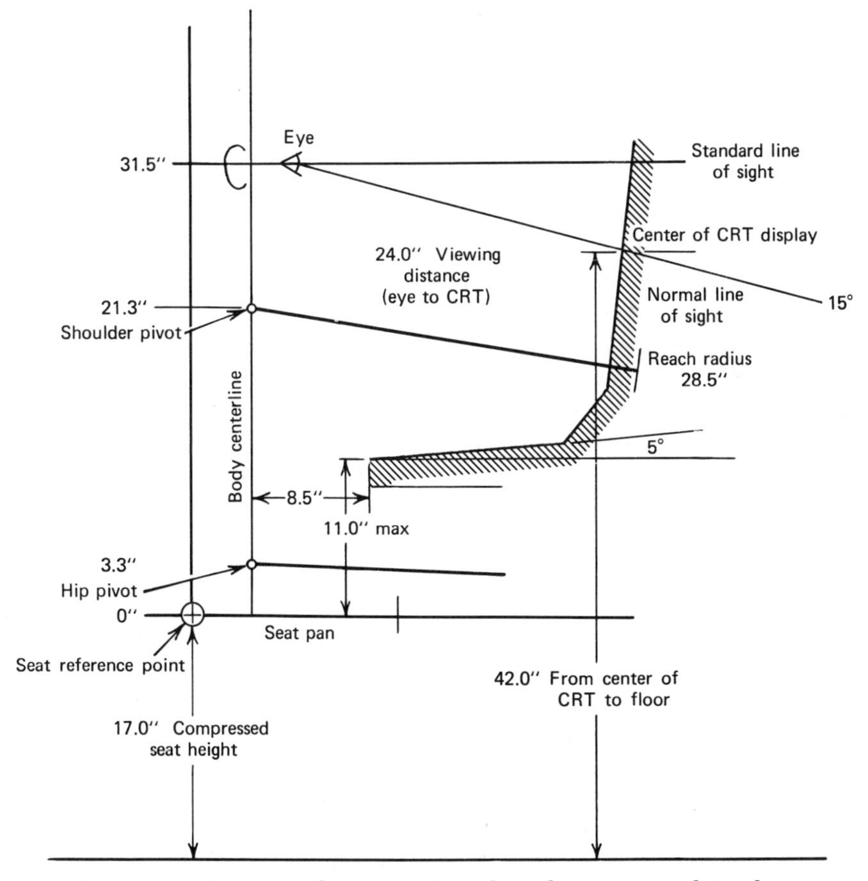

Figure 5-3 50 percentile operator/console anthropometric relationship.

be imagined) to justify the particular design being proposed. From the company's standpoint of selling a proposed design to a customer, the important point is to relate the response to the system design proposed. If the design is required by the functions and tasks demanded of system personnel, this is additional evidence for the necessity of that design. Almost certainly there will be other, purely equipment, reasons for the design, but the human factors rationale adds to the weight of the argument. Figure 5-2, which describes the design of a computer-aided display console described graphically in Figure 5-3, contains elements that rationalize the particular design selected.

Initial Equipment Concepts. Describe initial concepts of equipment layout with particular emphasis on control panel and workspace layouts, consoles, etc., and give the human factors reasons for including each design feature.

If the design proposed is one which is "off the shelf" or a series modification, it is wise to include statements that indicate that appropriate human engineering has been performed on earlier models of the system or similar types of equipment, that that human engineering experience can now be applied to the new equipment; for example: ". . . because the XYZ system is largely based on the design of earlier systems developed by the S company, extensive human factors design activities will not be required. Previous system development involved extensive human factors analysis of the CRT displays included in the new system; this analysis involved determination of appropriate alphanumeric size for maximum legibility, elimination of glare by shielding the CRT, and tests to measure the legibility of displayed material under various conditions of ambient lighting. However, an analysis of design requirements for the new system will be made to insure the applicability of previous human factors design solutions to the new equipment."

How H̲ Activities Will Be Performed. A substantial part of the H̲ response should be a description of how H̲ activities will actually be performed on the design project. Merely to say that human factors will be reflected in design is to utter a pious hope. The H̲ side of the customer's house is aware that there are sometimes formidable obstacles in the way of accomplishing H̲ goals. Consequently, without specifying a step-by-step means of achieving H̲ objectives and thus providing ways of monitoring their accomplishment, there may be little chance of satisfying either the objectives or the customer.

What the customer wants to learn is such things as will the contractor's H̲ staff

(a) participate in basic design concept planning;
(b) advise on the development of and review man-machine interface drawings and operating procedures, particularly with sign-off privileges;
(c) participate in formal design reviews and tests?

Mention should be made of such methods as periodic monitoring, special reports suggesting H̲ design characteristics, checklist evaluations, and periodic design review meetings. The tone in which this section is written should not imply that the company management is likely to fail to meet H̲ requirements but should present as an obviously desirable feature the measures taken to enforce the provisions of H̲ requirements.

As a way of ensuring that the contractor knows what is required of him to perform from an H̲ standpoint and also as a means of holding the contractor to a contractual commitment, the customer may require a complete description of the way in which the H̲ effort will be managed. It is, of course, impossible to specify each step in that process. However, the major steps should be described concretely in terms of the goals to be achieved and the outputs desired. A sample implementation plan is described in the Appendix to Chapter Six.

The H̲ activities to be performed should be tailored to the type of system development required. For example, if the system is to be composed of "off the shelf" components tied together in a new configuration, the H̲ requirement may be to ensure that the resulting new configuration is adequate but that no human engineering of the individual "off the shelf" components would be required. However, if the system is a significant advance on the state of the art and the role of personnel in the system is consequently ambiguous, much more attention will be paid to the delineation of those functions personnel should perform. If the RFP requires a formal testing program at a test site, it is appropriate to point out the part H̲ aspects will play in the testing. If a specific personnel requirement has been imposed (e.g., only four men will operate the system), the significance of that requirement should be indicated; that is, indicate that it will affect the design of the console in such and such a way, that it will require communications equipment, etc.

The H̲ specialist is well advised in the proposal not to build his role in system development much beyond what the nature of the system and customer demand. If he does, it will probably be viewed as presumptuous and excised by the proposal manager. There is a natural tendency for the H̲ specialist to build his program up if only because he expects it to be trimmed by the proposal manager or the customer during negotiations, but there is a greater chance of the H program being accepted by engineering and the customer if that program is relevant and concrete.

Personnel Requirements. If a new design concept is being proposed by the company, it would be useful to include in the H section the numbers and types of personnel needed and the role these personnel will play.

Reports. Describe any H-specific reports to be written. Periodic progress reports should be noted. Unless the content of these reports is immediately obvious, it is desirable to describe what the reports will contain. Any special reports recommending corrective design action should be described in detail.

Applicable Experience. Applicable human engineering experience on similar equipment should be noted.

Page Limitations. The total number of pages for the H section will depend on the complexity of the system and the extent of the H program anticipated. Often the customer enforces a page limitation, and this forces an allocation by the proposal manager upon the H specialist (as on everyone else). The general advice given to authors applies also to those writing the H response: Write only that which is needed to present necessary information.

HUMAN FACTORS IN PREDESIGN

Developing the Human Engineering Program Plan

"30.2 . . . The plan shall include but need not be limited to

(a) a description of the means by which the contractor will meet the requirements imposed by this specification and the procurement document;
(b) an implementation schedule in milestone chart form;
(c) description of the organizational structure and management relationship for accomplishing human factors functions, including provisions for control and approval. . . ." (Appendix to MIL-H-46855, Military Specification, Human Engineering Requirements for Military Systems, Equipment and Facilities.)

A program plan to describe the H effort represents the detailed planning of the project through all its phases. The contractor's proposal in responding to the RFP cannot (because of restrictions in preparation time and page space) describe all requirements in detail for completing the project. Therefore the H program plan assists in several ways:

1. It requires the human factors group to plan its efforts more intensively than it was able to during the proposal effort.
2. It informs the customer and the engineering management in more detail of what will be done in performing the H program. If there are any problems or objections, these can be verbalized.

Obviously a detailed program plan is suitable only for major projects. Some governmental contracts of a major nature may require such a plan within 30 or 45 days of contract go-ahead. Air Force contracts of sufficient complexity usually refer to AFSCM 375-5 (USAF, 1964) which specifies what such a plan must contain. Even when such a plan is not specified in the contract, however, it is useful in supporting a development project.

If the H activity is restricted to only one or two functions (e.g., helping to design a single console), it is pointless to create a program plan.

Program Plan Elements

The H program plan is an expanded H proposal response and contains the same elements minus those developed to persuade the customer to award the contract or to explain the system design concept. Much of the advice given in the previous section applies here as well. As a minimum the plan should contain the following elements:

1. *The purpose of the effort.* This may seem superfluous; at the general level the purpose of the H effort dovetails with that of the development project, which is to develop the most effective hardware with least cost. However, detailed H goals may not be identical with detailed equipment development goals. Specific H goals may include such things as: to describe operator tasks; to predict operator performance; to determine the frequency of human-initiated failures. These differ from, although they interact with, development goals like development of operating procedures, testing of hardware, etc. Each of the detailed H goals implies a specific activity, and to describe the activity we must first introduce the goal that the activity implements.
2. *What is to be done.* This is a listing of H activities similar to those listed in Table 5-2, which also indicates the duration of continuing efforts in each activity.
3. *How H activities are to be accomplished.* These are the detailed steps required to accomplish (2). For example, assisting in the development of control panel layouts may involve any or all of the following:

(a) consultation with engineers; (b) providing specialized information in the form of a checklist; (c) informal review of drawings; (d) formal reports evaluating drawings; (e) participation in formal design reviews; (f) sign-off. The procedures for implementing such detailed steps should be described. The greatest part of the H plan will involve this section.

Table 5-2 H Activities Initiated in Predesign

Activity	Activity Duration	
	Predesign	Detail design
1. Develop H program plan.	———	
2. Perform detailed human factors analyses		
(a) function/task analyses.	———	———
(b) training requirements analyses.	———	———
3. Perform or assist in performing human factors-related tradeoff studies.	———	
4. Predict personnel performance.	———	———
5. Human engineering of control-display, console, and workspace layout	———	———
(a) analyze equipment requirements and recommend equipment features to engineers.	———	———
(b) recommend control-display hardware components.	———	———
(c) review control panel, console, and workspace designs developed by engineers.	———	———
(d) participate in customer and government design reviews.	———	———

4. *Sequence of H activities.* This sequence should be described in the form of a milestone chart, if possible, which indicates major H activities on a time scale set by the developmental schedule (see Figure 5-4).

5. *H activities.* Since these will be largely written reports, it is desirable to describe them as concretely as possible, pointing out their utility to engineering. Many of the things an H specialist does have no concrete output in the sense in which an engineer has his drawing or a set of calculations as an output; for example, the specialist's most useful work may be consulting on an informal basis with and verbally influencing the design engineer. The former's contribution is often tied to a product he sometimes does not even get to sign off. To secure a proper evaluation of that contribution, and to insure that engineering knows what it is supposed to get for its money, it is useful to give these "hidden" outputs visibility.

6. *Information and cooperation.* What informational inputs does the H specialist expect from the engineer? How does he propose to work

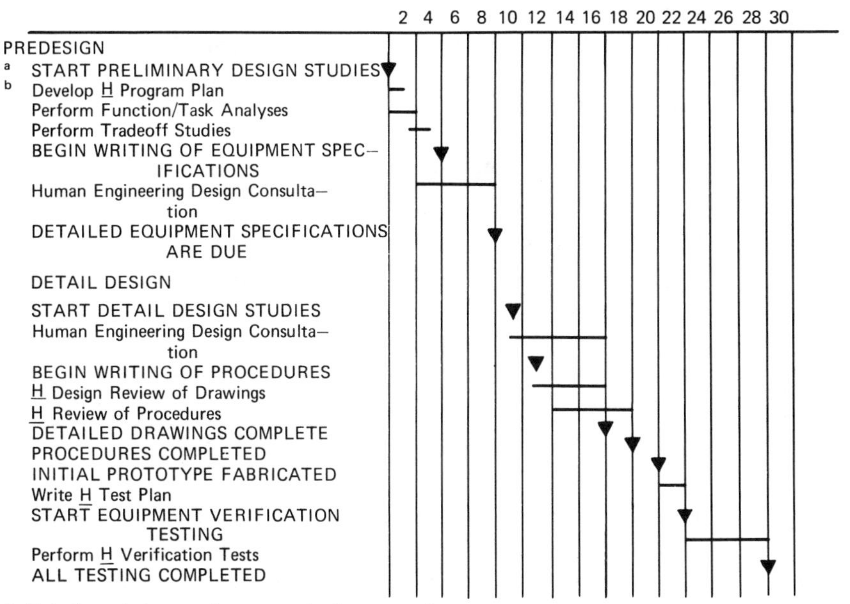

[a] Tasks in capitals are major system development tasks.
[b] Tasks in initial capitals are H tasks.

Figure 5-4 Sample milestone chart for H activities.

with the latter? For example, if drawings are to be reviewed, the specialist will need them from the engineer in sufficient time before they must be released to production. That time should be specified. The place of the H group in the project organization and the H group's charter (official responsibility for development activities) should also be indicated.

7. *Potential Problems.* Examples of some of these are: securing inputs from engineers on a timely basis; being alerted by engineering to changes in design concepts; invitations to participate in relevant engineering meetings such as design reviews. Listing these puts management on notice that the H specialist wants its help in overcoming these problems.

The H program plan should be distributed as widely as possible among engineers and engineering management (but only to those whose activities might be influenced by the plan or might affect its implementation). It is useful to request comments on the document. Few comments may be received, however; engineers are notoriously allergic to paperwork and the plan's intended recipients may not read it in any detail. The plan is useful, however, because it formalizes in the H staff's minds what they must do and how they must function. It also serves as a charter by putting management and the customer on notice about what the H program intends to do.

Performing Detailed Task Analyses

"***3.2.1.3 Analysis of Tasks.*** Human engineering principles and criteria shall be applied to analyses of equipment-oriented tasks. The analyses shall provide one of the bases for making design decisions, e.g., determining, before hardware fabrication, whether system performance requirements can be met by combinations of anticipated equipment and personnel, and assuring that human performance requirements do not exceed human capabilities . . ." (MIL-H-46855, Military Specification, Human Engineering Requirements for Military Systems, Equipment and Facilities).

The various human factors analyses performed in predesign are often referred to as "task analysis" as if they represented a single, homogeneous activity. The term, however, is far too simple. Task analysis includes the following:

1. The listing of the functions and tasks to be performed (e.g., detect and track aircraft, communicate with fighters, etc.).
2. Task description in terms of the behaviors involved (e.g., activate control, move joystick to track, examine status board to determine aircraft availability).
3. Analysis of tasks in terms of the interrelationships among task, equipment requirements, and operator performance (e.g., for the operator to read alphanumeric symbols on a dynamic TV display at a distance of ___ feet each symbol must be ___ inches high and ___ wide, subtending a minimal visual angle of 12 degrees).
4. The specification of the equipment mechanisms and characteristics required to implement tasks (e.g., the display required to perform the tasks listed above must have the following characteristics . . .
5. Specification of the number and type of personnel required by the design concept (e.g., one 5-level propulsion technician, two 7-level guidance specialists).

The similarity between these items of required information and those which the H section of the proposal was supposed to include is not at all coincidental. Task analyses in predesign (and to a certain extent also in detail design) are progressively more detailed elaborations of those developed during the proposal period. The H specialist therefore does not start his predesign task analyses with a blank sheet of paper; and the more detailed the proposal was, the further along he is. The proposal analyses, however, dealt with relatively large segments of man-machine activity (functions and gross tasks). To specify equipment details requires the specialist to describe system operations at a detailed task level.

For example, assume that in the proposal it was determined that the operators of an air tactical control system had to: (a) detect aircraft; (b) determine whether they are friendly or enemy; (c) determine where they are going; (d) communicate with friendly fighters; and (e) vector the latter within visual sighting distance of the enemy.

With no more detail than this it is difficult to say what special requirements will be imposed on the equipment configuration by operator tasks. (Actually the proposal should be more definitive than the example suggests.) Indeed, at this level of detail each of the above functions can be implemented in several different ways. A more refined analysis is therefore required to describe specifically *how* the operator is to perform these functions. This in turn will indicate how individual elements of the operator's equipment should be configured.

If it is performed correctly, the task analysis is a behavioral design concept because it describes how the system will work in terms of personnel behaviors. Consequently it should parallel and interact with the engineer's development of the equipment design concept.

Because of the swiftness with which the basic design concept is usually developed by engineers, task analyses must usually be performed very quickly and as part of the process of refining the engineer's design concept. On some major system development projects in the past, such as the Atlas, task analyses lagged far behind the design schedule, and vast amounts of paperwork were being generated even after the system had undergone testing. The cost of this paperwork was enormous, and its value was in inverse proportion. Major system development projects may call out task analysis as a specific development responsibility and include a block of time for it, but in most projects the analysis is done incidental to the development of equipment details. Consequently a formal task analysis with special analytic forms and reporting procedures may not be possible. This does not eliminate the need for the analysis, however, or the H specialist's responsibility for developing it.

To influence design (which is the major reason for the analysis) it may be necessary for the specialist to do his analyzing quite informally at breakneck speed while standing over the drawing board and conversing with the engineer. These may be less than desirable conditions under which to analyze a system, but it may be the only way the specialist can hope to make useful inputs to design. For the same reason, items (3) and (4) of the various task analysis outputs listed previously are of greatest significance because these outputs have the most direct bearing on the engineer's design.

We should not get the notion from this necessary informality that all we have to do to perform a task analysis is to list the task, as the engineer does when he develops an operating procedure. He has described the task in terms of observable events (e.g., turn control F33 to full on-position), but he has not described it in terms of human behaviors, much less analyzed the impact of these behaviors on the equipment configuration.

In this chapter we are less concerned with the analysis of functions into tasks and subtasks than with the questions to be asked and answered. The value of a task analysis lies in the answers to these questions, not in the task descriptions that result.

Unfortunately, because the number of questions which we may ask is very great (as well as being determined by the nature of the system under analysis), it is impossible to specify them as algorithms. (The factors which may induce excessive demands on the operator were implied in Table 3-5. These, as applicable to the system being developed, will suggest the questions to be asked.) This is admittedly an unsatisfactory response. Generally, however, the specialist is concerned about any factor that may require the operator to perform more *accurately* and respond more *swiftly* than he does ordinarily. Excessive operator accuracy and speed requirements cause performance breakdown.

Task analysis indicates the points at which the operator may be overly stressed by these speed or accuracy requirements. By examining the demands imposed on the operator and comparing these with the capabilities the operator has to satisfy these demands, the $\underline{H}$ specialist can determine whether a potential mismatch exists. These mismatches are points at which there is an unacceptable probability of error, degradation of performance, or equipment breakdown.

In addition, task analysis answers ensure that equipment and equipment characteristics for man-machine interfaces (e.g., controls and displays) are not overlooked, particularly those required for contingency functions and tasks.

Although task analysis seeks to determine whether the task, equipment, or environment make excessive demands on personnel, the demands referred to are not those which are obviously impossible to fulfill. Common sense avoids requirements such as asking the operator to lift 500 pounds unaided or survive without air. The demands that we say severely stress personnel are those which are capable of being fulfilled but at such cost to the operator that his potential for error or failure to meet requirements becomes unacceptably high. For example,

"an engineer designed a rotating radar antenna in which the maintenance mechanics were supposed to remain continually during their work sessions . . . A physiological psychologist who . . . happened to be looking over the engineer's shoulder noticed that the rate of rotation of this antenna would be optimum for producing nausea on the part of the mechanics . . ." (McCormick, 1969).

Task analysis helps to pinpoint such theoretically feasible, but actually undesirable, conditions by estimating the probability of error occurrence in such situations.

The analytic questions to be asked are of the following type: (a) is there an excessive requirement (given we know the operator's capability); (b) if so, what is likely to happen; (c) what equipment characteristic will minimize that demand?

Assume that temperature must be monitored. We may ask the following questions: What must the operator monitor? Must he report a precise value or simply when a cutoff point has been reached? (In the first case he needs a gauge; in the second, a simple indicator will do.) Is he expected to adjust the temperature when the heat reaches some point? (In that case he needs a control, and that control should be located underneath the gauge.) Are there temperature ranges which are safe, marginal, and critical? (In that case the scale on the meter should be color coded.) If the temperature becomes critical, how fast must he adjust the temperature downward? (If he must respond within five seconds, perhaps operator control is not appropriate and a device should be built into the equipment enabling it to shut itself off.

One of the major functions of task analysis is to elicit detailed human requirements against which the engineer's design can be evaluated. The answers to questions such as those raised in the temperature gauge example set standards for the engineer's design. They specify that either a meter or an indicator are required, that a control is or is not required, that the scale on the gauge should or should not be color coded, or that perhaps human control is not the best answer to the function of controlling temperature.

To the extent that developing a design concept requires one to assume certain personnel-design relationships, task analysis is inherent in design. Whether or not they are paid for by the customer, these analyses are performed. When they are not paid for and consequently are left to the engineer as one incidental part of his design effort, the analyses are performed unconsciously as assumptions implicit in design decisions or decisions often made without the engineer being aware of these assumptions; for example, if a decision is made to automate an equipment without conducting a deliberate (not necessarily formal) function analysis, the decision assumes (possibly without any evidence) that personnel cannot perform required system functions.

Task analysis as a conscious design process is the responsibility of both the design engineer and the H̲ specialist. (It is, however, more the specialist's responsibility than the engineer's because the former has been trained to perform the analysis.) Because it is another form of design analysis, task analysis, like all design, must be an iterative process. Because it is an interactive part of design and because basic design is very quickly completed, task analysis of major functions and tasks must be performed equally rapidly.

The difference between a formal task analysis—paperwork and formats—and the informal task analysis recommended in this section is that the former is paid for (as an activity distinct from other design activities), whereas the latter may not be. On this basis a formal task analysis is, of course, desirable because it ensures conscious consideration of personnel-design relationships; but it ensures this only so long as the formal task analysis does not divorce itself from on-going design.

One of the factors which determines whether or not a formal detailed task analysis is performed is the *size of the system*. When a system is small, requiring only one or two men, the likelihood of having the customer require detailed task analysis is much less than in a system that requires a substantially larger number of men. This is all wrong, of course; the small system may have many more severe human factors problems than the large one. The problem is that every aspect of system development, including its personnel elements and the analyses that relate to these personnel, is in scale with the over-all size of the system. Hence we may well get less task analysis in a small system simply because the system is small.

Another factor which is correlated with system size and which determines the extent of the analysis is the system price tag. All other things being equal, a system which is 1/10th the cost of another will get 1/10th or less the task analytic effort.

Human Factors Related Tradeoff Studies

"***3.2.2.1 Studies.*** The contractor shall identify and conduct experiments, laboratory tests . . . and studies required to resolve human engineering and life support problems specific to the system . . ." (MIL-H-46855, Military Specification, Human Engineering Requirements for Military

Systems, Equipment and Facilities).

All design is one tradeoff after another. Opposing considerations are balanced against each other, and a decision is reached in favor of one or the other.

When engineers in a study of design processes (see Chapter Seven) were required to design a system for moving missile fuel from a storage area to the launch pad, they had a choice of gravity drain methods (locating the storage tanks on high ground and letting fuel drain down to the pad) or of using pumps to perform the same function. Each method had advantages or disadvantages; for example, gravity drain may be cheaper but is slower and requires a particular geographical configuration. Pumps, on the other hand, are faster but more expensive and perhaps less reliable. Among the factors to be considered in the tradeoff is the effect of each configuration on control tasks required of the technician.

There are two reasons for the design engineer to consider human factors in tradeoffs:

1. The tradeoff, whatever its nature, may have serious consequences for personnel use of the system.
2. One of the factors to be traded off may involve personnel (e.g., training time, performance).

Although many design tradeoffs have no H implications, others do. For example, engineers in one study (Meister et al., 1969) were asked to decide between manual, semiautomatic, and automatic means of checking out electronics. The particular methods they selected resulted in different allocations of the number of personnel required, their skill level, and the tasks asked of them.

Even when human factors are considered in tradeoffs, they play a minor role in the engineer's final decisions. Table 5-3 lists the major tradeoff factors which engineers consider and their relative importance. Note that the human factors parameters (skill, availability, manpower cost, personnel, etc.) rank relatively low, after, in fact, all the major physical parameters. Essentially the same results were secured with three different samples of engineers who were asked to rank the factors they considered in making design tradeoffs.

Why then describe human factors tradeoffs? First, because human-factors tradeoffs may be critical to system effectiveness when they are performed, especially in advanced systems with high human involvement; second, because the reader should be aware of the human factors implications of even relatively routine tradeoffs.

A tradeoff, of course, is almost never between just two factors. Every factor has implications for other factors so that, if the tradeoff is to be properly analyzed, the latter must be brought into the analysis; for example, in choosing between two ways of performing an equipment function, the engineer may select one method because it has higher reliability, but he will have to balance the higher reliability against the additional cost of that reliability. Again, an engineer may prefer one method over another because the first requires less operator activity in using the equipment, but he will have to balance this consideration against higher cost and the fact that more automatic equipment generally requires more maintenance.

Table 5-3 Priorities Assigned by Engineers to Design Parameters *

Item	Rank
Equipment performance capability	1
Equipment reliability	2
Cost	3
Maintainability	4
Producibility	5
Development schedule	6–7
Personnel skill required by equipment design	(Tie)
Skill availability	8
Manpower life cycle cost	9
Quantity of personnel available	10
Length of training	11–12
Cost of training	(Tie)

* Taken from Meister et al., 1969.

There are, of course, no absolute tradeoff priorities. Every tradeoff is performed in the context of a specific equipment design problem. Although for experimental purposes we can develop abstract tradeoff problems, we have found that engineers need the equipment development context to assign an appropriate weighting to the factors. Priority number one in one design situation may be third or fourth in another.

Two conclusions can be derived:

1. In any tradeoff human factors parameters cannot be considered in isolation. The value of these parameters depends on the particular developmental context in which they are considered.

2. If the procuring agency wants human factors parameters to be considered in tradeoffs, it must specifically assign a high priority to these parameters. The customer's explicit requirements are the major deter-

minant in priority weighting. For example, if lowered cost is indicated by the customer as a primary factor, all other considerations will take second place to it. The author has also found that when considerable material on operator considerations is added to a procurement specification, regardless of other requirements, the engineer tends to give these considerations more weight in his decisions.

A tradeoff is a decision which involves criteria for weighting the various inputs that enter into the design decision. This decision can be reached analytically, consciously, and logically on the basis of objective data or intuitively, unconsciously and illogically on the basis of an acquired bias to one solution over another. The reader will see in Chapter Seven that both kinds of tradeoffs exist but that, unfortunately, many tradeoff decisions are made on the basis of or influenced by derived biases.

There is, of course, a kind of tradeoff in which the weight of the evidence so clearly favors one alternative over another that the decision in favor of that alternative is irresistible. An example of this is the decision between having a man lift unaided a 500-pound weight and having that weight lifted by automatic lift. This decision is obvious. For such a design decision a tradeoff study, whether formal or informal, is not necessary. A secondary tradeoff, to determine the *extent* of the operator's involvement in the lifting process, is a much more meaningful problem. Should the operator drive his fork-lift on to the elevator so that both he and his load can be transported to a higher level, or should the equipment be unloaded on the elevator and then reloaded at its destination? This is an illustration of the fact that top-level decisions are often made with relative ease; it is the more molecular decisions that may present difficulties.

It is only when the conclusions to be drawn from the available data are obscure or when they do not point logically to a solution or when data are insufficient, that tradeoff studies must be performed. Of course, the problem at issue must be important enough to warrant a study; the decision between two types of controls (e.g., a toggle switch or rotary) or their location on a panel would not ordinarily warrant a study.

Although relatively few design tradeoffs are centered around human factors variables, many of them have implications for personnel functioning and therefore require H inputs. How do we know when a problem requiring a tradeoff study requires such inputs? The following criteria should be applied:

1. When the decision reached or the solution selected will have significant effects on

(a) the number of personnel required to operate and/or maintain the equipment;
(b) the skill level of personnel required to operate and/or maintain the equipment;
(c) the amount of training these personnel would need;
(d) the manner in which they would be required to perform (i.e., the nature and difficulty of their tasks; this in turn would affect the efficiency with which personnel can perform their tasks);
(e) their safety.

2. When a major design requirement imposes a constraint on the number and type of personnel (e.g., the developer is required to design so that two men can operate four pieces of equipment). The question to be answered is, what effect will this constraint have on design?

3. When the problem involves the capability of personnel to perform a job and the question asked is, will personnel be overloaded by a particular design solution? For example, one design concept for an air-launched missile may require the pilot, using a television display, to guide the missile visually to the target over a five-mile distance. The question is, can pilots do this within a specified minimum error?

In most cases the tradeoff study will not be experimental in nature (i.e., require the gathering of laboratory-controlled data). Rather it will be a largely logical, systematic examination of alternatives using available data, an attempt to anticipate the potential consequences of design factors. Only when sufficient information on which to base a decision is lacking so that logic and empirical data do not suffice, will an experimental study be required. Hence in the tradeoff we accept the data we have; in the experimental study, we generate additional data.

The need for formal tradeoff studies (i.e., those consciously and deliberately performed) will therefore not be frequent. Such studies are performed relatively early in system development (e.g., in predesign) because major problems requiring tradeoff studies occur primarily in early design. At later stages the problems have been solved, or design has proceeded so far that only a major perturbation would require such studies.

Another factor which militates against formal tradeoff studies is speed, the necessity to make a decision rapidly in order to meet an explicit or implicit deadline. The informal tradeoff study is usually conducted over the drawing board or the engineer's desk. The fact that the engineer is usually working against time contributes to his predisposition to overlook human factors considerations. This is not to say that the informal tradeoff study does not consider opposing factors with evidence favoring one side or the other, but the opportunity for emotional bias to determine the decision is much greater than in a formal study.

There are no special techniques for the tradeoff study. The general procedure for performing any tradeoff study (regardless of its degree of human factors involvement) is to do the following:

1. Determine the goals of equipment design or what functions equipment must perform.
2. Determine the problems involved in meeting these goals or the problems which prevent an immediate decision.
3. Determine the tradeoff factors to be considered and the relative weight that should be assigned to each.
4. Determine what data are available to make a decision, their adequacy, and their implications.
5. Determine the alternative design solutions which permit achievement of system goals.
6. Determine the advantages/disadvantages of each design alternative.
7. If a given alternative is to be selected, anticipate the consequences (in terms of equipment and human factors) that follow.
8. Select the design alternative which most nearly achieves system goals.

From the standpoint of both the H specialist and the engineer, the tradeoff study is effective only to the degree that the consequences of each design parameter can be specified and contrasted. In making his inputs to the tradeoff study the H specialist must therefore phrase the consequences of his recommendation in concrete design terms.

Predicting Operator Performance

"3.2.1.1.2 . . . Estimates of processing capability in terms of load, accuracy, rate and time delay shall be prepared for each potential operator/maintainer . . . function . . ." (MIL-H-46855, Military Specification, Human Engineering Requirements for Military Systems, Equipment and Facilities).

Increasingly the requirement is being imposed by the customer on the developer of major systems to predict (quantitatively, of course) the adequacy with which system personnel—those who will eventually run the new system—will perform.

Although most equipment development projects (which are small to medium size, naturally) do not presently include this requirement, it is likely that as predictive methods become more accurate and easier to apply the requirement will be extended to all projects involving personnel activities.

Ideally, prediction involves two basic elements: (a) specification of what the system requires in terms of the performance measure selected; (b) knowledge of how well personnel will, on the average, be able to perform, again in terms of the selected performance measure. A comparison of the two will indicate whether operator performance will satisfy system requirements.

Since, as indicated in Chapters Three and Four, system specifications rarely include quantitative human performance requirements, it is rarely possible to accomplish this comparison. However, it is still possible to accomplish other prediction goals:

1. To suggest the manner in which the man-machine configuration should be designed. The operator performance prediction technique should indicate which functions the operator can and cannot perform; those which he cannot perform must be assigned to equipment (this is the function allocation use of the technique).
2. To aid in the choice of alternative man-machine configurations. The alternatives can be compared in terms of the performance efficiency that can be expected to result from the configuration, and, all other things being equal, the most efficient would be selected.
3. To predict operator performance when the system goes into operational usage. If we can predict what that performance will be under operational conditions and predicted operator performance is less than desirable (even though a quantitative system requirement does not exist), system design must be modified to improve that performance.
4. The predictive technique can actually serve as a model for the collection of human performance data in the operational environment. The definitions of measures to be employed and methods of integrating the performance of individual behaviors can be used as a strategy for observing and recording behavior in the real world.
5. To improve the man-machine configuration. If we can be reasonably sure that the operator of an equipment will make repeated errors in its operation and if we can pinpoint the probable cause of these errors (e.g., an inappropriate control or display arrangement), then something can be done to make the equipment easier to operate—a design change, a change in manning, in procedures, or even automation of the entire operator function, if nothing less will do.

Various measures may be applied to this prediction. These include the following:

1. Probability of error (how likely is the operator to make an error).
2. Time to respond (will the operator respond quickly enough for a

system input; will the time he takes to perform a task be within requirements).

3. Response consistency (will the operator's performance over repeated trials be consistently within bounds).

4. Response range (will his performance exceed maximum-minimum limits).

5. Frequency of response (will the number of responses the operator makes be sufficient).

6. Response accuracy (will the operator's performance be within specified accuracy requirements).

Any or all of these measures may be applied simultaneously (as appropriate) to the same behavior.

The most meaningful prediction measure is, in the author's opinion, the probability of task completion or, in the vernacular, that the job will be done correctly (within allowable error tolerances and in the time required). The author has a preference for a rather simple predictive metric defined as r_s/r_n, where r_s is the number of behavioral units which the operator successfully completes and r_n is the total number of such behavioral unit performances attempted. This is essentially equivalent to the measure of achieved equipment reliability.

There are other more complex metrics, however; for example, Swain, 1963, expresses total system or subsystem failure rate resulting from human error as

$$Q_T = 1 - \left[\prod_{K=1}^{n} (1 - Q_K) \right],$$

where "Q_T is the probability that one or more failure conditions will result from errors in at least one of n classes of errors, and the quantity in brackets represents $(1 - Q_1)$ $(1 - Q_2)$ ... $(1 - Q_n)$." This overall probability is ultimately derived (by a process which is too lengthy to discuss here) from determination of a predicted error rate using the Data Store (Munger et al., 1962) together with the determination of the probability that the error will result in system failure.

What we have been talking about are general predictive measures. They must be translated into system-specific terms such as bomb-aiming error or number of messages sent per unit time. The nature of the system outputs obviously determines the content of the measure.

A number of assumptions are involved in making a prediction: (a) that how well personnel function determines at least in part how well the total system will function; (b) that we can predict operator performance correctly or at least to an accuracy which permits meaningful conclusions; (c) that design or personnel changes are capable of modifying personnel performance in a positive manner. Assumptions (a) and (c), which underly the H discipline, were discussed in some detail in Chapter One, and empirical evidence was adduced in Chapter Two to support the assumptions. Assumption (b) is still in question although some suggestive evidence for the idea is available.

Underlying the requirement to predict operator performance is the further assumption that, since personnel performance interacts with equipment design (functioning) to produce system performance, neither the equipment nor the human aspect can be considered separately. If the reliability engineer predicts equipment reliability, the H specialist must also predict operator performance.

A number of possible operator performance prediction techniques are available, none of them more than crude at the present time. Methods of predicting error-probability are described in Siegel and Wolf, 1959; Payne and Altman, 1962; Swain, 1963; Meister, 1964; and Pickrel and McDonald, 1964. Whatever the particular method, however, there are four basic steps in the process:

1. Analysis of the system, subsystem, or equipment into discrete units for which the prediction of operator performance is to be made. The elements of these behavioral units are the function or task to be performed, the equipment component acted upon by the operator's performance of the function or task, and task or environmental conditions that modify the function or task. Such a behavioral unit might be "read temperature gauge quickly." "Read" is the function or task; "temperature gauge" is the equipment object; and "quickly" represents the modifying task condition.

2. Analysis of the behavioral unit in terms of the parameters that modify or influence the operator's performance of that unit. The accuracy with which the operator reads the gauge in (1) above also depends upon equipment, task, and environmental parameters such as the number of other displays on the control panel, the required accuracy of response, the ambient lighting in which the gauge must be read, etc. Not every parameter is significant for a particular task, but once it is determined that a parameter does influence performance, it must be taken into account in making the prediction.

3. Assignment of predictive values (such as the probability of successful task performance) to the behavioral unit being predicted. These values are presumably available from a store of already accumulated performance data describing this or similar behavioral units. The goal

of the prediction is to describe the behavioral unit in terms of the appropriate response measures listed previously.

4. The predictive values assigned to the individual behavioral units are combined to form a single value that represents the anticipated performance of all the behavioral units involved in accomplishing a given job. Behavioral unit predictions (usually formulated in terms of a probability value) are combined in accordance with task dependence/ independence relationships.

There is some evidence for the feasibility of all this. In a recent study (Meister et al., 1968) the author found it possible to distinguish among subsystem designs produced by engineers on the basis of predictions of operator performance in running these subsystems. Six missile engineers were asked to develop the conceptual design of a ground propellant transfer system for a space launch vehicle like Titan III. This was done under controlled conditions, using engineering and H inputs based on actual developmental data.

Each of the engineers produced a complete subsystem design: equipment descriptions, tolerances, schematics, operating procedures, drawings of the control equipment, and lists of components. To evaluate the adequacy of the subsystem designs they produced, a number of systems effectiveness measures were applied, of which one was equipment reliability and another operator performance reliability. When the human performance predictions were rank correlated with the equipment reliability predictions of the same set of designs, the correlation was significant at the .05 level of statistical significance. The prediction methodology was crude; the data base had many holes in it; but the results indicate the potential for use of the technique.

The prediction of personnel performance should not be considered a unique procedure. It is entirely parallel to reliability predictions of equipment design that are made early in design and for exactly the same reasons as those for personnel prediction. The operator performance prediction technique grew out of the equipment reliability prediction methods developed over the last 20 years and closely resembles these methods. Indeed, the methodology is often referred to as "human reliability" to indicate that what is being measured and predicted is an attribute of the system entirely parallel with that of equipment reliability.

In fact, the definition of human reliability as proposed by the author (Meister, 1966) is almost the same as that of equipment reliability. Human reliability is defined as "the probability that a job or task will be successfully completed by personnel at any required stage in system operation within a required minimum time." The definition of equipment reliability is "the probability of performing a function under specified conditions for which designed, for a specific period of time" (Zorger, 1966).

If we assume that personnel operation of equipment influences the reliability of that equipment to some degree, however slight, then that influence must be accounted for in predictions of equipment reliability. To ignore human performance in reliability estimates is to assume that operator performance is invariably perfect (i.e., 1.0). Since it is apparent that operator performance is rarely—if ever—perfect, equipment reliability estimates that fail to take account of human performance turn out to be grossly inflated.

Both equipment and human reliability predictions depend on a base of empirical data for their predictions. A data base has been prepared from the experimental literature by Altman and his co-workers (Munger et al., 1962), by Blanchard et al., 1966, and by the author (in Hornyak, 1967). These data bases are usable, but deficient. The problem seems to be not that there is insufficient data, but that much of the available data is inappropriate for predictive purposes. Lack of an adequate data base is one of the great failings in the predictive methodology as it is in the H̲ discipline as a whole. There are many difficulties in the application of personnel performance predictive techniques (see Altman, 1968, for an excellent review of these); however, if sufficient data describing what personnel will do under different equipment and task conditions were available, the difficulties referred to would be insignificant.

This is not to say that no personnel predictive data exist. It is safe to say that there is enough to apply one of the human reliability predictive techniques in at least a very crude manner. For example, Irwin and his co-workers, 1964, made predictions of certain maintenance tasks performed in engine maintenance of the Titan II. Predicted operator reliabilities based on the Data Store (Munger et al., 1962) were compared with actual observed reliabilities for 26 maintenance operations. The mean predicted reliability of the 26 operations was .9972; the mean observed (actual) reliability of the operator tasks was .9989. These techniques are therefore at least effective enough to enable the developer to differentiate between different design configurations on the basis of the operator's probability of correct performance although they cannot be used to make absolute predictions (i.e., personnel performance will be .9346 or .8788).

These techniques, it must be admitted also, are not easy to apply. Besides the lack of a large enough data base, there arises a problem of defining precisely what the behavioral unit (e.g., a task) is for which a prediction is to be made. Another problem is the basis for combining predictions for a series of such behavioral units particularly when tasks are interdependent. Interdependence poses severe difficulties. Another problem is that the techniques cannot readily handle continuous tasks (e.g., tracking), complex cognitive tasks (e.g., troubleshooting), or the performance of groups. The reader will note that the difficulties of task definition and integration are common to all human factors problems.

The existence of such difficulties should not, however, serve as an excuse for failing to attempt operator predictions, however crude. Not to make such predictions is to ignore the quantitative effect of the human element on system performance. If the author emphasizes the importance of quantitative operator performance predictive techniques, it is because he anticipates that eventually the refined methodology will become the cornerstone around which all sophisticated H work will be performed.

Obviously, the earlier in the design process the operator performance prediction is made, the less absolute confidence one can have in it. However, the results of the study by Meister et al. (1968) suggest that even at very early design stages (i.e., conceptual design) sufficient information is available to make comparative predictions with some confidence that they are at least "in the ball park."

The raw system data required to make human reliability predictions consist of two major items: detailed operating procedures and preliminary sketches of the control-display equipment used in conjunction with these procedures. These are often produced quite early in development. Even if they are not available in finished form, the information they contain can be extracted from the engineer by judicious questioning concerning the procedures that are to be employed and the nature of the equipment configuration.

There is a certain amount of serendipity in this questioning process. The necessity for the engineer to explain his operating procedures in detail may elicit equipment design details which he has overlooked.

If operating procedures and control panel layouts are available, the H specialist can make his predictions in relatively short order. The author found that he could make a prediction for a subsystem design in from 2 to 4 hours, using predictive tables like those in Munger et al. (1962). Hence the specialist and the engineer should not be put off by the fear that operator predictions will take an inordinate length of time or will be extremely costly.

Human Engineering in Predesign

Definition.

"3.2.1.2 The recommended design configuration shall reflect human engineering inputs . . . to insure that the equipment will meet the applicable criteria contained in MIL-STD 1472." (MIL-H-46855, Military Specification, Human Engineering Requirements for Military Systems, Equipment and Facilities.)

Human engineering is the application of human factors design criteria to the development of the equipment configuration. Although human engineering depends on function allocation and task analysis (see Figure 5–5) with a constant interaction among them as design iterates, human engineering can be distinguished from the earlier analytic activities by its emphasis on equipment details.

Human engineering is a—if not *the*—major H̲ activity in system development. Of all the things the H̲ specialist does, it is the most closely related to the detailed design of the ultimate equipment, and the activity most often performed from the initial proposal until the system is turned over to the customer.

In predesign it is—or should be—one of the inputs to the basic design concept; and, as long as problems arise that require redesign of the equipment, human engineering contributes to that redesign. Not only is human engineering performed throughout system development, but the manner of its performance does not change significantly, except that as overall design becomes more detailed human engineering deals with more molecular equipment units. Consequently, the following discussion applies with few, if any, modifications to detail design and test phases.

Since human engineering design must support human requirements, it follows from, and depends upon, earlier function and task analyses. If we do not know what the operator is supposed to do and how he must do it, we cannot understand how equipment should be designed for him. Human engineering as we sometimes see it being performed in control panel layout, without prior function allocation and task analysis, is like childbirth without gestation, at best a shuffling of controls and displays like counters around a child's game board.

Despite the common conception that human engineering is "control panel layout," "knobs and dials," it includes far more than that, and especially so in earlier developmental phases when fundamental design decisions must be made:

1. Analysis of proposed equipment configurations in terms of man-machine interfaces required by tasks.
2. Determination of man-machine interface characteristics.
3. Tradeoff between alternative design configurations (insofar as the aspects traded off relate to human factors).
4. Selection of controls and displays included in control panels.
5. Control panel layout, of course.

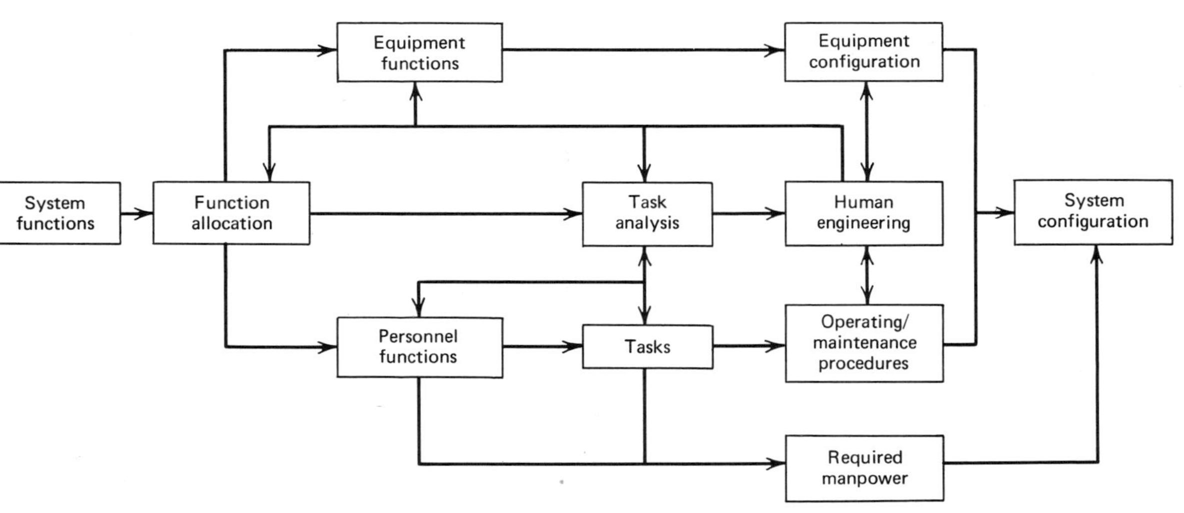

Figure 5-5. Human engineering (HE) takes task analytic information as well as equipment information to recommend appropriate man-machine interfaces for the equipment configuration and operating/maintenance procedures; HE feeds back to task analysis (TA) and function allocation (FA) when the man-machine interface presents such difficulties that tasks and functions have to be redefined.

6. Design of consoles, including arrangement of control panels within the console.
7. Design of the working environment (i.e., the arrangement of equipments within the work area and specification of lighting, temperature, and noise requirements).
8. Specification of human factors characteristics that make equipment maintainable (e.g., access spaces and location of test points).
9. Design of life support equipment (e.g., oxygen equipment).

Human Engineering Roles. In performing these activities the H̲ specialist serves in a number of roles:

1. *As an information resource*, he supplies detailed information about, and interprets the meaning and applicability of, human factors design principles included in design standards like MIL-STD 1472 (1969) or handbooks like Woodson and Conover (1964) or Morgan et al., 1963. The types of questions generally asked relate to

 (a) *operator capability*, for example, how much weight can the average man lift; what is the highest noise level tolerable before hearing damage results;
 (b) *design applicability*, for example, what is the minimum distance allowable between controls; what type of control should be used for a particular function;
 (c) *contractual applicability*, for example, when the engineer asks which section of MIL-STD 1472 applies to a particular design problem.

Although this information is available to the design engineer in documents such as those referenced above, the manner in which these documents are written—with considerable psychological terminology—tends to "turn off" the engineer. Moreover it is known that engineers prefer to receive their information from others with whom they can discuss that information. When faced with the necessity of analyzing H̲ documents, the engineer often has difficulty finding relevant material and often lacks the background to interpret the information (because he does not deal with these problems on a continuing basis and because human engineering standards are often qualitative and general).

The availability of the H̲ specialist as an information source helps also to create rapport with the engineer; this aids in the former's design consultations. The engineer's acceptance of the information and the H̲ specialists's role as an information source depends on the accuracy of the information transmitted, its applicability to specific design problems, the ease with which it is understood by the engineer, and the speed with which it is given. As will be seen in Chapter Seven, many engineers are inherently suspicious of workers in other disciplinary vineyards and prefer to secure their information from colleagues. They demand specific answers to specific questions (something which the H specialist may have difficulty providing because of the grossness of much of his data); and engineers want their information when they want it, without the necessity for literature reviews and lengthy analyses.

2. The H specialist assists in the *selection of control-display components* on the basis of equipment characteristics which have behavioral implications for helping or hindering operator performance; for example, toggle switches or pushbuttons have been found to be most useful for discrete control activation; rotary controls, when a large number of settings are required, etc. The H specialist assists in making such recommendations although in many cases the recommended application is relatively obvious. When the control or display task is complex, as in the selection of particular joysticks or types of CRT's, altimeters, etc., his advice is more significant.

Again, this kind of information is available in H reports, but it may be difficult for the nonspecialist to unearth despite efforts to make the material more available through handbooks like NASA's Biotechnology handbook (Webb, 1964).

In general, except for highly complex controls and displays, the H specialist contents himself with recommending component *types;* the design engineer selects the specific *part.* Many H workers, however, collect information such as manufacturers' catalogues on new control-display hardware and can supply valuable advice even in the specific part area.

3. The H specialist also plays a valuable role in *design reviews.* The company design review is a presentation and defense by the design engineer of his design concept against anyone who criticizes it. As do all other disciplinary representatives, the H specialist brings up questions concerning potential weak points that became apparent during his design analysis. He suggests possible solutions to these problems; when he has assisted in the development of some aspect of the design (e.g., control panel layout), he explains the human factors rationale behind that design.

Of lesser, but still substantial, importance is the customer-conducted design review. In major systems these may be contractually required at intervals such as every 90 days. The contractor's H specialist has an important role in that review in answering the customer's questions, explaining and rationalizing the human engineering features of the design, and, if all else fails and the customer is displeased with some human engineering design feature, assuring the customer that contractor specialists are available to make suitable modifications.

Of the two types, it is much more important for the H specialist to participate in the company review because here basic design decisions are made and unmade. The customer design review is important but is largely after the design fact. In the past it was often difficult for the H specialist to get permission to participate in company design reviews, but under pressure from the customer this procedure is becoming more common.

4. The H specialist's major function is to *consult* with design engineers on the design of equipment with major human factors features, most often control panels and consoles. (There are some occasions when the H specialist has the responsibility for designing these, but these opportunities are relatively few.)

This activity specifically includes making suggestions concerning desirable design configurations and reviewing the engineer's drawings but also subsumes the functions previously described in this section.

Human Engineering Ground Rules. Some of the specific methods by means of which human engineering is performed (with particular reference to control panel layout) were described in Chapter Three. The general method is for the H specialist to apply his knowledge of human engineering principles and data to a particular design problem. To do this effectively the H specialist must follow the rules laid down below:

1. *Know as much as possible about the purpose of the equipment to be designed, the specific operations to be performed with the equipment, the reasons for the engineer's design concept, and the engineering constraints on design.*

Proper human engineering cannot be performed unless the H specialist is fairly familiar with equipment functions. It is especially important to know the reasons why the engineer has decided to design his equipment in a particular way. Without this rationale the H specialist will not know the degree of flexibility he has in making his design recommendations. The H specialist must know, for example, if the engineer has decided upon a high degree of automaticity because he does not believe an operator can perform certain tasks without error or because of other reasons. This knowledge will permit the specialist to determine whether this design-rationale is justified, and if not, what design alternatives should be suggested.

2. *Volunteer his design concept rather than waiting for the design engineer to find him.*

Unless a high degree of rapport already exists between the engineer and the H̲ specialist or there is a contractual requirement requiring H̲ review of the design, it is unlikely that the engineer will seek out the specialist for advice or consultation concerning his design concept. The engineer's lack of knowledge of H̲ capabilities and his reluctance to submit his design concepts to criticism by other than engineers makes it unlikely that he will seek the specialist out. Consequently, in the interdisciplinary dance between H̲ and the equipment engineer, it is the H̲ specialist who must lead.

Some H̲ groups have been completely ineffective because they "stayed in their own corner" and insisted on the engineer coming to them for consultation. It is perhaps rash to make hard and fast statements, but the author believes that the H̲ specialist is not doing his job properly unless he spends 50% of his time contacting design engineers.

3. *Approach the engineer with a specific design concept in drawing form, well documented in terms of design implications and reasons for this concept.*

When the H̲ specialist does approach the engineer, it should be with a well thought out design concept. Easier said than done, perhaps, but at least the specialist can approach him with specific relevant questions concerning the rationale for particular design characteristics rather than with general offers to help.

If a design concept is presented by the specialist, it should be in drawing form (detailed drawings are not, however, required). The implications of the design concept being presented should be made clear to the engineer.

4. *Avoid behavioral "jargon" in communicating with the engineer and utilize wherever possible the engineer's own language.*

Above all, the analyses and recommendations that the specialist makes must be phrased in the design terms that the engineer can understand and apply. Nothing is more fatal to the effective implementation of the human engineering effort than for the specialist to talk to the engineer in psychological terms and to ignore the realities of the immediate design problem. It may be necessary for the specialist to provide a behavioral rationale for the design concept he advances, but he should emphasize the concept, not the rationale.

5. *Substantiate his recommendations with factual data and military requirements.*

The most effective rationale for an H̲ recommendation is a military requirement clearly spelled out. It is useful to review MIL-STD 1472 (1969) and pinpoint the items which call for a particular design recommendation. The next most effective rationale is the use of data and logic which demonstrate that the specialist's design recommendation is unavoidable or that the engineer's design concept can be improved on for behavioral reasons. If possible, the data presented should be quantitative and illustrated by concrete examples.

6. *Present the H design recommendation as quickly as possible after design is initiated and not later than the engineer's own design schedule requires.*

The engineer's design concept is developed to a great extent during the proposal and is completed (in outline, at least) shortly after the contract go-ahead is received. This may not be the case for control panels or consoles which are relatively molecular, but, since these depend on the basic design concept, it does not do to wait upon the engineer to present H requirements.

7. *Modify H design recommendations quickly when engineering constraints require this.*

The H specialist may not be aware of some factor which might invalidate his design recommendation when he presents it. He must be prepared for a continuing series of face to face confrontations with the engineer who will attack and demolish (if he can) the specialist's design rationale and who may raise points about engineering constraints that require some adroit maneuvering by the H specialist. To do this the specialist must be familiar with the engineer's design concept, the reasons for it, and the particular constraints the engineer is working under.

8. *Recognize the engineer's prime responsibility for design at all times.*

In most companies the engineer is directly responsible for the design of a particular component, assembly, or equipment. More recent military specifications or directives call for H signoff of relevant designs (e.g., control panels, consoles), and, when these directives are fully implemented, the specialist can exercise a veto over design. For many projects, however, these provisions are not contractual requirements. When they are not, the lever the H specialist has is the trust the engineer has in his judgment. Even when the signoff directive is in force, it has not been unheard of for the engineer to present his designs for H review at the last moment and to imply that any schedule delay will be the specialist's fault.

9. *Maintain continuing informal contact and possess good rapport with the engineer responsible for the design.*

The direct impact of human engineering on design forces a very close relationship between the H specialist and the engineer. All design is largely a communal effort performed in project teams whose members have subdivided the design responsibility among themselves. To be maximally effective, the H specialist as human engineer must be part of such a team. On those unfortunately too frequent occasions when he is not included in the project team, his efforts may be nullified.

When we discuss the organization of the H group, we shall see that the primary criterion of an effective H organization is one which promotes the closest possible contact between the specialist and the engineer. The project team offers this opportunity but only if the H specialist is given a meaningful task on that team.

Another perhaps even more important reason for his maintaining the closest contact with the engineer is the specialist's dependence on the latter for the basic engineering information which serves as inputs to his analysis. To "human engineer" a design is first to have a design to work on; the engineer creates design alternatives and communicates their details to others.

It is often difficult for the H specialist to maintain the desired intimacy with the engineer. Because the basic design concept is primarily the engineer's, the H specialist may appear—as he is to a certain extent—a critic of that design. Although the effective H specialist does not think of his activity as being negative, it is very easy for the designer to see him in that role.

10. *Perform human engineering functions in the engineer's design area, over the drawing board, if necessary.*

Everything which has been said previously implies that the H specialist must go to the design engineer rather than vice versa. This means working within the latter's customary environment.

11. *Monitor the engineer's design with concrete recommendations for any design features criticized.*

Since the engineer in most cases designs the equipment with which the H specialist is concerned, the latter is often forced into a monitoring role. Unless specific arrangements are made between the two of them for this monitoring function, the engineer's design effort will proceed apace without any consideration of the specialist's interest in that design.

Should the specialist have reason to comment unfavorably upon any of the engineer's design features, those comments should be phrased in specifics. General negatives, such as, "the ___ design feature is unsatisfactory," will only irritate the engineer unless he understands the reason for the complaint. (See Figure 5-6 for a sample human engineering report.)

12. *Discuss any negative comments on the design with the engineer before formally reporting them to engineering management.*

This is not only a matter of courtesy but also good tactics since it is possible that the criticism is founded upon a misapprehension of the design concept or the engineering data that led to it. In addition, the specialist risks the loss of rapport with the engineer unless conflicts can be ironed out directly between them. Only disagreements with major consequences for system operators should be carried to higher authority.

To: Head, Simulator Design Group
Subject: Human engineering evaluation, Launch Escape Tower System Simulator
Purpose: To verify that the design of the Launch Escape Tower System Simulator meets the requirements of Apollo Life System Design Criteria.
Results: The results of the evaluation showed the following discrepancies:

1. All pushbuttons on the Pyrotechnic Simulator were not concave or "non slip" and were smaller than the recommended diameter. (Reference paragraph 1.17.11, 1.17.15 and 1.17.21 of SRM 63-02, Designers Checklist.)
2. S1, S2, and S3 rotary switches on the Instrumentation Simulator had twenty four (24) positions instead of the recommended twelve (12). (Reference paragraph 1.18.5 of Designers Checklist.)
3. The green "squib" lights on the Pyrotechnic Simulator were labeled below the lights and not above as the remaining lights are labeled. (Reference paragraph 4.3 and 4.11 of Designers Checklist.)
4. On the Instrumentation Simulator S4, S5 and S6 are all three position toggle switches. On the Pyrotechnic Simulator S5 and S6 are three position toggle switches. These switches should be three position rotary type switches. (Reference paragraph 1.16.1 of Designer's Checklist.)

Recommendations:

1. Design of the pushbuttons should be changed to be a minimum of 0.5 inches in diameter and the pushbutton surface should be concave to fit the shape of the fingertips. (Reference pages 3-10 and 3-11 of SRM 63-02.)
2. The green "squib" lights on the Pyrotechnic Simulator should be labeled above the lights instead of below to conform with labeling of the other lights.
3. It is recommended that rotary switches be used when three or more tasks are to be performed. Toggle switches on the Instrumentation Simulator and Pyrotechnic Simulator should be replaced by rotary switches. Three position toggle switches are not recommended because they increase the possibility of error.

Concurrence: Mr. Farmer, in charge of the Simulator design, and with whom this memo has been discussed, concurs with the recommended changes.

Figure 5-6 Sample human engineering evaluation report.

13. *Compromise H recommendations when functional (e.g., electrical, mechanical) constraints on recommended design changes exist.*

Human engineering does not exist apart from other design aspects. To select an overly simple example to illustrate the point, it may be desirable to locate a particular control component at the edge of a control panel. However, the width of the mounting brackets required for the control may make it necessary to move the component away from the preferred location and toward the center of the panel. The H specialist must recognize that only after functional requirements are satisfied can we think about human engineering. However, this should not be used—as it sometimes is—as an excuse to waive all consideration of human engineering.

The effective H specialist also recognizes that design is a continuing compromise among many, often conflicting, demands, and that the demands of human factors are only one among the rest. There may be performance requirements that dictate one particular configuration, however much the H specialist would like to see another. The freedom to affect design is also constrained by the decisions, some of them hurried and poorly thought out, that were made in the proposal period. Design is like a tree whose trunk is the proposal configuration. The single bough the H specialist would like to bend may not grow in just the direction he wants because of the way the trunk grew.

14. *Eschew recognition for H design contributions except when required to report H progress.*

It was indicated in Chapter Two that human engineering design characteristics are considered by engineering management as merely one aspect of "good" design (in general). It is therefore difficult for the H specialist to emphasize individual design characteristics as representing special human engineering contributions. It is naturally desirable to point out these contributions, but engineers often feel it impossible to pinpoint any individual discipline as making more significant contributions than any other.

Continuation of Human Engineering Efforts. Because design proceeds continuously, the human engineering efforts described in this section are not confined to predesign but continue throughout development to the extent, of course, that they are required by different developmental phases. The same activities, although at a more molecular level, continue through detail design and into test. The attempt to confine human engineering to a specific time frame—to insist, for example, that human engineering review of control panels will be performed only after preliminary drawings are made but before final release—is fatal to the specialist's need for continuing contact with the engineer; and also tends to force on the H specialist the essentially negative role of over-the-shoulder design evaluator. If he is time-restricted, then design comes to the specialist already largely frozen.

Consequently the H group must insist that, if the company is to have any human engineering effort at all, it must continue through all those phases of system development which require design activity. If then, in discussing activities in the other developmental phases, we do not specifically describe human engineering, it should be understood that what we have said in this section also applies in the others.

REFERENCES

Altman, J. W., 1968. *Progress in Quantifying Human Error.* Paper presented at the Electronics Industries Association Systems Effectiveness Workshop, Chicago, Illinois, 18–20, September.

Blanchard, R. E., et al., 1966. *Likelihood-of-Accomplishment Scale for a Sample of Man–Machine Activities.* Technical Report, Contract Nonr-4314(00), Dunlap and Associates, Santa Monica, California, June. (AD 487 174)

Department of Defense, 1969. MIL-STD 1472, *Human Engineering Design Criteria for Military Systems, Equipment and Facilities,* 1 December.

Department of Defense, 1968. MIL-H-46855, *Military Specification: Human Engineering Requirements for Military Systems, Equipment and Facilities.*

Hornyak, S. J., 1967. *Effectiveness of Display Subsystem Measurement and Prediction Techniques.* Report RADC-TR-67-292, Rome Air Development Center, Griffiss AFB, New York, September.

Irwin, I. A., J. J. Levitz, and A. M. Freed, 1964. *Human Reliability in the Performance of Maintenance.* Aerojet-General Corporation, Sacramento, California, July.

McCormick, E. J., 1969. The Industrial Engineering/Human Factors Interface. *Human Factors,* **11** (2), 107–11.

Meister, D., 1964. Methods of Predicting Human Reliability in Man–Machine Systems. *Human Factors,* **6** (6), 621–646.

Meister, D., 1966. Human Factors in Reliability. Section 12. In W. G. Ireson (Ed.), *Reliability Handbook,* New York: McGraw-Hill.

Meister, D., D. J. Sullivan, and W. B. Askren, 1968. *The Impact of Manpower Requirements and Personnel Resources Data on System Design.* Report AMRL-TR-68-44, Aerospace Medical Research Laboratories, Wright-Patterson AFB, Ohio, September.

Meister, D., D. J. Sullivan, D. L. Finley, and W. B. Askren, 1969. *The Effect of Amount and Timing of Human Resources Data on Subsystem Design.* Report AFHRL-TR-69-22, Air Force Human Resources Laboratory, Wright-Patterson AFB, Ohio, October.

Morgan, C. T., et al., 1963. *Human Engineering Guide to Equipment Design.* New York: McGraw-Hill.

Munger, S. J., R. W. Smith, and D. Payne, 1962. *An Index of Electronic Equipment Operability: Data Store.* Report AIR-C43-1/62-RP(1), American Institute for Research, Pittsburgh, Pennsylvania, January.

Payne, D., and J. W. Altman, 1962. *An Index of Equipment Operability: Report of Development.* Report AIR-C43-1/62-FR, American Institute for Research, Pittsburgh, Pennsylvania, January.

Pickrel, E. W., and T. A. McDonald, 1964. Quantification of Human Performance in Large, Complex Systems. *Human Factors,* **6** (6), 647–662.

Siegel, A. I., and J. J. Wolf, 1959. *Techniques for Evaluating Operator Loading in Man–Machine Systems.* Applied Psychological Services, Wayne, Pennsylvania, February.

Swain, A. D., 1963. *A Method for Performing a Human Factors Reliability Analysis.* Report SCR-685, Sandia Corporation, Albuquerque, New Mexico, August.

Tessmer, R. J., 1967. *Criteria for Systems Tradeoffs, Vol. I.* Report RADC-TR-66-389, Rome Air Development Center, Griffiss AFB, New York, May.

United States Air Force, 1964. Report AFSCM 375-5. *Systems Engineering Management Procedures.* Air Force Systems Command, Andrews AFB, Maryland, 14 December.

Webb, P., 1964. *Bioastronautics Data Book.* Report NASA SP-3006, National Aeronautics and Space Administration, Washington, D.C.

Woodson, W. E., and D. W. Conover, 1964. *Human Engineering Guide for Equipment Designers* (2nd Ed.). Berkeley, California: University of California Press.

Zorger, P. H., 1966. Reliability Estimation, Section 5. In W. G. Ireson (Ed.), *Reliability Handbook*, New York: McGraw-Hill.

VI

HUMAN FACTORS IN DETAIL DESIGN

Although detail design is supposed to be a distinctive phase in the sequence of system development activities, there is rarely any official sign post that indicates to the developer that he is in a new stage of development.

In predesign the general shape of the system has been sketched; in detail design that skeleton is fleshed out. Alternatively, we can think of the predesign phase as being concerned with analysis of general functions and system operations, whereas in detail design the developer deals with individual equipments and equipment characteristics. All of these distinguishing characteristics are, however, over-simplifications since the developer may design in a detailed fashion in predesign and perform analyses in detail design. The only sensible way to look at system development is as a progressively more detailed elaboration of system characteristics that ends only when the system is turned over to the customer. The only objective landmarks are when the customer awards a contract to the contractor either for preliminary studies (e.g., predesign) or for full scale development of the system (e.g., detail design).

Table 6-1 describes the major H activities performed in detail design.

DEVELOPMENT OF OPERATING AND MAINTENANCE PROCEDURES

"3.2.3. Based upon human performance functions and tasks . . . the contractor shall apply human engineering principles and criteria to the development of procedures for operating, maintaining or otherwise using the system equipment . . . MIL-H-46855."

Two things determine what and how the equipment operator performs: the design configuration (equipment characteristics), and the procedures

Table 6-1 H̲ Activities Initiated in Detail Design

Activity	Activity Duration Design	Production	Test
1. Assist in development of operating and maintenance procedures	———		———
2. Perform human factors evaluation of system	———	———	———
(a) perform checklist design evaluations	———		
(b) perform mockup evaluations	———		
(c) evaluate initial hardware		———	———
(d) perform verification tests			
(1) develop H̲ test plan	———		
(2) conduct H̲ test			———

for operating and maintaining that equipment. Chapter Five emphasized equipment characteristics, but the latter also deserve attention.

Procedures are important because they determine the way in which the equipment will be controlled and what the operator will do. Equipment characteristics largely determine procedures, but not completely so. It is entirely conceivable, for example, that we could have two different ways of operating an equipment. One of these procedures may require controls to be activated in the sequence in which they are arranged on the control panel; the other may require control activation in a sequence differing from their arrangement. The difference in operator performance would be highly significant, the former procedure being easier to learn and performed with many fewer errors, than the latter. This is especially the case in maintenance, particularly for malfunction diagnosis, because the arrangement of internal components does not exercise as much influence on maintenance procedures as the arrangement of controls and displays does on operating procedures. Two maintenance procedures, one based on the split-half method (Miller et al., 1953) and the other based on the frequency of component failure, may be vastly different in terms of their effectiveness for malfunction isolation.

This does not mean in reality that any procedure can be adopted for any design layout. There must be a logical relationship between the function for which the equipment is designed and the way in which it can be operated. The procedure must, for example, account for all the functions to be performed in operating the equipment.

Note that in the last sentence the term "operated" was used and not "maintained." Everything said in this section about operating procedures should apply also to maintenance procedures, but this is true only when the maintenance is of the preventive type; that is, when the maintenance can be programmed in advance in a step-by-step manner. Procedures are difficult to analyze and modify when they cannot be programmed in advance; this is the case with troubleshooting.

Theoretically the procedure should be developed at the same time the design is developed since the procedure is often implicit in that design. In fact the engineer develops at least rudimentary procedures at the time of his design, if only by implication. Unfortunately the engineer, in his preoccupation with hardware, often waits until design is completed before elaborating or formalizing his procedures. This is unfortunate from the $\underline{H}$ standpoint because there are some occasions—fortunately relatively few in number—when a procedural inadequacy really reflects a design inadequacy.

In any event, this delay in developing *fully detailed* procedures is why the author has referred the human engineering analysis of procedures to the detail design phase although logically it belongs in predesign. Nevertheless the analysis methodology is the same; the process described in this section applies equally well to the same activity when performed in predesign.

The $\underline{H}$ specialist's role in the development of operating and maintenance procedures is similar to his role in the human engineering of design. Procedural design, however, depends to an even greater extent on the equipment's internal functional requirements than does the design of the man-machine interface. The $\underline{H}$ specialist often lacks technical information about these internal functions. Hence, although he is concerned with ensuring that procedures have no built-in flaws that could lead to operator error, he tends to wait for the engineer to construct these procedures.

Obviously the possibility of an operator error in performing a procedure cannot be completely avoided. The $\underline{H}$ specialist, however, looks for the conditions which *predispose* the operator to make a procedural error and attempts to eliminate these conditions in advance. As in the case of equipment design, the procedure imposes certain demands upon the operator. As these demands increase, the probability of error increases. Hence the job of the $\underline{H}$ specialist is to reduce these demands as much as possible.

The $\underline{H}$ specialist analyzes a procedure to secure answers to the following questions:

1. Is there a discrepancy between equipment functions and arrangement and the procedure which could lead to error? If the procedure requires one sequence of operation, while the equipment must be operated in a different sequence, there will obviously be an error. Does

the procedure include all the operations that must be performed? All too frequently in complex operations the written procedure is found to overlook certain operating requirements.

2. Is the operator given all the information he needs to perform correctly; in other words, are all potential ambiguities clarified? When information is lacking, the operator is faced with a number of possible alternatives to the correct response; this in turn increases error likelihood. If all necessary information has been specified in the procedure, theoretically the number of alternatives to the correct response decreases to zero, allowing only the correct response to be made. This is, of course, only theoretical. Including all "necessary" information could result in a procedure the length of this book.

It may appear as if there is little to choose between "turn power on" and "turn power on by throwing the POWER ON-OFF SWITCH to the ON position." When there are many controls, however, the novice operator, or even the experienced operator performing under stress, may forget which is the power control. Perhaps the equipment may require that power be turned on in two distinct stages, first by throwing a toggle switch and then—after 15 seconds—by adjusting the voltage to a particular value. If the procedure merely specifies "Turn on power," the chance that the operator will fail to adjust the voltage at the correct time to the required value is increased.

Procedural ambiguities often arise because the design engineer is quite familiar with required operations and may assume that the eventual operator will be equally familiar with those operations. The fact that the H specialist is not a skilled operator or engineer and can therefore adopt a "naive" attitude toward the procedure can help him uncover such ambiguities.

3. Does the procedure make excessive demands on the operator in terms of

(a) strength (e.g., move a lever requiring 75 pounds of force);
(b) speed of response (e.g., type a 30 digit coded message in 3 seconds);
(c) frequency of response (e.g., adjust a rotary control continuously to match a digit readout whose numbers change every two seconds);
(d) perceptual ability (e.g., track a just perceptible radar pip);
(e) accuracy/precision of response (e.g., adjust a vernier scale graduated in tenths halfway between 8.56 and 8.57);
(f) memory (e.g., adjust a 10 digit thumbwheel indicator to correspond to a number received over headphones).

If the system imposes these demands, then it may be impossible to change the procedure to eliminate them. Under these circumstances it may be necessary to modify equipment design. On the other hand, equipment design may permit several different ways of performing an action, in which case it is reasonable to think in terms of a procedural change.

4. Does the procedure supply all the feedback information necessary for the operator to know that he is performing correctly? If an astronaut must initiate the second stage of space vehicle thrust only after the first stage has built up to full power, he cannot do so unless the procedure reflects an indication that first stage full power has been achieved.

All of the above questions demand behavioral answers except possibly item (1). To know when a procedural demand is excessive requires that we know what the operator is or is not capable of. The determination of what information is required by, or should be fed back to, the operator requires an analysis of informational needs.

All of the above examples of procedural inadequacies have been deliberately exaggerated to illustrate the point, yet comparable problems, although not perhaps of such a striking nature, are often to be found in procedures.

Implied in these inadequacies may be design characteristics which led to the procedural problems. If the operator of the two-stage rocket referred to previously is required to estimate when the second stage should be fired, it may be because the engineer did not include a display of first stage completion in his cockpit. Evaluation of procedural problems often evaluates *by implication* the design that may have led to these problems.

Procedural inadequacies may therefore result from two causes:

1. Design is inadequate; hence the procedure that follows that design is inadequate.
2. Design is adequate, but the procedure does not describe required operations satisfactorily.

Case 1 requires that design be modified. This presents more problems than does case 2 because it is much more difficult to modify hardware (even in drawing form) than to modify a procedure.

A procedure can always be revised as long as the revision does not demand that hardware be redesigned. Indeed it is always easier to make a system change which affects only procedures than one which affects hardware. When an $\underline{H}$ problem arises after design is completed, the specialist would do well to look first and very hard at the procedures as a possible means of solving his problem before he recommends equipment modifications.

Unfortunately, even though procedures represent a potentially lucrative payoff for his efforts, the H specialist often does not pay enough attention to them. The reason is twofold:

1. Even in a small system there are many procedures to review.
2. To understand a procedure fully requires the H specialist to learn almost as much about equipment operations as the design engineer (because the procedure is constrained by internal equipment operations more than are controls and displays).

There are two solutions to this problem. With regard to the number of procedures, the H specialist should concern himself only with *critical* ones. These are procedures whose incorrect performance is likely to lead to serious equipment, personnel, or system consequences, or they are procedures that are required to accomplish major system functions or achieve system goals. Moreover, even though maintenance procedures may have just as much impact on system performance as operating procedures, it is as well if the H specialist confines himself to critical operating procedures. This is because the system determinants of maintenance procedures are less well understood than the system factors directing operating procedures.

The second solution flows from the first. Since only a relatively few critical procedures are involved, it is easier for the H specialist to learn the specific equipment operations which underly these procedures.

What does the H specialist wish to discover in his analyses of critical operating procedures? Since he is concerned largely with preventing error and ensuring that the task will be performed on time, he looks for places in the procedure where the opportunity for error and delay is greatest. To uncover these sore spots he looks for situations which could produce these error opportunities.

These situations are the same as those described earlier when the specialist analyzed the procedure to determine:

1. Discrepancies between procedure and equipment design.
2. Failure to include all equipment-required operations in the procedure.
3. Informational ambiguities which may result in misinterpretation of procedural requirements.
4. Excessive demands on the operator.
5. Lack of required feedback information.

One way of determining the potential for error is to systematically analyze the various ways in which error could occur:

1. List the types of errors most likely to occur and those most critical in their effect on accomplishment of the system function.
2. Determine what characteristics of the procedures (and/or its accompanying design)

(a) encourage the commission of an error;
(b) could be modified in some way to prevent the error.

3. Determine if there is any other way in which a procedural step could be performed which would reduce the probability of error in performing the step.

An abbreviated example of what such an analysis would look like when documented is shown below:

REQUIRED PROCEDURE	POSSIBLE ERRORS	QUESTIONS TO BE ASKED
Place propellant system in stand-by condition.	1. Failure to initiate purge of propellants from system.	1. Is procedure clear about when and how the purge is to be initiated; are the controls for indicating purge indicated?
	2. Failure to complete purge of all propellants from system.	2. Information may be lacking about purge completion; has display or other feedback indication been provided? Does procedure indicate how operator is to know when purge is complete?

This analysis will not tell one that an error *will* inevitably occur, but it will suggest where major errors *could* occur. From this we can deduce what procedural features could cause the error, and these features can then be revised. The questions in the example above do not imply that the procedure is actually inadequate or at fault. The questions are asked simply to verify that the procedure covers major error contingencies. Note also that the error analysis is just as appropriate for design as it is for procedural analysis. We can ask exactly the same questions about a projected hardware configuration.

Ordinarily an analysis of every conceivable error would be very tedious and time-consuming, but this burden is reduced if we eliminate at the outset those error possibilities which are either very remote (improbable) or inconsequential (have insignificant effects on system functioning). This of course presupposes that we can determine which error possibilities are improbable or inconsequential. In all likelihood, once the specialist has listed all the error possibilities that he can think of, he will have to consult with the engineer concerning their frequency and criticality.

We cannot, however, eliminate *infrequent* error possibilities if their error effects are major. Sometimes error possibilities are disregarded because the action which could result in the error is very infrequently required of the operator. Even under these circumstances, if the error effect is significant, precautions should be taken against its occurrence.

Procedural analysis must be performed by the H specialist in consort with the engineer. The specialist asks the engineer to review the procedure with him by answering the questions listed earlier in this section. Ideally the analysis is a dialogue: the specialist asks the questions and the engineer supplies the answers. The specialist examines the answers for their procedural implications and makes recommendations, after which the engineer considers the recommendations to determine their feasibility. What the H specialist is doing is requiring the designer to *verify* his procedure by exposing it to human performance criteria.

Obviously this analysis is qualitative only. This is, however, not necessarily a disadvantage. After all, the procedure under examination is qualitative too. It is of course possible to apply to the procedure the operator performance prediction techniques described in the previous chapter. Using the prediction technique described earlier, the specialist assigns an error probability to each step. This permits us to determine those steps in which error probability is highest, and it is these which may require procedural changes. The major disadvantage of these prediction techniques—inadequate data—has been noted before. In addition, the predictive techniques do not indicate the *type* of error which may occur; this information, which must still be inferred from the error-prone situation, is necessary for recommending procedural improvements.

Assuming that we find a potential weak spot in a procedure (one which could conceivably lead to an error with significant consequences), what do we do? Obviously a different way of performing the procedure, one which will eliminate the error possibility, is required. The H specialist should recommend several alternative methods of performing the procedure and let the engineer choose the one that fits most readily into his design. If the specialist cannot recommend specific procedural alternatives but contents himself with merely pointing out a potential problem, the chances of his securing a change are very considerably reduced. However, the ability to recommend an alternative procedural step requires considerable background knowledge of the equipment operations reflected by the procedure.

Procedural changes can, of course, be made only within the constraints of the equipment configuration. The freedom available for modifying a procedure is not unlimited. The opportunity for making procedural changes is much greater when the procedure can be analyzed early in design, particularly when it is developed concurrently with hardware drawings.

HUMAN ENGINEERING EVALUATION OF DESIGN

Design Criteria

"Human engineering principles and criteria shall be applied, during detail design, to equipment drawings . . . to insure that the equipment can be efficiently, reliably and safely operated and maintained . . ." (3.2.2.2, MIL-H-46855).

Evaluation is inherent in design to the extent that the designer compares, evaluates, and trades off one design concept with another. This is even more true of human engineering, because the H̲ specialist usually does not design equipment on his own; much of his activity involves analysis of the designer's outputs to ensure that these meet H̲ standards. It is therefore worthwhile to examine the different ways in which the H̲ design evaluation can be performed.

The goal in H̲ evaluation is not to secure the best possible design but rather one that offers the greatest likelihood of accomplishing system functions and the least possibility of operator error. Even if it could be defined, we cannot achieve design perfection. Many human factors compromises with parameters such as performance requirements, cost, and reliability are usually required before design is concluded. As a result, the H̲ specialist is usually willing to accept a design if it does not violate any critical H̲ criteria.

For the H̲ specialist the basic question is how he can determine when a design is—or is not—acceptable from a human factors standpoint. The two—is or is not—are not identical. It is much easier to discover when a design violates H̲ criteria than when it meets these criteria; first, because, as was pointed out in Chapter Two, many H̲ criteria are indistinguishable from criteria of "good" design in general; for example, "failure of the display . . . shall be immediately apparent to the operator" (5.2.2.6, MIL-STD 1472). Second, it is easier because H̲ criteria usually contain a range of acceptable values; any number of design alternatives may satisfy a given criterion; for example, ". . . microphones . . . shall be designed to respond optimally to that part of the speech spectrum most essential to intelligibility (i.e., 150 to 4800 Hz) . . ." (5.3.7.2.1, MIL-STD 1472).

Two interrelated factors must be considered in the determination of acceptable H design: (a) the criteria of acceptable design, and (b) the methods used to apply these criteria.

A human factors design criterion is a rule for selection of appropriate design characteristics. It should indicate when, under what circumstances, and how, particular characteristics should be included in design. Because design characteristics imply some operator performance in response to these characteristics, the effective criterion must be defined at least partially in terms of operator performance. Unfortunately too few H criteria are defined in this way.

Criteria of acceptable human factors design are to be found in two sources: (a) military standards such as MIL-STD 1472 and (b) human engineering textbooks or handbooks such as Woodson and Conover, 1964, and Morgan et al., 1963. Military standards provide the binding regulations that constrain the human engineering characteristics of government-procured design; human engineering handbooks provide much the same material but with more discursive and detailed background explanation. Chapter Nine reviews the current human engineering standards and discusses their applicability to design problems.

Although some H standards are quantitative (e.g., recommended console dimensions, minimum distances between control centers, anthropometric data, etc.), many are qualitative, which reduces their usefulness; for example, "The precision required of control manipulation shall be consistent with the precision required by the display . . ." (5.1.1.4, MIL-STD 1472). Nevertheless H standards are useful in providing limiting (cut-off) points in design, that is, values beyond which design presents a definite danger to operator efficiency.

When the criterion is qualitative and general and when a range of values is permissible, the H specialist and the engineer face a problem in determining which explicit or implied value of an applicable design parameter will most effectively promote operator efficiency.

Because no weighting values have been explicitly suggested for its criteria, MIL-STD 1472 implies that each criterion has an equal value in in terms of the particular man-machine characteristics to be designed. (This situation is characteristic, however, of all military standards.) Since equipment design is an effort to organize design characteristics into a meaningful whole, the criteria that describe them may assume different values as a function of the total equipment configuration. As for choosing between the requirement for simplicity of control panel design ("the equipment shall represent the simplest design possible" [4.6, MIL-STD 1472]) and that for maximum feedback ("feedback on control response adequacy shall be provided" [5.1.1.5, MIL-STD 1472]), the applicable principle depends on the total control panel configuration.

Naturally, during design the engineer does not act as if these criteria were of equal value. He picks and chooses, but, since he lacks adequate guidelines for selection, he may do so erratically. He may not consider all the relevant criteria because he does not think of all the operational conditions that could affect equipment operation.

In some cases the task of selecting a human factors criterion is quite simple. Provisions for the selection of control levers, for example, can be ignored if the operator has no requirement for using a lever. Many other principles are of a more general nature and therefore are not so exclusive; for example, the requirement to provide to the operator only that information which is necessary for his task applies across the board. However, such general requirements are usually ignored because they lack any substantive force.

The relativistic nature of H̲ criteria also involves some inherent contradictions. If we are to guard every critical control on a panel (5.4.1.7.1, MIL-STD 1472), the indiscriminate use of switch guards for safety can severely hamper operation of these controls. Thus the H̲ criterion can be applied in full effect only when a technique exists for analyzing the conditions under which the criterion applies.

There is also no one-to-one relationship between a design feature and the operational requirement that elicits its use. A control may be located on the periphery of a panel because it is used only infrequently, but knowledge that a control operation will be performed infrequently does not necessarily and invariably lead to positioning that control in the periphery.

The criteria generally found in H̲ standards deal largely with required operations, that is, those needed to accomplish a task. Such criteria are generally inadequate for dealing with most contingent operations such as potential malfunctions or emergencies. Unfortunately, contingent operations are the ones in which the design configuration most affects operator performance.

Although it may be relatively easy to discover an individual design feature that does not meet human factors standards, it is much more difficult to determine the *significance* of that feature in terms of the equipment as a whole. How important is one control or display which deviates from human engineering standards when the total equipment includes 20 controls and displays? If the component is critical, then even one human factors discrepancy requires revision, of course. Most man-machine interfaces, however, are not that critical. Presumably all controls and displays are required for operation, otherwise they should not be included in design (there have been situations in which nonfunctional control-display elements have been included on control panels, but fortunately rather few), but one control or display may not be critical in terms of exercising a greater effect on operator performance than another control or display. The point is that, unless the specialist can differentiate one human engineering discrepancy from another on the basis of operator performance, all discrepancies—large and small—assume the same value, and then a sort of human engineering "Gresham's law" applies in which all discrepancies tend to be downgraded.

The difficulty of assessing the significance of the individual human engineering discrepancy also presents severe methodological problems to the H specialist when he attempts to evaluate the operability or maintainability of the equipment as a whole (preferably in terms of a figure-of-merit). Are two individual discrepancies twice as bad as one; is one critical discrepancy more important than four noncritical ones? All such judgments tend to be a matter of opinion rather than of empirical evidence.

It is obvious that if we were to assign priorities for human factors research, among the highest would be the strengthening of design criteria. These criteria are standards derived from extensive research and the consensus of H specialist opinion. To the extent that they are deficient, they reflect on the adequacy of the research specialist's performance. They also reflect poorly on the research direction provided by the governmental agencies for which specialists perform research.

Checklist Evaluations

The tool used most often by the specialist to evaluate design is the checklist. The checklist is a series of written statements that describe the individual characteristics which an equipment ought to have to be properly "human engineered."

The over-all checklist is usually broken up for convenience into sections corresponding to major equipment characteristics such as scales, counters, etc. Some checklists such as the sample below (Krumm and Kirchner, 1956) have spaces to enable the evaluator to check off the acceptability of the individual item.

III. WARNING LIGHTS

	Acceptable	*Not Acceptable*	*Not Applicable*
1. Contrast with dark, dull-finish background.	________	________	________
2. Mounted within 30° of visual axis	________	________	________
3. Critical lights isolated from those less critical	________	________	________
4. Labeled	________	________	________
5. Color red used where dark adaption necessary	________	________	________
6. If flashing lights used, rate is 3 to 10 per second and of at least .05″ duration	________	________	________

Two things about evaluative checklists are important:

1. The items found in the checklist represent selections from the total population of equipment characteristics that might affect performance, therefore judgment (with the possibility of error) was involved in their development.

2. The items are selected on the basis of their presumed effect on operator performance. A checklist item that presumably has no effect on operator performance is not included. However, if data are lacking to support the presumed relationship between equipment characteristics and performance, the checklist is essentially based on opinion.

The items selected for inclusion in any one checklist therefore represent the developer's concept of those equipment characteristics that are important to operator performance. Discrepancies between alternative checklist forms for the same types of equipment reflect each developer's differing concept of what characteristics are important. In general, most checklists contain the same items, despite minor variations in wording. There may, however, be differences in the number of items in the checklists, indicating differences in the developer's concept of the detail required to describe an equipment.

A procedural checklist and an evaluative checklist are not identical. A procedural checklist describes actions to be performed. An evaluative checklist is a statement of a desired quality that is inherent in equipment, procedures, or environment.

Since most checklists have provisions only for indicating the presence or absence of a desired attribute, the attribute is assumed to be binary in nature. This is an artifact of the checklist methodology; almost all design attributes are continuous.

It may appear from the preceding as if the checklist's shortcomings would severely limit its use. The solution would appear to be more effective research to develop better checklists.

There are more sophisticated checklist methods such as Siegel's Design Evaluation Index (Siegel et al., 1964). A design may also be evaluated on the basis of the operator error associated with those design characteristics (Payne and Altman, 1962). However, these methods are presently too cumbersome and make too many demands for data to be used in a tight developmental schedule. The disadvantage of the checklist's imprecision is balanced by the overwhelming advantage of its quick and easy use.

The checklist is used in evaluating two types of design products—the drawing and the completed hardware. The difference in dimensions between the verbal checklist and the physical drawing and the hardware and the fact that the evaluator is using a binary attribute scale make it difficult to perform a valid evaluation of either one. The checklist judgment inevitably requires a very considerable amount of subjectivity. A study comparing five specialists' evaluations of nine control panel layouts (Meister and Farr, 1966) revealed low consistency among the experts' evaluations.

Evaluation by means of a checklist will therefore be less than completely satisfactory to the methodological purist, yet some form of checklist is used by all design personnel either implicitly (e.g., when the engineer reviews his own drawing with certain categories in mind as evaluative criteria) or explicitly in a checklist device. Indeed it is not unknown for human engineers to write their own special checklists to perform special evaluation tasks. Often human engineers at the start of a major development project abstract particularly relevant standards from reports and supply these to designers as guidelines. This permits engineers to perform their own human factors evaluation.

The utility of the human engineering checklist becomes most evident when we consider what the results would be if even this rudimentary tool were not available. The human factors discrepancies found in systems developed in the past usually represent situations in which even something as unsophisticated as a checklist would have been highly useful, had it been applied.

Moreover the simplicity of the checklist has one overwhelming advantage: standards considered of prime importance can be (and usually are) memorized by the specialist and used on a highly informal basis, thus avoiding the need for elaborate testing situations.

The standard against which any new design is evaluated should be the "ideal" fully configured equipment, but this is obviously impossible because no one knows what the ideal is and because any equipment is a complex of many individual elements such as controls and displays, access spaces, and cabling, each of which has its peculiar characteristics. It is difficult therefore to evaluate design as a package. It is necessary to consider each design characteristic separately and to abstract certain desired design characteristics that are presumed to represent "good" human engineering. These (in the form of written checklist descriptions) are compared with the new design.

It is in the abstraction of the relevant design characteristics to serve as an evaluation standard that we encounter trouble. Although a large number of specialized checklists have been designed (for controls, for displays, for maintainability, for transport vehicles, etc.), the task of developing an effective checklist is a difficult one.

One reason for the difficulty lies in the hierarchy of abstraction that the human engineering standard or the checklist can represent. Take the general attribute, "operability." At this level of abstraction (e.g., is this equipment operable?), it is impossible to make a meaningful evaluation since operability is a function of a number of specific characteristics such as the arrangement of controls and displays by function or procedure, color coding, labelling, etc. Obviously the use of an evaluative criterion, which merely asks if the control panel is operable, does not permit us to zero in on the factors responsible for operability.

One order of abstraction lower is the use of an evaluative criterion which asks a question phrased in terms of a standard (e.g. all displays operated by controls should be adjacent to these controls [section 5.1.1.2, MIL-STD 1472]). This is more effective but still difficult to use. Although we can determine which displays and controls are interrelated, the definition of what adjacent means precisely is troublesome. What is the maximum distance that should separate interrelated controls and displays? Ideally the human engineering standard should specify that 2, 3, or 4 inches (or some empirically derived value) is the maximum separation permitted between interrelated controls and displays. MIL-STD 1472 is not that specific at present.

Since the design to be evaluated is either a drawing or an actual equipment, the method of comparison between the actual design and the standard should ideally be graphic also. The goal is difficult to achieve because the same human engineering characteristic can be reflected in design in somewhat different ways. For this reason it is probably impossible altogether to avoid verbal statements, but these should be as design-specific as possible and should be supplemented by one or more graphic examples of the verbal standard.

In evaluating designs the H specialist concentrates on those which are critical; that is, equipment with which operators interact frequently and whose inadequate operation by personnel would seriously jeopardize system performance. Of those designs involving considerable man-machine interaction, only a few will have significant consequences for the satisfactory performance of system functions. If, for example, the equipment has a single control that must be activated only at the beginning of equipment operation, no elaborate human engineering evaluation of that control will be required. Unless the H specialist is concerned specifically with maintenance, the most detailed man-machine interface which is to be evaluated is usually at the control panel level, occasionally the individual control or display, although a limited number of maintainability characteristics (e.g., access spaces, test points, connectors, cabling) may also be evaluated. H evaluation of internal component design is not usually performed because of the difficulty in specifying observable H criteria to evaluate that design.

In practice the H specialist evaluates primarily control panels, console configurations, layout of groups of equipment, and environmental conditions (e.g., noise, temperature, lighting). To a lesser extent he evaluates maintainability characteristics that are relatively obvious such as accessibility, test point arrangement, cabling, color-coding, etc.

Performance Evaluations

"*Mockups and models* . . . Three dimensional full scale mockups of equipment involving critical human performance . . . shall be constructed . . . These . . . shall provide a basis for resolving access, work space and related human engineering problems and incorporating these solutions into system design . . ." (3.2.2.1.1, MIL-H-46855).

The only H evaluation tool considered so far has been the checklist. This section will describe *performance evaluations of drawings and mockup tests.* Evaluation tests performed on hardware are described in the section on human factors in production.

The kinds of performance evaluations we can perform with a drawing are necessarily crude. However, they can help reduce the number of highly subjective judgments by the evaluator. We can record such indices as performance or reaction time and number and type of errors. If a subject is used instead of the evaluator, we can secure his personal opinions, which in many cases are more valuable.

One method of accomplishing such an evaluation is by means of a "walk-through" or simulation of the equipment operating procedure. Such a simulation can be performed in a number of ways: mentally, in which the evaluator imagines ("pretends") that he is carrying through the procedural steps involved in "operating" the equipment represented by the drawing and attempts to anticipate the operating problems he might encounter. Or the simulation can be partially realistic; the evaluator (or a test operator) reproduces the physical act involved in the procedure, but he does so abstractly by touching the component symbol which represents the control to be activated or a display to be read. This permits the specialist to develop a more concrete idea of the sequencing of the operating procedure as it relates to the arrangement of components and the confusions that may be involved in locating a control/display.

Alternately, we can glue a full scale drawing of a control panel layout (or cutouts of the individual controls and displays) to a piece of cardboard and have an operator indicate (e.g., by tapping) each simulated control-display with a pencil as the operating procedure is read to him.

What can we measure under these circumstances? If the operator consistently has difficulty finding a control or display or touches the wrong one, operational personnel may have the same difficulty. If we assume, as is reasonable, that the operator progressively learns the positions of controls and displays in relation to an operating sequence, then any excessive delay in learning these positions or repetitive errors in indicating them may suggest an inappropriate control-display location. It is even possible to compare alternative control panel layouts by repeating the test with each layout and comparing results.

This method is admittedly quite crude and permits evaluation only of control panel design. Obviously it is not feasible to do the same things with the design of internal components. However, it has some value since it makes the operating procedure less purely verbal and more concrete. Any difficulty revealed by the "walk-through" procedure deserves further consideration.

The performance evaluation of a drawing does not replace the checklist evaluation but supplements it. The checklist can deal with a larger number of design attributes than the performance evaluation can. Hence both should be utilized. The more realistic the representation of an equipment design, the more realistic the simulation can be. Obviously we can do much more in a mockup and even more in a simulator.

When the design is critical and the need to insure efficient human performance is very great, the H specialist may wish to perform a *mockup test*. An extensive discussion of different types of mockups is beyond the scope of this section (although see Meister and Rabideau,

1965, for a more detailed explanation). Generally, however, mockups range in complexity and sophistication from relatively simple and inexpensive cardboard, plywood, or styrofoam constructions (which only partially reproduce and represent the operational equipment) to very complex, costly, and computer-controlled mockups which approximate full scale simulators. However, the latter type will be provided for the H staff only when the human factors questions that need answering are critical to system success. The Apollo system is an example of a system development project in which, for obvious reasons connected with the necessity for preserving astronaut safety, the money became available to build elaborate mockups. Of course, if the company has developed mockups or simulators for other engineering purposes, the H specialist also can make use of them.

To the extent that the mockup faithfully simulates the eventual operational equipment, it can be used to study most questions which might arise about that operational equipment (except for maintenance). *Static* mockups (i.e., no moving components except possibly switches, no functional displays, no dynamic event presentation) permit us to study in a more sophisticated fashion the control panel layout studied in the walkthrough. In addition, anthropometric suitability (e.g., reach capability) and simple maintainability features (e.g., size and location of access spaces) can be examined. Because of its relative simplicity, this is the most common type of mockup. Static mockup subjects may be asked to perform various physical actions representative of those they would have to perform operationally such as reaching for and grasping objects, climbing up or down, entering confined spaces, lifting, removing/replacing components, connecting plugs, etc. When the mockup is controlled environmentally, they can be asked to perform all these under varying light, temperature, noise, or atmospheric conditions.

Dynamic mockups (controls and displays can be activated and read and events can be presented although perhaps not with complete operational fidelity) permit us to evaluate both design and operating procedures as well as the same maintainability aspects studied in the static mockup. *Programmable mockups* (in which operational events can be represented and subjects can make alternative responses to stimuli) permit us to study the full range of human factors design features.

An example of a static mockup test is available from Apollo developmental history. This test was performed to determine the feasibility of a proposed fixed couch position in the Apollo Command Module. The subject performed in a fully pressurized space suit. He walked upstairs in a crouched position while unpressurized and entered the couch with a rolling motion to his left. After the subject was settled in the couch, his suit was pressurized to 3.5 psi and reach capability to various control panels was observed. The subject was able to reach all positions except the upper inboard corner of the module. He then demonstrated egress movements, first moving his right leg to get it out of the couch and then rolling to the right to get his left leg out. The test was concluded at this point, since the limited mobility of the suit would have made further motion unsafe.

It was concluded that the relationship of the couch and panels tested permitted the crew in pressurized suits to enter and leave the couch in a weightless environment. However, specific interferences of headrest, docking window, couch support struts, and other couch mechanisms were observed; these would require redesign. The space for couch access was so limited that the crewman had to position himself parallel to the couch and then slide into it.

Note that the data were secured only by means of observation of the subject's movements and that quantitative analysis was restricted to a simple frequency count. The subject's opinions on the difficulty of performing required actions were secured following the test.

HUMAN FACTORS IN PRODUCTION

H Evaluation of Initial Hardware

A special type of performance evaluation is that performed on the initial operational hardware. The fabrication of the prototype or initial production item of the new system is a milestone event. The prototype serves as the initial test vehicle although others later in the production series may also be used for testing.

The evaluation of the human factors adequacy of this prototype (sometimes called the first article inspection) is important for both the H specialist and the engineer because it is their first opportunity to deal with operational hardware. This evaluation is especially important for the specialist if the equipment requires a high degree of man-machine interaction. Nevertheless, because this evaluation necessarily occurs only once, too much emphasis should not be placed on it.

The equipment in this evaluation is operational hardware but not necessarily functioning in an operational environment or in an operational manner. In general, first article inspection is performed within the manufacturing area. When equipment such as a typewriter, radio, or television set can be operationally utilized in any environment, the first article inspection need be no different from the operational tests described in the next section. Other equipment such as rocket engines may require special operational facilities (e.g., a test range) for full functioning.

When the installation of this equipment in the test area will require a long period of time or when the H specialist will not have the opportunity to participate in operational equipment tests, the equipment in its production environment offers an opportunity to perform interim tests. Since they are largely static and emphasize checklist type observations, we can look at first article inspection tests as merely highly sophisticated static mockup tests.

Operational tests are always preferable to first article inspections; the latter are useful only if the opportunity to perform the former is missing or will be substantially delayed.

Needless to say, in performing this evaluation the human engineer takes potluck. Since the equipment being inspected was built primarily as an engineering test vehicle and since it is the first of its type, many engineers will seek to make use of it. If the H specialist participates in the first article inspection, he is unlikely to have the prototype to himself; the author remembers one such inspection on the first Atlas series E missile during which at least 20 engineers were walking around and into the Atlas engine area at any given time.

The following H evaluation program for the first article inspection is recommended. The extent to which we conduct such a program depends on whether a full fledged operational test program is to be carried out, how soon, and whether H has a part in that program. There is little point in performing first article inspection tests if the opportunity for a much more valid evaluation will be available soon:

1. Checklist evaluation of human engineering characteristics, performed in exactly the same manner in which the specialist conducted his drawing evaluations.

2. Examination of the prototype from the standpoint of the static aspects of maintainability (e.g., availability of required access spaces, standardization of fasteners, efficient routing of cabling, noninterference of extrusions with operating equipment, keying of connectors, etc.).

3. Dynamic evaluation of the anthropometric aspects of the operator's job. For example, can the operator reach controls from the sitting position as measured by his actual performance of the reaching action? Can he reach into the engine compartment to unfasten certain bolts?

4. Determination of the time required to perform certain actions, particularly with relation to maintenance activities. When the procuring agency has imposed a maximum or mean-time-to-repair (MTTR), it is useful to determine how long it takes to remove covers and components. A demonstration of compliance to maintainability requirements may be called out in the statement of work. Engineering may therefore permit

the controlled removal/replacement of components so that the time required can be recorded; or if not, we can still observe any required removal/replacement activities and collect time data on these. The performance of such simple actions as fastening cables together may also indicate if the technician is likely to experience undue difficulty because of hardware characteristics. For such measurements simple stop watches are all the instrumentation required, and subjects can be members of the H staff.

The preceding four types of tests do not require the equipment to be "fired up." If the equipment can be activated in the factory area, it can be used for partial simulation of operational sequences. The extent to which such tests can simulate operational conditions will depend on the degree to which operational inputs can be provided; for example, an air traffic control console may present "canned" targets to permit the operator to track, thus providing a reasonable approximation of operational tasks; or a programmed checkout set may have a special deck of cards which, when inserted into the set, will permit a technician to simulate a malfunction diagnosis.

5. Where partial simulation of an operational sequence is possible, an investigation of the adequacy of task performance can be conducted; for example, the operator's ability to track the canned targets described earlier can be ascertained.

The measures we can secure under these circumstances include the following:

1. The *time* it takes to perform a particular task. This is important when the task is time-limited (i.e., must be performed in some specified maximum time if the mission is to be accomplished). This factor is particularly important in maintenance functions where a MTTR may be imposed by contract. We can then determine for at least some of the operational tasks whether the time requirements will be met. If they cannot, we can then look at the equipment or the procedure to see what modifications are necessary and possible.

2. *Errors.* Although maximum error requirements are not generally imposed in contracts, it is important to know whether an excessive number of errors is being made in a given task. The reason for such errors may suggest sources of difficulty which require redesign or, more probably, a modification of the procedure or some job aids to be supplied to the operator.

3. *Subject opinions.* If the man performing the simulated task is interviewed (debriefed) following task performance, he can provide much useful information concerning sources of difficulty he encounters. The

author has found that operator and technician subjects are usually quite aware of at least the most pressing factors that influence their performance and can, when questioned, report these.

HUMAN FACTORS IN TESTING

"*Human Engineering in Test and Evaluation.* The contractor shall establish and conduct . . . a test and evaluation program to assure the fulfillment of the requirements herein . . . The human engineering design verification program, contained in the approved system test plans, shall be implemented by the contractor . . ." (MIL-H-46855, 3.2.4 and 3.2.4.2)

Types of Testing

Earlier it was pointed out that on occasion formal H tests (as contrasted with continuing informal design evaluations) are performed during predesign and early development. Such tests, which have been termed *exploratory* and *resolution* tests (Shapero and Erickson, 1961), may be performed when a question arises concerning the following:

1. The capability of personnel to perform a particular function (e.g., can space pilots dock their vehicles using only radar data)—*exploratory test.*
2. The selection of the most effective system configuration from a number of alternatives (e.g., does configuration A produce more efficient operator performance than configuration B)—*resolution test.*

Such formal tests are relatively infrequent and are performed only if a system design problem arises that cannot be solved either logically or by recourse to available data. Information about personnel capabilities (exploratory testing), although not very great, is considered by engineers —although perhaps not by specialists—sufficient to answer most system development questions; more importantly, most actions required of personnel are not so extreme that questions of capability often arise. The selection of alternative man-machine configurations is almost always a matter of logic and tradeoff of design parameters although mockup evaluations involving formal subject testing can be considered a form of resolution testing. However, in most cases the engineer is confident of his ability to make a proper selection without formal resolution testing. Consequently we find exploratory and resolution tests more characteristic of highly advanced systems (e.g., Apollo) which stress the state of the design art than of "routine" system development.

Tests which verify that the completed system will perform in accordance with system requirements are, however, an integral part of

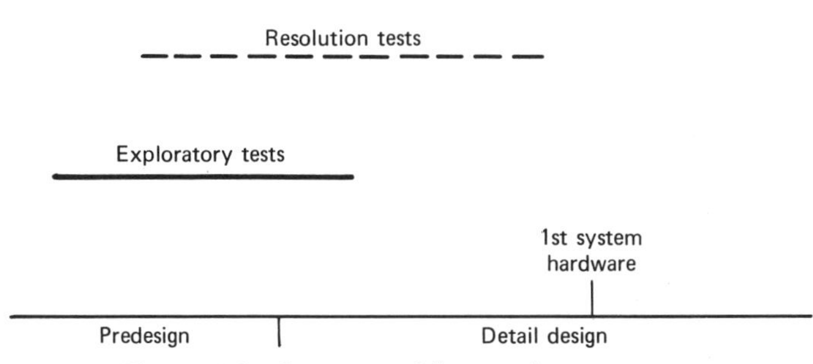

Figure 6-1 Sequence of human factors testing.

system development. They are called *verification* tests and are carried out during the testing phase of detail design. First article inspections can be considered a form of verification testing. The three types of test—exploratory, resolution, and verification—may overlap in time, depending on the idiosyncracies of system development (shown in Figure 6-1).

Verification tests are performed to demonstrate that the system will work; consequently they require developed hardware (even though the hardware may be a prototype or the first model). These tests are the ones we are concerned with in this section.

Verification tests relate personnel performance to a particular design configuration. The question is not whether a personnel function can be performed (thus requiring an exploratory test) but rather whether that function can be performed to particular system requirements (e.g., accuracy/time) with a specific system configuration. Verification tests have a much more specific system context than exploratory tests and hence are much more common in system development projects.

Verification Test Requirements

Before H verification tests can be performed properly, however, three preconditions (other than that hardware exists) must be satisfied:

1. The actions required of operational personnel must be explicitly described.
2. The system must be exercised during the test in an operational manner.
3. There is a human performance requirement (preferably quantified) with which operator test performance can be compared.

This does not mean that verification tests cannot be conducted without first satisfying these preconditions; but the results will be somewhat dubious.

Most systems describe required personnel actions fairly clearly before equipment must be tested, and during the test the system is usually exercised in a reasonable approximation of the operational manner unless functional problems (requiring design modifications) arise.

The third precondition—the availability of a quantitative human performance requirement—is, however, most difficult to satisfy because, as has been pointed out previously, formal operator requirements are rarely included in procurement documents.[1] It is, however, possible to infer such requirements with regard to individual tasks from an analysis of these tasks and in particular from the outputs required of task performance. If the output can be quantized, then a performance standard for the operator can be established; for example, if the bombardment system aboard an aircraft must place 8 bombs out of 10 within 20 feet of a target from 20,000 feet, this becomes the operator performance standard for the bomb-aiming task. The human performance requirement represents what the bombardier and his equipment must do to satisfy system requirements. Of course, the operator behavior to be evaluated must be directly linked with the system output (i.e., system output must depend upon operator performance).

In the example above, bomb release is the terminal output of the bombardment system. Just as the internal components of system equipment must perform satisfactorily for some period of time prior to bomb release, so the bombardier must perform adequately in intermediate tasks before dropping the bombs. Human performance requirements must therefore be specified for all bombardier functions involved in bombing activity even though the system developer may be primarily interested in the terminal output.

However, if the operator and his equipment are so closely related, is it necessary to measure human performance in a manner distinct from that of the over-all system? Should we not merely measure the end product of the operator/equipment performance—bombs dropped within the acceptable target radius—and let it go at that?

It may appear that it would be sufficient to demonstrate electronically that the bombardment subsystem will permit dropping projectiles with the required accuracy. However, unless the system is completely auto-

[1] For relatively simple discrete (step-by-step) tasks the required operating procedure can be used as a form of operator-performance standard. If the operator performs the procedure correctly, he has satisfied the task requirement. Simple discrete tasks, however, are often the exception rather than the rule in most systems.

matic, the bombardier is an integral part of the system and must be tested along with the equipment. Suppose the sighting mechanism is so complex that it does not permit the bombardier to line up the target properly, with the result that the bombs fall short on 50% of test runs. Can the bombardment subsystem be said then to be functioning according to requirements? If we cannot test the operator apart from his equipment, it is equally pointless to test the equipment without its operator.

H̲ verification tests distinct from (although not conducted apart from) other engineering performance tests are necessary because the developer is—or should be—interested not only in the terminal output of the system but also in the performance that led to that terminal output. Suppose the sighting mechanism on the bomb equipment requires calibration immediately before the bombs are dropped. Suppose further that this calibration depends on radar returns which the bombardier has difficulty differentiating. If the bombs fall consistently to one side of the target because calibration is difficult, the developer will never know this unless measures are taken not just of bomb drop error but also of the bombardier actions prior to the bomb drop. If all goes well during the engineering test, then specific human performance measures may appear to be redundant. Almost never, however, does all go well during engineering verification tests; human performance measures, therefore, help to determine the cause of system deficiencies, thus permitting remedial action to be taken.

The measures we take in verification testing are measures of task output. If we are to verify that system goals have been achieved, it is necessary to ensure that appropriate measures of those goals are being recorded. Task output is determined by the goal to be achieved in performing the task. Goal achievement is defined in terms of some equipment or system function that is required if the system mission is to be completed satisfactorily. Take a teletype communication task as part of a system function to be accomplished. It can be analyzed in the following manner:

Task: teletype communication.
Task goal: to transmit N messages per unit time.
Task output: messages transmitted.
Task measure: number of messages entered into the teletype by the operator.

Obviously the teletype operator may do things other than entering messages via his keyboard; he files messages; he may answer inquiries over the phone, etc. These subtasks, while essential to his job, do not directly affect the goal of the equipment, which is to transmit messages; hence, from the standpoint of measuring system performance, they can be safely ignored except when they influence the performance of the primary task.

Again it should be noted that in the teletype example above it is impossible to secure a measure of operator performance apart from the equipment that he was using. In fact, human- and equipment-performance testing must be performed simultaneously; in the former, however, the measures *are derived from the operator or involve his operation.*

Take the measurement of radar set performance. We can secure a measure of the equipment's performance by electrically picking off signals from individual circuits or modules or by measuring output voltage. These do not involve the operator, but, on the other hand, they do not tell us anything about how well the system (as opposed to the equipment alone) performs the functions of detection and tracking. Measurement of radar set performance in terms of the over-all system goal—detection and tracking—is measured by the number of targets detected and accuracy of tracking. These latter measures can only be secured by the operator's verbal report or by measurement of his motor responses. To the extent that we wish to measure system performance, therefore, we must include the operator in the measurement. A major part of any system test plan must therefore include a section dealing with human performance measures.

Since verification testing is performed only after design has been completed, it is almost always performed with operational gear even if that equipment is a prototype or first model. (It is conceivable that verification testing could be performed with mockups or simulators, but, since operational hardware will eventually be available, there is not much point to it.) For this reason, because of the cost involved in operating hardware and because the equipment engineer also wishes to verify equipment functioning, it is likely that any $\underline{H}$ verification test will be primarily an equipment verification test with human factors objectives as one subsystem testing goal. In other words, the $\underline{H}$ verification test will "piggy-back" the engineering test.

Equipment verification tests must be differentiated from equipment development tests. The latter, like exploratory testing, are designed to solve an engineering problem; consequently, operational hardware may not be available (breadboards or engineering development equipment models are extensively used) and operational procedures may be neither available nor followed. The common feature of all equipment verification tests, whether they involve a single equipment or a system of equipment, is that *equipment is made to function in a manner as close as possible to its ultimate operational use.* This is an advantage for $\underline{H}$ testing since a prerequisite for $\underline{H}$ testing is operational usage. Since operational use conditions require operator performance, it is at this point that H verification testing becomes not only possible but necessary. Unfortunately, unless the customer specifically requires H testing or the specialist fights for it, it is unlikely that engineering will automatically include provisions for H verification testing as part of its over-all test planning.

The fact that the equipment must be exercised in order to perform engineering tests, whether or not H data are collected, means that the cost of an H verification test when it is performed as part of an engineering test is actually quite low. Since the H test requires merely that H personnel participate in the test as data collectors (and since the operating personnel would be there in any event) the only thing that must be paid for is their time.

Since the H verification test assumes that the test system will be exercised in an operational manner, no modification of the test situation for experimental H purposes need be made. The H verification test usually does not demand separate test facilities or test performance. Consequently the participation of the H specialist in an operational test will not significantly influence the manner in which that test is conducted, nor will H testing influence over-all test scheduling because the H test schedule is completely dependent on engineering test factors and is predetermined by the same system constraints which influence the way in which the equipment works.

The advantages of performing an H test in conjunction with equipment testing are therefore obvious: (a) verification that system personnel will be able to perform to requirements; (b) at little cost; (c) with no negative influence on the equipment test.

What the H test does require is a special set of measures distinct from those ordinarily collected by engineering. These can be collected at the same time other engineering data are collected.

Because the H verification test is so closely tied to the equipment test, the suggestion is sometimes made that the test engineer should collect the necessary human performance data as an incidental part of his major equipment testing activity. This is theoretically feasible but practically impossible because few if any test engineers have the necessary time, specialized skills, and interest to collect such data.

The H Test Plan

The first task in performing verification testing (as in any systematic testing activity), is to develop a detailed test plan. The H test plan has two major purposes: (1) it requires the specialist to determine precisely what must be done in the test, thereby avoiding obscurities which may later embarrass his data collection; (2) it communicates H test objectives and procedures to engineering management. Since the H test is only one part of the over-all equipment verification test, it is important to make any special test requirements known clearly and early to engineering.

The test plan should contain the sections described in Table 6-2. A more complete description of the plan is provided in Meister and Rabideau, 1965, and the reader who wishes to pursue verification testing in detail should refer to the series of papers included in Kincaid and Potempa, 1969.

Several of the items in the test plan are obvious, others are not. General objectives (1.1) are usually so general that they must be broken down into specifics (1.2) which supply the reasons why particular measures are taken and data collection procedures employed. That the objective of H testing is to verify the adequacy of man-machine interfaces is an interesting statement but not very illuminating. It is much more meaningful to say, for example, that specific objectives are to (a) evaluate the adequacy of interior lighting; (b) determine crew response time for pre-launch checkout, etc.

Since performance data will be collected on specific tasks (2.3), performed with specific equipments (2.1) and tests (2.2) using specific procedures (2.4), it is necessary to list these. When many tasks are performed concurrently, it may be impossible to collect data on them all (because of time constraints or lack of personnel); hence it is necessary to select those that are most critical to test. Altman (1969) describes methods of selecting such tasks and evaluating the cost of testing them.

If the system is to be exercised in several operational modes (e.g., preventive maintenance vs. troubleshooting, prelaunch vs. launch, etc.) or if there are factors inherent in system operations which may influence or reflect differences in the operator's performance, those factors should be studied. For this reason the factors to be studied should be listed (3.0) if we intend to collect data that will permit performance comparisons among them; for example, if it is possible that the particular shifts on which a crew works make a difference in its performance, we may classify or segregate the performance data by shift. The data collector records the workshift during which the data are recorded; if shift is not relevant, he simply ignores work shift as an item of data. The factors/conditions which the tester intends to compare will, together with the specific objectives, determine the measures to be taken and the method of data collection.

The measures taken (4.1) stem from the test objectives and the factors to be compared. If, for example, one of the test objectives is to record all human-initiated failures that occur during the test, then a

Table 6-2 H̲ Test Plan Contents

1.0 Purpose of the H̲ test.

- 1.1 general objectives (e.g., verify that system personnel can perform required tasks).
- 1.2 specific objectives (e.g., collect data on human errors).

2.0 Description of system being evaluated.

- 2.1 list of equipments on which human performance data are to be secured.
- 2.2 list of equipment tests during which H̲ data will be collected (if H̲ data will not be collected during every test).
- 2.3 list of operating/maintenance tasks for which H̲ data will be secured (if H̲ data will not be collected on all tasks).
- 2.4 applicable T.O.'s or other operating/maintenance procedures.

3.0 Factors for which operator performance will be compared (e.g, length of work cycle, size of checkout crew).

4.0 Test criteria and measures.

- 4.1 H̲ measures to be recorded (e.g., start/stop time, errors, discrepancies, operator actions, and use of T.O. procedures).
- 4.2 criteria of H̲ test accomplishment (e.g.., two satisfactory repetitions of every task measured).

5.0 Data collection methodology.

- 5.1 number of data collectors, tasks to be performed, assigned to which test locations.
- 5.2 training of data collection personnel (if required).
- 5.3 detailed data collection procedures in scenario form.
- 5.4 data collection tools (e.g., questionnaires, interview forms) and how they are to be used.
- 5.5 required instrumentation (e.g., tape recorders, cameras, biosensors).
- 5.6 facilities required (e.g., test stand 3).

6.0 Subjects.

- 6.1 number.
- 6.2 type (e.g., characteristics, background, if pertinent).

7.0 Data analysis.

- 7.1 relationships which are to be tested statistically (e.g., performance on day shifts is more effective than on night shifts).
- 7.2 statistical design to test relationships (e.g., Chi-square, regression analysis).

8.0 H̲ test schedule and sequence. (Describe in relation to equipment test schedule.)

9.0 Corrective action and reporting procedures.

- 9.1 description and schedule of reports to be written (e.g., bi-weekly progress reports).
- 9.2 procedures for recommending design changes.

record must be maintained of all equipment malfunctions resulting from human error. Task outputs (e.g., messages per unit time) also determine performance measures.

It is assumed, of course, that all measures to be taken are *valid* and *relevant* indicators of the operator performance the specialist desires to examine; by this we mean that the measure selected should reflect those operator actions directly associated with or responsible for the terminal output of the task. Although there are many theoretical issues involved, for most practical purposes the tester determines validity and relevance on the basis of his familiarity with the system operations to be studied and prior analysis of the tasks involved.

It is also important to note how much data are to be collected (4.2). It is usually not possible in verification tests to collect all desired data. Because many engineering tests are, from the specialist's point of view, fragmentary, the problem is usually to secure the required minimum of data. Hence the value specified is that minimum of data acceptable as an adequate sample.

Data collection methods (5.3) and data collection forms and how they are to be used (5.4) should be described in great detail to serve as a procedural guide for the data collector. When several testing locations may be used, the selected location should be indicated (5.6). If data collection personnel require special training (5.2), an outline of that training should be provided.

If we have a choice of subjects (6.0) or must select from the subject population the particular individuals to be tested, the number required and the criteria for their selection should be indicated.

In order to develop the statistical analyses to be performed on the data to be collected, it is necessary to specify as precisely as we can the relationships among the system factors to be compared (7.1); those relationships assume some judgment as to the factors responsible for variations in operator performance. For example, if the analyst intends to compare task duration in flightline maintenance with task duration in shop maintenance, it is necessary to say so specifically and to indicate why; otherwise the rationale for the statistical technique required to perform this comparison (7.2) will be obscure (even to the statistician). The reason for specifying these relationships in advance of the analysis is above all to make them visible to the test planner so that he can examine the data he is collecting to ensure that they will permit the required statistical analyses.

When the H test schedule (8.0) differs from the over-all engineering test schedule, it is necessary to indicate the discrepancies. Any special reporting procedures (9.0) anticipated by the H specialist should also be indicated.

Conducting the H Test

A detailed description of how to conduct the H test is beyond the scope of this brief overview. The objectives, measures, and tools to be employed largely determine how the test will be conducted. For example, if one of the goals of the test is to determine equipment design deficiencies from an operator standpoint; if it is decided that these will be gathered subjectively by interviewing operators; if the interview has been developed; and if the timing and location of the interviews have been decided upon; then in fact the test has been described and all that is required is to carry out the interview schedule.

The conduct of the test—once this is properly planned—should therefore be fairly routine. (For further details see Meister and Rabideau, 1965, and Kincaid and Potempa, 1969.) That this is often not a "smooth" process results from perturbations in the over-all engineering test procedure (discussed in the next section).

Testing Problems

The major problems which the H specialist may encounter in testing and which should be avoided, if possible, include the following:

1. Tests may not be run according to anticipated operational procedures. Obviously procedures may be revised as a result of testing, but testing should begin as much as possible in accordance with procedures to be employed by the operational system. A different operating procedure constitutes a different mode of operation for the system. If an agreed-upon set of operational procedures does not exist or is not used by engineers or if engineers deviate from them at will, the test verifies only that the equipment will work in some mode which may or may not be the operational mode. Since the operating or maintenance procedure is a major determinant of what personnel do, obviously a nonoperational procedure elicits and tests nonoperational performance. We can therefore set up the following rule: *an H verification test is valid to the extent that it reproduces the conditions and procedures which will be employed operationally.*

2. The personnel performing as operators and maintenance men in the test may not resemble the personnel who will eventually exercise the system. It is much simpler to use company engineers or technicians to act the role of system personnel in the verification tests even though these may be army privates or seamen. To do so, however, is to ignore the fact that company engineers are much more familiar with the system that they personally developed than are enlisted men; hence their performance should be, if not superior to, then at least qualitatively different

from that of enlisted men or other system consumers. In verification tests for major system development projects, the eventual operational user may be brought in to run the system (and have his performance tested) after the initial system bugs have been ironed out. A second rule: *an H verification test is a valid test to the extent that it utilizes personnel who either are or resemble the personnel who will eventually perform operational tasks.*

Less critical but still a significant problem is the following:

3. Unanticipated difficulties may constrict the test schedule and cause equipment tests with lower priority to be eliminated or reduced in terms of number of replications. The equipment test may be performed only once, and, since the H test depends on the equipment's operating cycles, only one instance of the operator's performance may be secured. If we could repeat the test several times, the operator performance recorded might well turn out to be atypical. Human variability is such that often it requires substantially more replications than one equipment test if we are to have confidence in the validity of the human performance data. A third rule can be advanced: *an individual H verification test is acceptable to the extent that it is repeated at least once and produces essentially the same data.*

It is also sometimes difficult to get engineers to replicate tests in precisely the same way each time because in an R&D program they are interested not so much in statistical significance as in progressing through developmental milestones. Under these circumstances considerable data extrapolation may be necessary.

It is also essential to remember that a test in which the operator fails for some reason or other to complete a task or completes it in a nonoperational way (e.g., using an unauthorized procedure) or in which some of the H data is missing has primarily a diagnostic, rather than a verification, value; that is, it may indicate what factors caused the operator to fail the test or task. However, it does not demonstrate that the procedure or design configuration being verified will in fact permit the system to accomplish its mission. A "no" is not a "yes." Such "dud" tests should be repeated. The problem is that the equipment test may provide the functional data which the engineer wishes but may not provide satisfactory H data. If the engineer proceeds then to the next equipment test in the series, it does not help the specialist who is left with incomplete data.

Despite all this H verification testing does achieve many of its goals. Since the major question asked in this type of testing is fairly simplistic (did the operator complete or fail to complete his task), the H̲ specialist can secure data even in an environment in which precise control over variables and measures is difficult. For other questions (e.g., why a performance deficiency occurred or the effect of system operational factors on performance) testing may be more or less successful. In any event, the specialist is not attempting to perform a laboratory study or test the significance of differences between experimental variables; mainly he wants to determine that satisfactory performance does or does not occur. He is, for example, concerned not with all errors but only with those that are sufficiently great to prevent task completion.

The specific techniques involved in conducting the H̲ test (as differentiated from planning the test) are therefore not overly sophisticated. The major reason for this is that the precision controls the specialist may wish to exercise over the data collection process cannot usually be exercised in the operational or quasi-operational environment; hence sophisticated techniques are largely out of place. Nevertheless, simply to overcome the difficulties resulting from lack of situational control, great care must be exercised in performing the test.

It may be asked, why, in view of the relative lack of behavioral sophistication in this type of testing, cannot the engineer himself conduct the test and collect the necessary H̲ data? The answer is that many of the techniques used in H̲ data collection (primarily observational and interview) are highly subjective. These are methods in which the data collector himself is the measurement instrument. Because of this the data collector's orientation toward, and his background in, the measurement of human performance is a vital element in that measurement.

Both in observation and interview, considerable judgment is required. In observation of task performance, the data collector looks for indices of difficulty, for error-causative factors. As was pointed out in Chapter Four, the recording of task completion, of response duration, and of error frequency data will indicate how well personnel perform but not necessarily what may be responsible for inadequate performance. It is a matter of subtle cues, admittedly hard to define objectively; and for that very reason the engineer who is not cued to error-causative factors will have difficulty in collecting the necessary data.

The interview requires considerable experience in establishing rapport, in anticipating where the interviewee is leading the interviewer (what the interviewee means by partial hints), and the ability to open up new channels of communication. This does not mean that the engineer cannot be trained to collect such data, but it does mean that he cannot do so *without* training.

On the other hand, the H̲ specialist suffers from one disability that the engineer does not, and this is his (generally) less than optimal knowledge of the hardware aspects of the system. If the specialist's data collectors are nonengineers, he may find it necessary to put them through an intensive equipment indoctrination program.

APPENDIX: SELECTIONS FROM A HUMAN FACTORS PROGRAM FOR THE XYZ SYSTEM

The human factors program plan for the XYZ system generally will follow MIL-H-46855 and MIL-STD 1472. Previous development efforts, such as those dealing with the AN/SQZ-43 Visual Analysis Subsystem for SAC, have involved extensive human factors activity, particularly in the area of human engineering of console design. The experience gained on this and other related programs will be directly applicable to the XYZ program.

This program will include all the necessary analytical, experimental, design, testing, and evaluation activities required to provide for human components of the system. The elements of the program will consist of the following:

1. *Personnel-equipment analysis.* The collection and analysis of data during the effort will result in a central file that will be used to support other human factors activities and engineering in general.

2. *Human engineering.* The human engineering effort will be accomplished in accordance with the provisions of MIL-STD 1472. Generally, this effort will involve

   (a) participation in the system functional analysis to ascertain function and task performance requirements for personnel;
   (b) translation of the technical requirements into detailed design of equipment;
   (c) evaluation of design to support personnel task performance and the specifying of test objectives required for such evaluations in Category I and II test programs.

3. *Personnel requirements.* The personnel requirements analysis will identify positions to be filled by operator and maintenance personnel.

4. *Personnel subsystem test and evaluation.* This activity will be designed to ensure that the entire system can be operated, maintained, and controlled by the assigned personnel. In addition, this activity will focus on the identification and resolution of problem areas and deficiencies that could degrade total system effectiveness.

The relative emphasis on the above aspects will be dictated by the nature of the equipment components to which the H program is applied;

for example, the work performed on the data processing set will emphasize equipment maintenance characteristics and tasks because this equipment is automatic and involves only a few routine operator tasks (such as card punching and magnetic tape handling). With regard to the console unit, greater emphasis will be placed on the human engineering characteristics (e.g., arrangements of controls and displays) because the console interfaces directly with operational planners. The development of the large-screen projection equipment will involve greater emphasis on the visual and perceptual parameters such as brightness, contrast, resolution, and color qualities. System characteristics, such as the arrangement of major system units in relation to each other and their intercabling, will be examined to ensure ease of operation and maintenance . . .

Concurrently, but based partially on the preceding task analysis, a human engineering program will be established to provide guidance in the detailed design stage of the program. Coordination will be maintained between human factors personnel and design engineers responsible for the development of major equipment components. Human factors specialists will review and approve all design drawings that affect operator and maintenance performance. Human engineering specialists will participate in all equipment design reviews, and recommendations for design modifications will be submitted in writing to the responsible design engineer. Documentation describing recommended modifications, together with the action taken, will be included in a central file.

Human engineers will assist engineering design groups and maintenance specialists in the development of procedures for the operation and maintenance of XYZ equipment. Operating and maintenance procedures will be reviewed as they are generated. The purpose of this review will be to identify and modify any procedural steps that have a significant probability of causing erroneous operator or maintenance operation. As in the case of engineering drawings, no operator/maintenance procedures will be released without approval of the cognizant human factors procedures specialists.

Following the development of hardware a human factors test program will be instituted to ensure that the completed system is operable and maintainable. Human factors testing will emphasize two major areas:

1. Evaluation of the legibility of symbols displayed on the large-screen projection equipment; this will be accomplished by means of special test procedures described in the next section.
2. Evaluation of technician capability to perform maintenance tasks in accordance with maintainability requirements. This will be accomplished by observations made during Category I and II operational tests.

Observations will be made of all tasks to determine the adequacy with which these are performed. All deviations will be recorded and analyzed to determine their causes.

In addition, interviews will be held with operational personnel to determine any difficulties experienced. All engineering test, accident, and safety reports will be reviewed to ascertain information on errors made by test personnel, deviations from procedures and technical manuals . . .

REFERENCES

Altman, J. W., 1969. *Choosing Human Performance Tests and Measures.* In J. P. Kincaid and K. W. Potempa (Eds.), Proceedings of the Human Factors Testing Conference 1–2 October 1968, Report AFHRL-TR-69-6, Air Force Human Resources Laboratory, Wright-Patterson AFB, Ohio, October.

Kincaid, J. P., and K. W. Potempa, 1969. (Eds.), *Proceedings of the Human Factors Testing Conference 1–2 October 1968,* Report AFHRL-TR-69-6, Air Force Human Resources Laboratory, Wright-Patterson AFB, Ohio, October.

Krumm, R. L., and W. K. Kirchner, 1956. *Human Factors Checklists for Test Equipment, Visual Displays and Ground Support Equipment.* Report AFSWC-TN-56-12, Air Force Special Weapons Center, Kirtland AFB, New Mexico, February.

Meister, D., and G. F. Rabideau, 1965. *Human Factors Evaluation in System Development.* New York: Wiley.

Meister, D., and D. E. Farr, 1966. *The Methodology of Control Panel Design.* Report AMRL-TR-66-28, Aerospace Medical Research Laboratories, Wright-Patterson AFB, Ohio, September.

Miller, R. B., et al., 1953. *Systematic Troubleshooting and the Half-Split Technique.* Report No. 53-21, Human Resources Research Center, Lackland AFB, Texas, July.

Morgan, C., et al., 1963. *Human Engineering Guide to Equipment Design,* New York: McGraw-Hill.

Payne, D., and J. W. Altman, 1962. *An Index of Electronic Equipment Operability: Report of Development.* Report AIR-C43-1/62-FR, American Institute for Research, Pittsburgh, Pennsylvania, January.

Shapero, A., and C. J. Erickson, 1961. *Human Factors Testing in Weapon System Development,* Presented at the ARS Missile and Space Vehicle Testing Conference, Los Angeles, California, March 13–16.

Siegel, A. I., W. Miehle, and F. Federman, 1964. The Display Evaluative Index Technique for Evaluating Equipment Systems from an Information Transfer Point of View. *Human Factors,* 6 (3), 279–286.

Woodson, W. E., and D. Conover, 1964. *Human Engineering Guide for Equipment Designers.* Berkeley, California: University of California Press (2nd ed.).

VII

THE ENGINEER AND HUMAN FACTORS

RATIONALE FOR STUDYING THE ENGINEER

No analysis of H and the way it is applied in system development would be complete without considering the engineer, his design processes, and his attitudes toward human factors. It may seem strange—yet curiously appropriate for a behavioral discipline—that the efficiency of H as a factor in system development should depend largely on the engineer's attitude toward human factors.

The reason is that the engineer's role in system development is focal. Like it or not, he has the ultimate responsibility for design. The H specialist—or for that matter any other specialist, reliability, maintainability, weights, etc.—may propose, but the design engineer disposes. Just how efficient the H specialist is therefore depends in the long run (among other factors, of course) on the engineer's willingness to use the data and the recommendations the former supplies.

That willingness has not always been in evidence. Anecdotes of how H specialists were ignored or of how their recommendations were rejected during system development are common topics of conversation at professional meetings of specialists. The most extreme case the author heard of—fortunately very uncommon—was the chief design engineer who would bodily eject from his office any human factors engineer who had the temerity to appear with a request. We can only hope that the anecdote is apocryphal.

In any event specialists take it for granted that they will encounter a certain amount of suspicion and resistance from design engineers (and from company management as well) at least until they "prove themselves." It is not enough, however, merely to accept this somewhat uneasy relationship. It is necessary to dissect it and to determine why it exists and what can be done about it. Given the irrefutable fact that, to be effective, specialists must interact with engineering and engineers, what can be done to improve that interaction?

Since the relationship bulks so large in the specialist's consciousness, a great deal has been written and said about the necessity of improving communication between the specialist and the engineer. Much has been made of the necessity for $\underline{H}$ personnel to communicate their inputs to the engineer in the latter's language (whatever that language consists of). Before we can communicate effectively, however, it is necessary to learn something about the characteristics of the communicators, their reactions to human factors and to $\underline{H}$ as a discipline.

(It is unnecessary to discuss the specialist's attitude toward the engineer. This is ordinarily determined by the degree of acceptance he achieves among the engineers of his company. Those whose efforts are accepted feel warmly toward the engineer; those whose efforts are not are apt to consider the engineer as an intellectual barbarian.)

A word of caution, however; the author has discussed this topic before a number of audiences, some of whom included design engineers. Invariably there is a certain sensitivity to the subject; it suggests a certain criticism of engineers and engineering generally. Nothing was or is further from our intention. Whatever the relationship may be between the specialist and the engineer, it is necessary to clarify it before any substantial understanding can be achieved.

Although there is considerable anecdotal evidence to support the statements made in this chapter, it is best to base them on experimental evidence. This chapter reflects what has been learned from studies (Meister and Farr, 1966; Meister and Sullivan, 1967; Meister et al., 1968 and 1969) performed by the author in collaboration with several colleagues.[1]

Questions to be Answered

These studies were performed to answer a number of questions:

1. *How engineers design*

(a) what kinds of information does the engineer ordinarily use as the basis for his design decisions;

[1] Special acknowledgement must be made to Dennis J. Sullivan who, except for the first study in the series, collected all the data reported. The first study was performed with the assistance of Donald E. Farr. Mrs. Dorothy Finley was extremely helpful in statistically analyzing the data gathered in the later studies. The research was supported by the Engineering Psychology Branch, Office of Naval Research (for which the author is grateful to Drs. J. W. Miller and M. Farr) and by the Air Force Human Resources Laboratory (with the aid of Dr. W. B. Askren). The author is, however, solely responsible for the conclusions drawn and they do not reflect the opinions of either the Navy or the Air Force.

(b) how does he perform his design analyses?

2. *Engineering attitudes toward human factors*

(a) what are the designer's attitudes toward H̲, human factors information, and the role of the operator in design;
(b) to what extent does he routinely include human factors in his design analyses?

3. *Engineering use of H̲ inputs*

(a) how efficiently does the design engineer utilize H̲ inputs;
(b) what inputs does he feel are most valuable to him;
(c) what design implications does he draw from these inputs;
(d) in what developmental sequence will these inputs have their greatest effect?

4. *Effect of H̲ inputs*

(a) what is the effect on hardware design of these inputs and in what form;
(b) how effective are personnel requirements included in system design specifications, and how do they influence design?

5. *Utility of human engineering standards*

(a) how useful are human engineering standards for directing design;
(b) how much influence do they have on design?

6. *Adequacy of human engineering guides*

(a) how adequate for the engineer's informational needs are popular human engineering design guides (e.g., Morgan et al., 1963)?

Study Method

The method used to secure answers to these questions assumes that how engineers analyze design problems determines (a) how they use human factors information and (b) how they include operator considerations in that design. In other words, information has value to the engineer only to the extent that he can relate it to his design task.

Consequently, if we wish to secure data about the usefulness of human factors inputs to design, it is necessary to place the engineer in a situation that requires him to design. H̲ inputs can then be systematically supplied, and we can observe whether or not the engineer uses them.

Previous experience has shown that it is useless to ask designers to verbalize their design methods. The answers received are vague generalities, or there are no answers at all. Because it is creative, much design activity cannot be observed; and like most people the engineer may even be unaware of some of his mental processes or unable to describe them.

Consequently in our studies we rejected any methodology which depended on *formal* interviews or *questionnaires* (especially the latter); we also discarded any technique which was not based on, or could not be incorporated into, concrete design situations. The method developed by some trial and error over five years of research was to do the following:

1. Present the engineer with a series of realistic design problems either representative of those he ordinarily encounters or based directly upon actual design situations.
2. Provide him with information describing both equipment and human factors aspects of these problems.
3. Require him to solve the problems.
4. Observe how (or if) he used the experimental inputs in the problem solution.
5. Following problem solution, review with the engineer how he achieved his design solution and the value to that solution of the inputs provided.

Two types of design test have been used to implement this research strategy:

1. Something we have termed the *Design Product Test* (DPT) in which the design solution or product of the design process is the actual conception of an equipment to satisfy the problem requirements. Such a test measures design "longitudinally," that is, the entire process from receipt of the problem by the subject to its completion. The method is one which requires that each engineer subject be tested over a period of 8–10 hours at the least, and preferably considerably longer. This is because of the time required for the subject to assimilate the engineering details of the problem. It may be possible to shorten this time period, but the results of such an abbreviated procedure would be dubious.

As an illustration, Appendix A contains the first design problem presented; subsequent problems became more elaborate and complex, as will be seen.

A variety of problems has been used; for example, the first DPT required the layout of a command/control console aboard a missile frigate (Figure 7-1); the second, the design of a circuit-board tester (Figure 7-2). The figures illustrate the level of detail to which subjects designed in these studies. The third called for the design of an air traffic control radar monitoring console; the fourth, for the design of an entire propellant transfer and pressurization subsystem for a space launch vehicle. The last required the design of the test equipment needed to check the electronics subsystem of an air-to-ground missile.

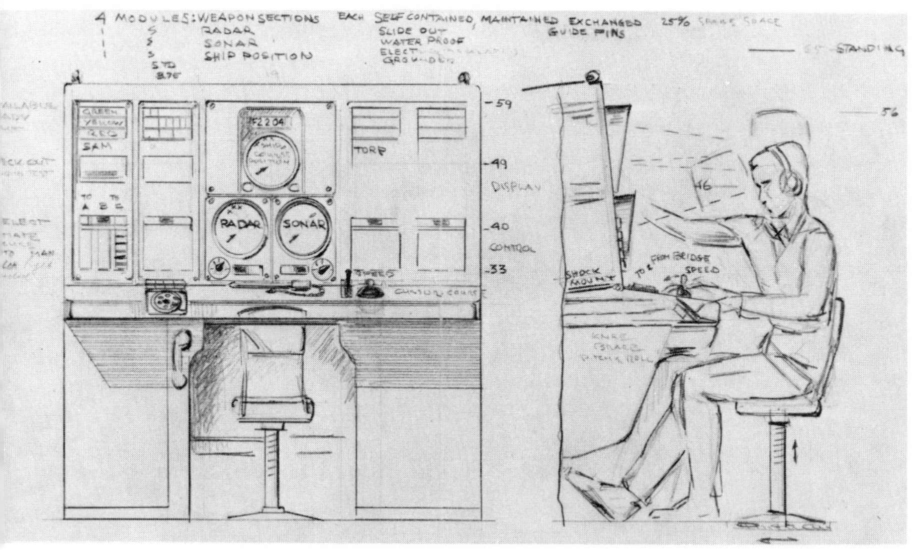

Figure 7-1 Designers sketch for DPT I—Weapons Assignment and Control Station (WACS).

In all these problems design requirements were described in terms of the design specification commonly used in military procurements. The format of the design problem is illustrated in Appendix A. A description of the characteristics of the design problems and the engineering subjects who solved them is provided in Table 7-1.

The type of equipment layout required of our subjects was something between an artist's rendition and a fully detailed design; it is characteristic of what we might find in a contractor's proposal. These abbreviated designs were necessary because of the comparatively short time for which subjects were available. We also wished to restrict the engineer-subjects in terms of the level of detail they had to provide since we were interested in having them concentrate their attention on parameters affecting the over-all configuration rather than on molecular details of individual components. The actual studies proved that it was often necessary to prevent the subjects from becoming absorbed in the minute details of the equipment they were creating.

2. Because we wished to investigate reactions to individual aspects of the design situation, a second type of test (the *Design Input Test* or DIT) was also employed in studies I and II. In this an equipment layout was not required of the engineer, but he was asked to analyze the problem verbally in terms of specific types of human factors inputs. We

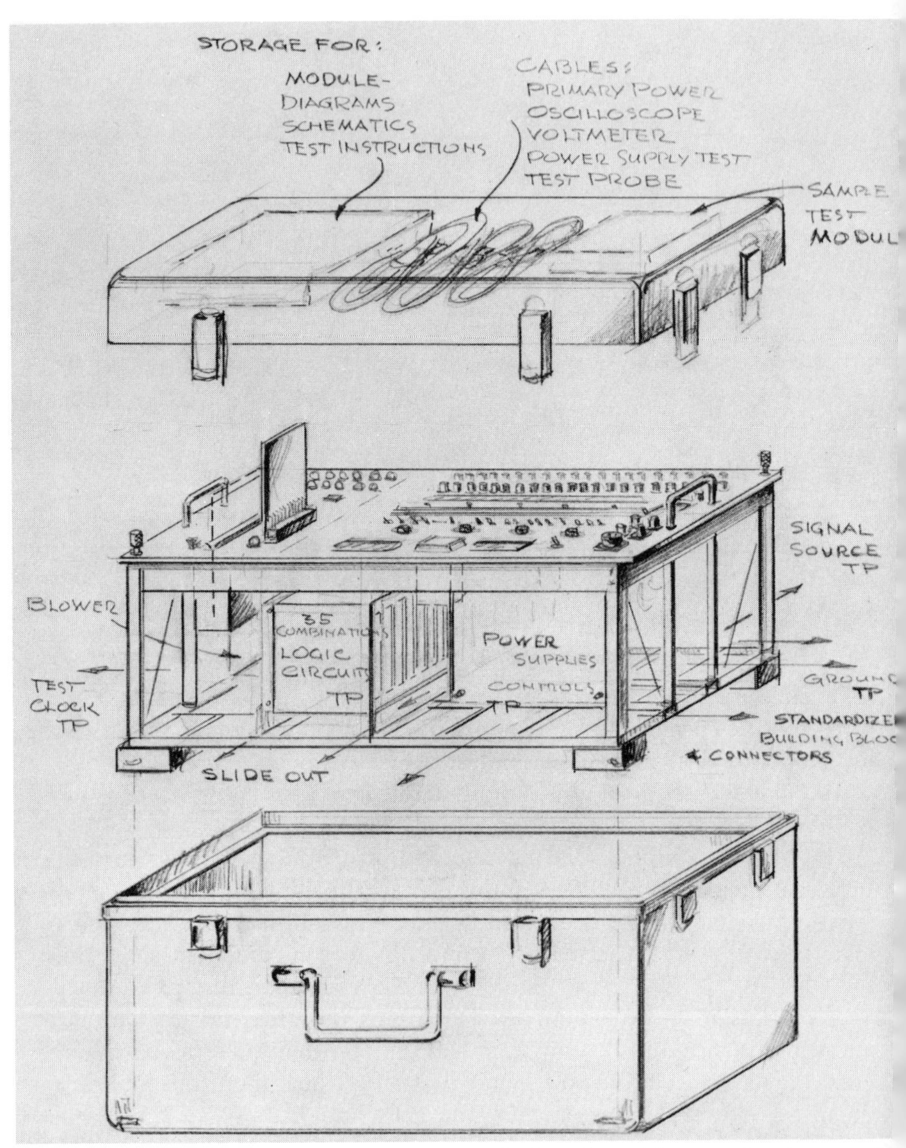

Figure 7-2 Designers sketch for DPT II—Shipboard Module Test Set (structural elements).

Table 7-1 Summary of Test Characteristics

Study	Subject Experience Level (years)	Number of Subjects	Educational Level	Type of Design Required	Time Allotted (hr per problem)	Type of $\underline{H}$ Input	Output Required
I–II	14.5	30	65% BS degrees	Console	4–6 (Total 12)	Control/display	Control panel/console drawings
References Meister & Farr, 1966 Meister & Sullivan, 1967							
III–IV	16.5	14	100% BS degrees	Subsystem	16–32	Control/display plus function/task analytic information	Control panel drawings Function flow diagrams, schematics Manpower estimates Equipment descriptions Bill of materials
References Meister et al., 1968 Meister et al., 1969							

were attempting to isolate details of the analytic process, much as a pathologist would excise and observe a tissue section under the microscope.

In one DIT item, for example, the engineer was presented with the problem of designing a shipboard equipment and asked to list the human factors inputs he would need to solve the problem. The assumption was that the number and type of inputs listed would indicate how the engineer conceptualized the problem. Appendix B presents a complete list of DIT items.

DIT situations were necessarily somewhat more abstract than DPT situations because they dealt with the analyses leading to design solutions rather than with the solutions themselves (i.e., drawings). The difference between these two types of test situations was, however, one of degree only; in each case a design problem was explicitly or implicitly presented.

Both test situations were so constructed that they demanded analysis of operator factors if the problems were to be correctly solved. In other words, human factors were an integral part of the design solution; for example, the design specification for the first DPT problem suggested that a decision be made between single versus multioperator use of the equipment, a decision which would have significant implications for design.

We attempted to make the test problems highly realistic since designers react negatively to situations in which technical details are incorrect or inappropriate. To ensure this realism experienced senior design engineers (not the subjects, of course) reviewed the tests and made any required modifications before the tests were presented to subjects.

Realism was further assured by modeling the problem, particularly in studies I and II, after a system which had been actually designed or produced. The same equipment and human factors inputs used in the original design were incorporated in the problem (with only minor modifications). When designers were questioned concerning the realism of the tests, they considered them, with very few qualifications, representative of design problems ordinarily encountered.

In the first two studies subjects were informed that the purpose of the tests was to discover techniques typically employed by experienced designers in solving design problems; these methods would then presumably be incorporated in engineering curricula. The purpose of this bit of theatrical business was to conceal our interest in human factors, thus eliminating as much as possible any bias subjects might show toward the discipline. It is, after all, not uncommon for subjects to try, half consciously, to give investigators what the latter presumably want. As it turned out, however, our fears that engineers would be biased toward human factors, if they guessed the purpose of the study, were completely unwarranted.

In the first place it was impossible to conceal our underlying interest since the questions asked dealt mostly with human factors; but more important, it soon became evident from the frankness of the interview responses that the engineer is relatively impervious to the contamination of psychological influences at least as they relate to human factors. Engineers are unusually outspoken concerning their likes and dislikes. In subsequent studies we discarded the notion of concealing our interests and decided to make a virtue out of necessity. Hence we made no bones of our concern for human factors and deliberately encouraged an attitude of interest toward our subject matter. Our reasoning was that, if engineers were negative toward $\underline{H}$ inputs even under the inducement of "playing along" with the investigators, these inputs could not be more effective in "real life." If we induced any positive bias in our later subjects it had not the slightest apparent effect on the indomitable honesty of our engineers; they know what they like and don't like.

Test Administration

Before taking the DPT, test subjects were given the design specification (problem) to read and study for several days. This is because the engineer does not typically proceed to design as soon as he receives the problem but goes through a gestation stage in which he more or less consciously gears himself to his problem.

A highly informal test atmosphere was encouraged. Testing was performed individually in an office in the engineer's own work area. During the test the engineer was entirely unconstrained; he could work as fast or as slowly as he wished; he could consult any reference sources or other engineers (other than fellow subjects); he could talk to the experimenter or not. The experimenter did not volunteer information or comments but would respond if the subject initiated a question or a discussion.

The subject had available to him equipment for drawing and copies of all military specifications noted as applicable in the design problem statement in addition to "standard" human factors handbooks. The designer was informed that he would be observed during the session but that he was free either to ignore the experimenter or to interact with him as he wished. A tape recorder was available to record the designer's verbal responses.

Originally, we had hoped to be able to note fairly minute aspects of the designer's behavior such as the number of times he referred to documents, the length of time he took to meditate, the number and types of changes he made to his drawings, etc. This turned out to be impossible; the engineer's behavior was either completely covert (e.g., reading the design specification) or much too molecular (e.g., manual drawing movements). Actually, the more abstract/creative a subject's activity is, the less relevant his molecular behavior is likely to be. To make any sense, therefore, of what he was doing and particularly why he did it, the engineer was intensively interviewed as soon as he indicated that his drawing was completed. The questioning emphasized the rationale for the design concept as well as the utility of the design inputs provided to the subject.

During the DIT the investigator interacted more directly with the subject, and the test items can be considered as a special form of interview. Because of the unstructured manner in which DIT questions were asked, the test situation had an almost clinical atmosphere.

The debriefing following the engineer's completion of the design was extremely "loose." Although a series of standard questions was asked by the investigator,[2] the subject's responses were followed up to secure greater detail so that in effect the engineer determined where the discussion led. At the same time the investigator did not content himself with the initial response to a question but continued to probe intensively, requiring the engineer to explain his answers in more detail until the latter's ability to respond was exhausted. This procedure was followed to get beyond the stereotyped responses engineers typically give to questions about their design processes. Discussions were sufficiently probing that a few subjects became somewhat emotional in their replies.

Subjects

The engineer-subjects for these studies represent, with one qualification, a cross section of the engineering population ordinarily found in the industrial environment (see Table 7-1). The qualification was that subjects must have had a reasonable amount of experience in actually

[2] Sample questions included the following: (1) Did the specification contain enough information for you to design what you would consider a satisfactory equipment? (2) Did it lack any information that you felt you needed? If so, what was lacking? (3) What would you consider to be the major problems you had in designing this equipment? (4) What factors (design parameters) did you consider most important in designing this equipment? (5) Did you find the human factors information useful in performing your task? These questions were considered only as models. The investigator was free to modify them (and did) in terms of the sequence and content of the subject's responses. In addition, the subject was asked to explain his design behavior. (Example: I see that you located this bank of toggle switches at the extreme lower left of your control panel. Why?)

designing equipment. Draftsmen and junior engineers, who ordinarily merely fill in the details of a drawing after the basic concept has been created by others, were excluded.

In the initial study engineers were largely mechanical engineers (packaging specialists) from an electronics company. Their responsibilities included design of control panels and of the external chassis and the packaging of the total equipment for maintainability. These were detail design responsibilities; in other words, although they made design decisions, these decisions dealt with molecular component characteristics.

Subjects in study II, primarily electrical engineers, came from a major aerospace company. Their range of design specialization was substantially broader than that of the first subject group: crew compartment and cockpit design, escape and life support systems, aircraft interiors, and electronics packaging.

The subjects for studies III and IV were all highly experienced propellant and missile design engineers. These were the most sophisticated of the three sets of designers. Their duties required taking a project all the way from initial design through test of the fabricated system. They were accustomed to making fundamental design decisions on their own. The contacts that all these engineers had had with H specialists were minimal although the discipline was not unknown to them.

Inputs and Outputs

In the first two studies no effort was made to simulate the progressive development of subsystem design. Design problems were solved in a single 3- or 4-hour session. In the second two studies subsystem development was simulated by scheduling each subject individually for a minimum of 10 weekly 3-to-4-hour sessions (depending on his design speed). In each session the subject received increments of data corresponding to those that he would ordinarily receive as system design progressed; for example, in the first session he had available to him the design statement of work and a list of personnel functions; by the fourth session he had been given function flow diagrams, equipment descriptions, and a preliminary task analysis of operations; by the sixth session a time-line analysis, etc. Appendix C lists the H inputs provided.

Naturally the subject had available to him at each successive test session all the accumulated data and his previous design outputs from preceding sessions. At each session the subject was asked to supply certain design outputs which the investigators hypothesized should be affected by the H input for that session; for example, in session 6 the engineer was asked to sketch the control panels needed to operate the subsystem; it was hypothesized that their layout might be influenced by time-line analyses supplied in the same session.

One of the ways in which the effect of H requirements was tested was to modify them drastically late in the simulated developmental cycle. For example, if up to this time the subject had been required to design for a maximum of two men, he now had that maximum reduced —to one man—or expanded—to four men. Subjects were required to analyze their previous design configurations in terms of the new requirements. If significant design changes resulted, these would suggest that H requirements do influence design.

We mentioned earlier that equipment and personnel inputs presented to subjects (the "scenario") were based on inputs derived from an already designed or operational subsystem; for example, in study III the operational subsystem selected as a model was the Titan III ground propellant transfer and pressurization subsystem. This subsystem, whose functions were described in Chapter Three, was responsible for receiving rocket propellants from railroad cars, for transferring fuel from these cars to ready storage vessels, for storing propellants for a period of 30 days, and then for transferring the stored fuel to the missile tanks. In study IV the subsystem selected for simulation was the aerospace ground equipment of the AGM-69A, which is an air to ground missile designed to be launched from the B-52 bomber. The specific equipment to be designed by subjects was the electronic test equipment used to check out the status of the missile prior to its loading on the aircraft. In both these studies the original documentation for the system was researched and the desired inputs extracted. Design engineers from the company producing the system checked the experimental inputs to determine their correctness.

To determine the effect of a particular input on the design task, it was presented at the same time at which subjects had to perform the design task to which the input was presumably relevant. At the start of any session the engineer was told his design task, the inputs available to him were described, and he was asked to examine them (in the event he had not reviewed them since he was first handed them at the close of the previous session). The subject then performed his design task. About a half hour before the end of the session (unless he obviously was not finished, in which case the session would be continued to the following week), the subject was informed that his work was to be reviewed. His output then was reviewed with him by the investigator to elicit additional information, particularly the reasons why distinctive design features were incorporated. At the same time, the subject was questioned to determine whether: (a) he thought the input was useful; (b) the input was understandable and meaningful; (c) he used the input in deriving his design product; (d) the format of the input was satisfactory; (e) the timing of the input was appropriate; and (f) any additional information was needed.

At the close of the session he was handed the inputs for the next session and was asked to study them before reporting for the next session.

RESULTS

It would be tedious and repetitious to report the results of the four studies individually. The results of each study, although adding some new facet of information to the total picture, build on the data supplied by those preceding. Our conclusions are therefore described in terms of the questions on pages 240–241 to which the studies as a whole were directed.

How Engineers Design

The general impression, fostered by reports and articles such as Marples, 1961, and Coryell, 1967, is that design is a logical, systematic, step-by-step process. Design as we see it under controlled study conditions is indeed logical in the sense of "if-then" deductions, but it is not completely systematic if by systematic we mean the consideration and trading off of all the factors relevant to design; moreover, it is only partially a step-by-step process.

Beyond that our results show that to quote Mitroff, 1967:

"engineering design *must* be described in behavioral terms, and not just in technical terms . . . There is no more basic behavioral variable than the personality of the engineer . . . our point is that any model which only conceives of design as a process of optimizing technical variables is dangerously incomplete . . ." (pp. 11, v, and vi).

All of the engineer-subjects studied exhibited the following characteristics:

1. Almost immediately after examining the problem, they started to design hardware without performing much deliberate, systematic analysis. In other words, they exhibited what can almost be characterized as an *obsession with hardware*.

2. They relied overwhelmingly on design solutions which they had used before—presumably satisfactorily. This prevented them from considering novel approaches to design.

3. Their use of information was determined almost exclusively by the design requirements and constraints in their SOW; they interpreted in-

formation in terms of the implications of that information for those design requirements.

4. They preferred to design with a minimum of advice and/or constraint by others, including the customer.

5. They were reluctant to consider operator or human factors in making decisions fundamental to system configurations.

The remainder of this chapter presents evidence to support these conclusions.

Obsession with Hardware. In studies I and II we observed that our engineer-subjects, when presented with the Design Product Test problems, proceeded almost immediately to the layout of the equipment drawing without any "systematic" analysis of the design problem. As we saw in Chapter Three and as is generally assumed by human factors specialists (e.g., Van Cott and Altman, 1956, and Rabideau, et al., 1961), system development presumably proceeds sequentially along the following analytic stages:

1. Determination of mission functions.
2. Determination of system functions.
3. The allocation of system functions between equipment and personnel.
4. The specification of equipment design requirements (as distinct from system requirements).
5. The interrelationship between equipment and personnel functions.
6. The specification of equipment and personnel characteristics to satisfy equipment and personnel requirements and functions.

It may well be that this sequence is solely an H invention. Since it is logical, the specialist may believe that it is actually the way in which system development proceeds. The question is whether engineers in their everyday design activities apply that logic.

If some such step-by-step conceptual process occurs in design, it is difficult to observe overtly. It is possible with some difficulty to elicit these analytic steps from the engineer, but they must be specifically requested and emphasized. When we can detect the process, it appears highly compressed because the engineer relies on his experience with previous systems and equipment to make the decisions that are required by these steps.

This was particularly marked in studies I and II which involved consoles rather than systems. Since the test problems listed a series of control and display functions, the engineer's first reaction was to extract these functions from the design SOW and to specify the precise controls and displays needed to perform the functions. There was almost no consideration of how these functions were interrelated or what their purpose was in serving the over-all goal of equipment functioning. As a matter of fact, designer behavior in these studies was most distinguished by what the engineer did *not* question or comment on. The result was an almost immediate resort to the drawing board and the equipment layout.

At the end of the first study we made the following statements:

"The first thing the designer does, as would be expected, is to review the design specification or statement of design requirements. He identifies certain key words as clues to particular work activities (e.g., 'drawings,' 'lists,' and 'parts' would be noted by the designer responsible for drawing, creating lists or selecting parts).

"Following this he lists the equipment components required (if the specification notes these) or the functions to be performed (if equipment components are not described). It is significant that he fails to distinguish any equipment component or function in the specification as being more important than any other, and that he fails to ask questions (when he has the opportunity) concerning the manner in which equipment components would be used operationally. Such questioning is, we assume, essential if one analyzes design at a more than superficial level. . . .

"After the initial stage of listing components is completed, the designer prepares a quick rough sketch of the over-all (i.e., external) configuration. The designer then fits the components he had previously listed into the over-all configuration on the basis of largest controls/displays first, etc. The initial rough drawing may be repeated, this time in finer detail and much more care. There appears to be relatively little indecision about positioning controls and displays, which presumably reflects the absence of any tradeoff conflict" (Meister and Farr, 1966).

This lack of analysis might, we thought, be explained on the basis of two hypotheses:

1. Because our design SOW was fairly specific in describing control and display functions, it may have eliminated the necessity for more sophisticated analysis, which presumably precedes the determination of equipment details. In other words, the behavior of our engineer-subjects may have been constrained unduly by the relatively concrete nature of our design SOW.

2. The engineers we had as subjects in these first studies were detail engineers; in other words, engineers who were perhaps not used to making conceptual design decisions and whose experience was primarily

in the detailed development of equipment after these basic decisions had been made by more analytically oriented (i.e., system) engineers.

Both possible explanations have some plausibility. Engineers can be divided (roughly) into two categories: those, like the subjects of studies III and IV, who are essentially system engineers responsible for analysis of system requirements to determine basic design concepts and characteristics, and those, like the subjects of studies I and II, who are concerned with elaborating the details of individual equipments after more fundamental decisions have been made by system engineers.

Perhaps we had the wrong kind of engineers as subjects for the first two studies. (To say so is perhaps intellectual arrogance. It implies that the only design activity which is really meaningful—from a specialist's standpoint—is that of a fundamental, decision-making nature, like deciding whether or not a system will perform a particular function or whether a system will be automatic or manual. The hypothetical analytical sequence listed previously also embodies a value judgment that only critical decisions are meaningful. But such critical decisions may be quite infrequent in design.)

(It is possible that many H specialists are not really interested in molecular/minor details of engineering design; they want to participate in the "big" decisions, and, if they are left out of them, they feel frustrated; but what if the "big" decisions are few in number and difficult to pinpoint?)

In any event, when we examine the results of studies III and IV, we find that the explanation supplied for the lack of analysis in studies I and II no longer apply. Our engineer-subjects in the later studies were analytically oriented (from an equipment standpoint). The design SOW's for the second two studies were, moreover, not highly specific and left to the subject the primary decisions for the amount of automation and the determination of how the system was to function. Overall system requirements (e.g., in study III the nature of the propellants to be transferred, the time constraints within which transfer was to take place) were specific, but the manner in which these were to be implemented was left to the designer.

Moreover, the experimental approach in these latter two studies explicitly called for the subjects to make analytical decisions; for example, the first session in study III called for subjects to elaborate the very simple subsystem functions provided in the SOW (e.g., receive propellants, store propellants, provide blanket pressure) and to break them out between equipment and personnel responsibilities. We asked specifically for a *subsystem* functional block diagram which is even less detailed than an equipment functional block diagram.

To our surprise all of the subjects responded by producing a fairly detailed schematic diagram (see Figure 7-3) which included the following features:

1. Explicit equipments needed (e.g., heat exchanger).
2. Piping lines between equipments and even geographical (site) arrangements.
3. Valving required to operate the equipment.
4. Determination of which valves would be remotely (automatically) operated and which manually operated.
5. Equipment tolerances.

If we apply, as a purely arbitrary yardstick, a six months period for an average concept formulation phase, then in the first three or four hour session (corresponding perhaps to a week or two of actual development time) the engineer-subjects had accomplished just about everything which we would expect them to perform in the first two of the six months. (Moreover, they were not compelled to hurry; they knew they had 40 hours to complete the design.) What remained then for these engineers to do? The subsequent three sessions (corresponding to the next four months) involved the detailed elaboration of the subsystem whose configuration they had decided on in the initial session. This detail elaboration (e.g., addition or deletion of valves or other equipment assemblies, specification of equipment characteristics), which is of course much slower, did not change except in very minor scale the earlier decisions. In effect, in the first session the engineers had made the fundamental decisions for which we presumably have a concept formulation phase. Everything that followed was elaboration.

It is difficult to call these decisions into question since they provided each of the subjects with a design concept which was feasible, if perhaps not optimal; however, there is evidence, volunteered by the subjects themselves, that they had overlooked certain factors that the H inputs eventually called to their attention. However, the essential point is that, whether or not their decisions were justified, they were made almost immediately after they received the SOW.

We could say, of course, that our engineer-subjects were presented with no problems which their experience could not solve. Their test situation did not require any breakthroughs in the state of the art, nor were they presented with functional requirements which were highly stressful in a design sense; but then, with the exception of a relatively few system development projects (e.g., the development of the first radar or sonar, the development of the first ICBM, the first manned

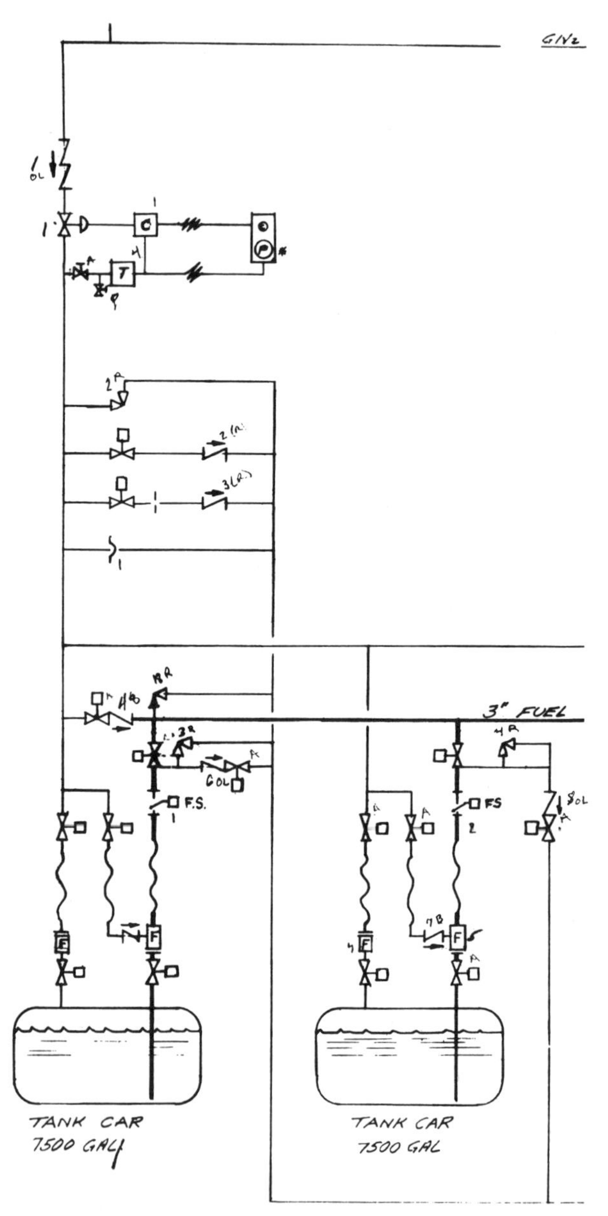

Figure 7-3 Schematic diagram—transfer from RR shipping cars to ready storage fuel tanks.

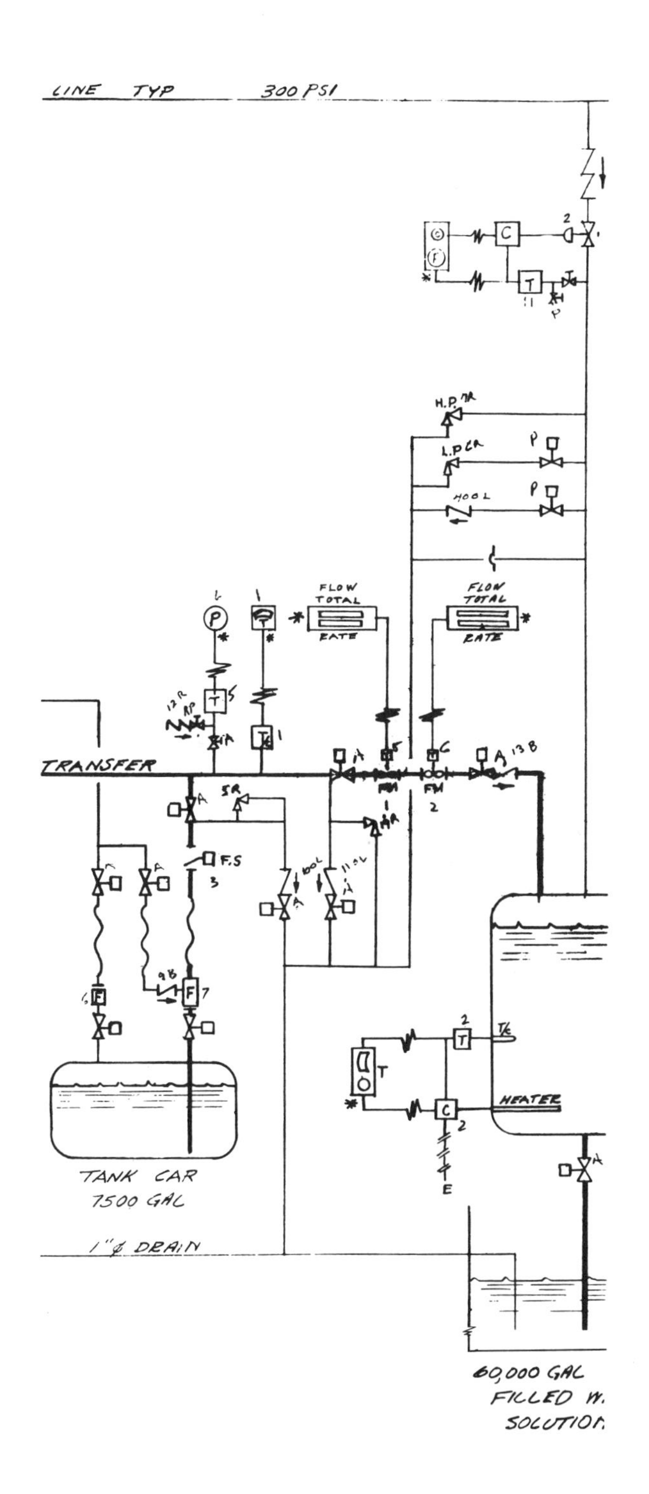

LINE TYP 300 PSI
FLOW TOTAL
RATE
FLOW TOTAL
RATE
H.P.
L.P.
TRANSFER
F.S
FM
FM
HEATER
TANK CAR
7500 GAL
1"ϕ DRAIN
60,000 GAL
FILLED W.
SOLUTION

space vehicle), most developmental projects represent only a progressive advance over an already familiar state of the art. From that standpoint the test situation was not atypical.

There is other evidence of the engineer's continuing concern for hardware. In studies I and II subjects were required to develop control panel/console configurations in which (by our standards) human engineering considerations should have been paramount. Typically, however, the subjects explained that the basic decisions for control panel layout were made by electrical engineers responsible for developing the internal circuitry behind the panel. In other words, by specifying certain internal components (e.g., amplifiers) these engineers implicitly (and sometimes explicitly) set the arrangement of the control panel face. Many of the subjects said that typically they relied on the electrical engineering group to tell them how they should arrange the control panel. We call this "inside-out design" since its starting point is internal componentry. The human factors specialist, of course, is oriented toward "outside-in design," in which the arrangement of the control panel is based on factors depending on operational usage (e.g., operational functions to be performed, priority and frequency of use, etc.).

Of course, it is possible that this hardware tendency—or at least its prominence—may be more characteristic of detail, than system, engineers.

There are certain implications of this "flight to hardware" and the concomitant swiftness of basic design. One of the hypotheses which was gradually developed as the studies progressed was that the reason many human factors recommendations are not implemented is that they are made too late in system development. If the swiftness of equipment design is true, this tardiness would be fatal for the $\underline{H}$ input.

It is possible that $\underline{H}$ inputs are delayed because the specialist, who often lacks all but a minimal engineering background, depends on the design information provided to him by the engineer as the basis for his design contributions. The specialist is reactive rather than active. Because of this dependency then, his inputs tend to follow rather than precede fundamental design decisions.

If designers make fundamental decisions immediately after confronting the design problem, $\underline{H}$ analyses and inputs, if they are to be effective, must also be available at the very start of design. Specialists cannot wait two or three months or, as in the case of detailed task analysis, until the end of the concept definition phase to submit them to engineers. To be in synchrony with design, such analyses must be available at the very start of the design problem.

Since the engineer uses as his basic informational tool the design SOW, it would seem necessary that the human factors analyses (in the form of design requirements) should be included in that SOW. The SOW, which contains the design requirements, is particularly important because "a statement of the design requirements cannot be completely removed from a description of the design process as a whole" (Mitroff, 1967, p. 43). In other words, to have developed a set of design requirements means that much of the basic design concept must have been anticipated or implied during the process. The progressive, incremental elaboration of H̲ inputs will therefore no longer suffice.

Naturally, to provide human factors inputs to the engineer at the time he starts his design effort means that much of the H̲ analysis that is ordinarily performed by the contractor after he receives his go-ahead must be performed by the customer prior to that time. Needless to say, the customer (e.g., the Air Force, the Navy, the Army or other governmental agencies) does not usually perform these human factors analyses.[3]

If these analyses are available when the design SOW is delivered to the contractor, the job of the H̲ specialist will be not to attempt to perform the human factors analysis himself but to interpret its meaning in design terms. In other words, if as a result of such analyses it is decided (and specified in the SOW) that the maximum number of personnel available for running the system is 10 men of the following specialities to perform the following tasks, it will be the job of the specialist to indicate what this requirement should mean to the designer in terms of *design alternatives.* Of course, some of the paper (e.g., task) analysis created presently by specialists in an attempt to affect the design configuration will no longer be necessary and will presumably be eliminated; and naturally it will require that the H̲ personnel who presently create that paper work become more attuned to hardware and the meaning of human factors requirements for hardware. Documentation such as the Air Force's Quantitative and Qualitative Personnel Requirements Information (QQPRI) may not be eliminated, but it will certainly have to be revised to make it an instrument for interpretation of human factors in design terms.

It is to be expected that the specification of such human factors inputs as design requirements may elicit opposition from engineers and company management. As we shall see, the engineer resents any restrictions on his design freedom, and the consequence of this precontract analysis will be to constrain design from a human factors standpoint.

[3] The customer will object that he often does not know enough about the ultimate design concept to be able to specify personnel requirements to the contractor. This objection is sometimes valid; however, the author feels that the customer usually can perform much more detailed personnel analyses than he actually tries to perform.

Such a concept will work only if the H specialists analyzing a system requirement prior to the issuance of a design SOW are highly qualified and if they have time to analyze that requirement not only in human factors terms but also in terms of their hardware consequences and their relationship to cost, schedules, etc.

Reliance on Experience. Obviously, to bypass analysis as much as he does the engineer must rely heavily on his experience. There is nothing inherently wrong about that; we all do it. However, design, if it is to be creative, must examine many novel approaches. When his reliance on experience prevents the engineer from considering novel design solutions, it is reasonable to conclude that this experiential factor makes him somewhat inflexible. There are a number of details we can cite which describe this process.

In study I the design specification for DPT I required the subject to decide whether one man or two should operate the console. The number of functions required to be performed was fairly extensive; for example, sonar and radar detection and tracking, own ship navigation during an attack, monitoring of weapon status, and firing orders to weapons modules. Hence the requirement to consider one versus two man operation was a reasonable requirement. Yet all subjects opted for a single operator *because most of the consoles they had designed previously had been single operator consoles.* They failed to analyze the information in the design SOW as it related to the problem of one versus two man operation. There was in fact no consideration of the requirement at all despite the fact that the SOW specifically asked that such an analysis be made. The excessive reliance on experience suggests the possibility that customer desires may be rejected or ignored when their requirements conflict with the engineer's convenience or with his experiential stereotypes. The rationale given is that the designer "knows best," the cost will be less, etc.

Another illustration—during study III one engineer, after examining the SOW, suggested that he had in his files an already created design which would precisely satisfy the requirements of the problem. He offered to go and submit it to the investigators. Naturally, his offer was politely but firmly rejected.

The experience factor reveals itself also in the designer's use of stereotypes. A design stereotype is a consistent preference, all other factors being equal, for one design solution over others. These configurations are repeated in successive designs unless some feature of the design problem prevents the engineer from applying his stereotype. One stereotype we found in study III was a consistent tendency, all other things being equal (and sometimes unequal), to prefer an automatic design to one requiring more manual inputs.

Another stereotype found in study IV was to consider that certain design solutions are more appropriate for skilled personnel, whereas others are more suitable for the relatively unskilled. In this line of reasoning it is more appropriate to design for the unskilled with qualitative, go-no/go displays, step-by-step procedures, throw-away modules for maintenance purposes, and higher level (e.g., subsystem) as opposed to lower level (component) troubleshooting checks. Skilled personnel can be given quantitative displays and required to troubleshoot to the circuit level. Given information concerning the skill level of the equipment maintenance men for whom they were designing, engineers tended to include the above characteristics in their design configuration.

The point of the foregoing is not to denigrate the particular engineering stereotypes, which may in fact be highly reasonable but rather to point out that they exist and that the engineer brings to design various attitudes that influence that design.

The same point is made by Mitroff (1967):

"Design, in other words, is not merely a game governed solely by technical equations. . . . It is the engineer and not the equations who does the designing. Either the engineer learns to manipulate his equations . . . or he is manipulated unconsciously by equations he does not understand for purposes he does not intend to be accomplished. . . . It (engineering design) is a technical activity directed by a behavioral creature. . . ." (p. 19)

If then we look askance at the engineer's reliance on experience and stereotypes, it is because this tends to perpetuate inadequate design solutions, and these in turn may be responsible for much of the poor human engineering we find.

Design Requirements and Constraints. We were also interested in determining the primary source of information on which the design engineer relies for his design decisions. We reasoned that if we could attach our inputs to that source the likelihood of the engineer's using these inputs would be increased. Naively, we were surprised to find that the primary informational source for the design engineer is the SOW which contains his design requirements. It seems perfectly understandable that the engineer should be concerned about his design criteria; and yet his concentration on these surprised us in view of the fact that a torrent of inputs, mostly equipment-related but also some dealing with human factors, flow to the engineer throughout development.

The concern for design requirements, to the point of almost ignoring other data sources, manifests itself in engineers of widely different backgrounds and for design problems which range from simple console layout to full scale subsystem development.

In study III, for example, the sources of information provided in the first three sessions consisted of the SOW, lists of personnel tasks, equipment and personnel flow diagrams, equipment requirements descriptions, and control-display recommendations. Of these, the SOW, consisting of rather general requirements, presented the least specific information. Although subjects examined each informational source carefully when it was first presented, the relative amounts of time spent in reviewing and referring back to each source heavily favored the SOW; none of the other informational sources approached its attraction. Characteristically, as the subject expanded on his initial design concept by proliferating the amount of equipment detail, he constantly checked the validity of the detail against the original criteria in the SOW.

As a consequence, we can conclude that the engineer looks at each new informational input in terms of whether it is a design requirement (and hence constrains him) or is not (in which case it can be blithely ignored).

The emphasis on design requirements cannot be understood without considering their significance as the other side of the coin—as design constraints. Every design requirement reduces the number of degrees of freedom within which the engineer can function. Manifestly, if there are few design requirements, the number of choices the engineer can make is quite wide. The larger the number of design requirements, the fewer choices he has. The engineer is quite aware of this and interprets each item of information which he receives in terms of two criteria:

1. Does the information represent a requirement levied on my design?
2. If so, to what degree does it constrain or reduce my design alternatives?

Information failing to meet these criteria will fall into the "nice to know, but so what?" category.

Most engineers naturally prefer as few design constraints as possible. A few engineers become quite resistant when the number of design constraints increases beyond the minimum they will tolerate. They may also reject information when they interpret that information as constraining their design. One of the subjects of study III had to be replaced after the initial session because he considered that the personnel inputs for the second session were so detailed as to demean him (e.g., "an experienced engineer doesn't need all this information; if you have to tell an engineer all these things, you may as well get someone off the street, etc."). This rejection of design constraints has certain negative implications for H inputs when they are interpreted as constraining design.

The utility of H inputs is profoundly affected by the designer's concentration on design requirements and constraints. The most technically sophisticated human factors analysis will be of little significance to the engineer unless it is phrased in such a way that he *interprets* the inputs resulting from that analysis as a requirement or constraint on his design or, less probably, an improvement on that design.

This means that to be maximally influential the input should be described in *prescriptive* terms and especially in such a way that equipment characteristics inevitably flow from the input. In other words, the inputs should be of the following form, explicitly or implicitly: "equipment will be designed to accommodate the following human factors requirements. . . ." The emphasis on the phrase "interprets" in the preceding paragraph is important; no input, as we shall see, is unequivocal. To avoid the problem of interpretation as much as possible, the input should be phrased so that certain design consequences inevitably flow from the input. The H input must be mandatory, and it must be interpretable (either by the engineer or the H specialist) in terms of these design implications.

This last point is most important. Because he has difficulty in handling behavioral inputs, the engineer may not be able to perceive the design implications of an H input; these must be given him by the specialist. Unfortunately, there are many H inputs which, even if expressed in mandatory terms, cannot be interpreted even by the specialist in terms of design implications. The most unfortunate example of this is the description of desired personnel characteristics which is provided as part of Air Force QQPRI. Unless such inputs do have discernible design requirements and implications, they will be largely ignored by the engineer.

It would be a mistake, however, to assume that the design requirement inevitably imposes a particular type of design solution. The requirement that a design be highly automatized, for example, does not inevitably lead to a particular design solution because there are various ways in which an equipment can be automatized. If the H specialist often can discern no direct relationship between an input he has made and the eventual configuration produced, it is because requirements and inputs act only to restrict the number of design alternatives but not to suggest a particular one.

In fact, there is reason to believe that design requirements are used by the engineer mostly as criteria for evaluating the adequacy of the configurations he designs. (This would account for the heavy emphasis on the SOW in their design activity.) If an H requirement is imposed such that the equipment must be designed for one-man operation, for example, this means only that after the preliminary design is completed, the engineer compares that design against the requirement to ensure that in fact it can be operated by one man.

This applies not only to H̲ inputs but also to equipment inputs. If the H̲ specialist wishes to control design (from the human factors standpoint only, of course), he can do so only in three ways:

1. By getting specific H̲ requirements included in the SOW.
2. By pointing out to the engineer during design the design implications of these requirements.
3. By ensuring that engineers evaluate the adequacy of their designs in accordance with H̲ requirements.

Inflexibility of the Design Concept. The speed with which the engineer develops his design concept (often by the end of our first test session; in actuality often during the proposal period) is matched by his relative inflexibility in modifying that concept later, except in minor details.

In order to determine the effect of personnel requirements on design, these requirements were drastically changed in the middle or latter part of the testing sequence. For example, the subject might be told that he could have only one or two men to operate his subsystem after he had informed us earlier that his subsystem could only be operated if he had four men. Manifestly the subject's original subsystem design was no longer applicable after the requirements had changed.

However, the number of changes he was willing to make in his original design to accommodate the new demands was minimal (e.g., minor changes in controls and displays, the addition of a loudspeaker system, changes in technical procedures). No changes were permitted in the basic design concept. This design inflexibility was characteristic of all subjects tested.

This applies of course only to the basic design concept. The engineer is perfectly willing to modify the molecular details by means of which the major concept is to be implemented.

The implications for fundamental H̲ inputs like function allocations are obvious: they must be supplied at the time the basic design concept is developed, or, once the design concept has been completed, they are irrelevant. Again, this does not apply to human engineering recommendations of the "knobs and dials" variety which the engineer is usually willing to consider later in design.

Need for Information. One of the clearest tendencies which the engineer manifests is an apparently inexhaustible desire for information bearing on his design. The design process is a search for possible solutions. Information must be collected about each possible solution in order to decide whether that solution is feasible or not. The engineer rarely feels that he gets enough information of either the equipment or human factors variety.

When he receives that information he evaluates it in terms of its impact upon the alternative design possibilities he is considering. He does this by working out the implications of the new information mentally and then by checking those implications against SOW requirements.

Most of the information he receives he rejects as being irrelevant, but nevertheless he cherishes the opportunity to examine it. Since it is always possible that some new datum or consideration will negate a possible design concept, it is necessary for him to analyze that datum or consideration merely to reject it. This is the reason also for his recurrent reference to the SOW, to compare the new information against the SOW design criteria.

In fact, much of the information provided to the engineer (both equipment and human factor-related) has little effect on his choice of a design concept. However, the H specialist can capitalize on the engineer's informational need in putting forth his own recommendations.

Variability in Design Solution. At the outset of our studies we were worried that all of our subjects would produce very similar designs in response to our experimental design specifications. Such uniformity would have made it difficult to determine the influence on design of the experimental variables we were studying.

We need have had no cause to be concerned. One of the most striking things about design is the great variability which we encounter among engineers who are designing to the same specification. There are of course similarities among these designs—how could it be otherwise—but the differences are even more striking. For example, the same requirements will elicit two designs, one a highly automatized system (e.g., two computers, one being backup for the other), the other largely manual.

This observation is well known in engineering circles: ". . . The restrictions on engineering work imposed by the need or the engineering dimensions are seldom so exacting that only one solution is possible . . . The task . . . can often be performed by several alternative products . . ." (Ranstrom and Rhenman, 1965).

One reason for this variability lies in the engineer's interpretation of design requirements. Any set of requirements, whether these be cost, reliability, or human factors, must be filtered through the engineer's concept of what these requirements mean; for example, is a reliability requirement of .9950 (study III) so inordinately high that it demands—as one subject assumed—a completely automatic system with two computers; or can it be achieved, as the other subjects felt, with less Draconian measures? The engineer's individual "design style," probably determined by his successful and unsuccessful design experiences, obviously channels the interpretation he places on the requirement.

From a human factors standpoint the fact that several design solutions are possible and that one is selected by the engineer after trading off various parameters is mildly satisfying. It means that a solution which is desirable from a human factors standpoint stands as much chance as any other, provided we can phrase the human factors design requirement in such a way that the engineer interprets it as of sufficient importance.

Engineering Attitudes toward Human Factors

One of the most consistent and hence significant findings of our research, confirmed in each study with all subjects regardless of sophistication, years of experience, or type of design problem, is that the typical design engineer does *not* consider human factors in his design. Despite what the engineer claims, observation of how he actually designs reveals an almost complete lack of interest in the behavioral implications of his design.

One of the subtests in our first study was a series of 42 statements concerning the engineer's attitudes toward human factors and his consideration of them in design. The engineer had to respond to each statement by indicating his agreement or disagreement. When we examined the engineer's responses to this subtest, we found that they indicated a strong, we could almost say profound, interest in human factors and a dedication to their consideration in design. After we had compared these paper and pencil responses with the engineer's actual design behavior, we found they were 180 degrees at variance with his actual behavior in solving design problems.

In his actual performance it was impossible to detect any but the most cursory consideration of human factors. Over a number of years we have come to believe that we cannot take at face value what most engineers say they do or feel when they refer to human factors in a positive manner (on the other hand, we can believe their negative statements). They are sufficiently astute to recognize what others—their company management, customer representatives—think they should say.

It would be unfair to say that the engineer is deliberately lying. By his lights, he does consider human factors in design. This is because he views what we would consider human factors problems as essentially a subset of the general problem of selecting and arranging hardware components. There are no human factors problems because all problems reduce to equipment parameters.

Thus, to the average detail designer, the layout of controls and displays on a panel face is largely a problem in having enough physical space in which to arrange these components. To him it is not a question of which principle of control-display arrangement to apply to his design but rather a problem in physical dimensions; whereas the specialist considers control-display arrangement, for example, as a problem to be solved in terms of its effect on the operator or in terms of the anticipated performance of the operator, most engineers do not.

How do we know this? We deduce it from that fact that none of our subjects considered his design in terms of the *operational use* of the equipment. In studies I and II engineers failed to ask any questions concerning that operational use. This is less true of the subjects in studies III and IV who can be considered system engineers, but their behavior was not significantly different from the others. If we assume that it is impossible to consider human factors without considering the operational use of the equipment, then failure to question this use represents lack of interest in behavioral factors.

Are we doing engineers in general an injustice in what we say? The answer is "yes" for that *rara avis,* the behaviorally oriented engineer. The answer is "no" for the great majority of other engineers. For the statements we have made the evidence is unfortunately largely negative. It is difficult to prove a negative, but the uniformity of engineer responses compels belief. None of the principles which H specialists consider to be human factors principles was given by engineers as the reasons for their design. Thus subjects did not rationalize their designs in terms of such factors as function allocation, the relative criticality of controls and displays, principles of control panel arrangement, etc. We did not, of course, ask subjects to explain their actions in behavioral terms. We did, however, look for explanations which could be interpreted in terms of a behavioral rationale.

We may ask then on what basis the control panel layouts required in the first two studies were performed. In actual system development much of the design would, as pointed out, have been determined by requirements imposed by the engineers designing the internal components of the control panel. This factor played no part in our studies, however. Then there is the experiential factor which bulks so large in design. The detail design engineer comes to design with an inventory of most favored configurations which he applies rather routinely until or unless he runs into a problem. He applies certain general rules from past experience (e.g., the power on-off switch is commonly located in the lower left or right hand corner of the control panel).

Another basis upon which the detail engineer designs (and which may also account for his feeling that he considers the operator in his design) is the fact that he typically puts himself in the place of the equipment operator. He says to himself, in effect, "If I were the operator, how would I prefer the design to be laid out?" We could conceivably call this a form of H̲ analysis, but, if it is, it is extremely primitive.

In practice the detail designer utilizes a few of what we consider the more basic human engineering criteria: controls should be as close to the operator's hands as possible, hence should be located as much as possible at the bottom of the panel space; displays should be visible as much as possible, hence should be placed toward the center of the panel area; controls and displays used together belong together, hence should be grouped as modules, etc.

These are rules of thumb to be applied in all design situations, rather than principles to be applied after detailed analysis. We emphasize the analysis because without analysis the consideration of human factors becomes a matter of rote rules. From that standpoint there is nothing like a human factors analysis to be found in the average engineer's design behavior. Since the engineer does not examine the conditions of operation under which his equipment will be exercised, his design is likely to be quite satisfactory only so long as there are no special circumstances that may affect the application of these rules of thumb. When this happens, his design is inadequate.

What about our more highly sophisticated subjects of studies III and IV? Despite their sophistication they did not differ significantly from their country cousins in human factors attitudes. We have already pointed out their obsession with hardware to the extent that they bypassed the analytic phases which are essential to human factors considerations.

Additional evidence that our system engineer-subjects were only minimally concerned about human factors is that, as we shall see, personnel requirements, changes in personnel requirements, and human factors information in general had relatively little impact upon their designs (although it would be untrue to say that they had no effect).

Paradoxically, the logical result of the engineer's attitude toward human factors is to reinforce the need for the H̲ specialist. Many engineers, some company managements, and even some military customers have questioned the usefulness of H̲ specialists and the human factors requirement in design. They might be justified in their scepticism if the engineer did in fact perform the same service as the human engineer, if he did consistently incorporate human factors in design.

However, the engineer's lack of interest in human factors makes it mandatory that someone be available to prod the management into considering this parameter. Although it is reasonable to trust the design engineer to consider relevant equipment factors in his design, it is obvious that we cannot trust him to do the same for behavioral or operator factors. The governmental customer and the industrial manager who think they lose nothing by eliminating H activity on a particular system development project are in reality fooling themselves as well as doing a disservice to the personnel who have to operate and maintain that system. All the evidence we have gathered suggests that reasonable consideration of human factors can be accomplished only by encouraging (perhaps even forcing) H participation in design.

Engineering Use of H Inputs

There is evidence that engineers will make use of H inputs if these are emphasized. For example, approximately half of the subjects in the second two studies felt that H inputs were of some value. Because the engineer develops the hardware details of his design so rapidly and because he is responsive primarily to design requirements included in the system statement of work, H inputs have their greatest impact on the engineer when they are presented early in design, preferably in the SOW.

Curiously enough, this contradicts the feeling engineers have that H activity properly belongs to the later phases of system development (detail design). Perhaps this feeling exists because even sophisticated engineers have the concept that human factors is no more than "knobs and dials." Certainly it is easier for them to understand human engineering with its concrete, design-oriented parameters.

In general, the kind of H inputs the engineer understands best (even though he may not appreciate them) are those that constrain his design or that deal with concrete system operations. In addition to control-display inputs they include the following:

1. The maximum number of personnel for which equipment is to be designed.
2. The skill level of personnel who will run the system.
3. A description of the jobs and tasks personnel will have to perform.

The first two of these represent personnel requirements levied as constraints on system design. The third provides information which can be transformed into operating procedures.

Some H specialists believe that every H input should be of value to the engineer simply because it is an input and regardless of whether or not that input has meaningful design implications. There is a tendency on the part of these specialists to feel that, because they can conceptualize a potential input, that input must be critical to the design of the system.

Engineers are, however, relatively indifferent to abstract H inputs (e.g., those that describe training requirements, availability and experience level of personnel, and the cost of training and supporting personnel). They do not see the design relevance of such inputs. They may be of value to other H specialists or to the development manager, but they are so much noise in the system to the engineer. The more abstract a behavioral input is, the more difficulty the engineer has in deriving design implications from it. If he is unable to draw meaningful conclusions from the input, he rejects it as useless.

Even when the engineer does not find the individual H input to be valuable, however, there is some evidence that he views the total amount of H information supplied as part of the design specification as an indicator of the customer's interest in human factors. In study IV we included in the SOW given to our experimental groups as much information on personnel and personnel operations as we could muster. The results of the debriefing interviews with these subjects indicated that there is "attention-value" in any H input even when the engineer says that he cannot or did not use the input in his design. This, of course, refers only to information supplied as part of the SOW, not to inputs supplied incrementally during system development.

The utility of these inputs is directly related to their sequencing in system development. Although it conflicted with the subjects' feeling that H information belongs in detail design, they also felt that, if personnel requirements were to be imposed on design, these would have considerably greater impact if they were made available either in conjunction with or soon after the SOW. Subjects emphasized again and again that in order for H inputs to be effective in influencing design, they must be provided early enough to assist in forming the design concept. This accords with the engineer's desire for as much information as rapidly as possible. More molecular human engineering inputs (e.g., recommendations for specific control-display arrangements) are appropriate to detail design; however, these too must be provided at the same time the engineer is designing at that level.

Effect of H Inputs

The crucial questions to be answered are, of course: what impact, if any, do H inputs have upon the design concept, and what impact *could* they have? As a consequence of the engineer's general indifference to human factors as a design parameter, the influence of H inputs on design is less than desirable. Purely informational inputs have minimal influence. H inputs which are phrased as design requirements and which are included in the initial SOW do have a measurable influence on design but less than they perhaps deserve. The evidence for these conclusions requires some discussion.

In study III two groups of subjects were given varying personnel requirements: one group was asked to design for a small number of highly skilled personnel, whereas the other group was to design for a larger number of low skilled personnel. There were significant differences between the groups in terms of the amount of control-display hardware that they considered necessary for their systems (but whether the nature of these differences was logically related to the particular H requirements is difficult to assess). Where H inputs were supplied incrementally through the ten sessions, half the engineers reported that these would have had a significant impact on their design if they had been included as requirements in the SOW.

Subjects were also asked to predict the influence upon design of a number of hypothetical personnel requirements. Those input requirements which were concrete, precise, and quantitative were predicted to be highly influential; those formulated in more general, less specifically design-relevant terms were rated as having slight or no effect upon a design.

Study IV contrasted a situation in which H inputs *as a unit* were presented to one group of subjects as part of the SOW following design go-ahead; in the contrasting situation the inputs were presented to a second group incrementally. It was hypothesized that "early" inputs included in the SOW would have a significantly greater impact on design than those presented incrementally. Major differences were found between the designs produced by the subjects. Three-fourths of all subjects reported that inputs phrased as requirements did affect their design decisions.

Ideally, therefore, a SOW should indicate requirements such as the following: that the number of personnel to be employed in the subsystem will be between 6 and 8, but no more; that the following specialists (e.g., petroleum handler, computer operator, described in such-and-such manner) will be required; that operational personnel will have such-and-such functions, performed in the following sequence; that the system concept will involve a specified amount of automation such as a central control station with valves operated remotely once operations have begun; that system operation will be preceded by a checkout, etc., etc. Even under these circumstances, however, H inputs will not be responded to in isolation. H inputs alone are not important enough to dictate major design decisions but are considered along with other factors (e.g., performance requirements, reliability, cost, time constraints) to produce the design concept ultimately adopted.

Even when H inputs are not phrased as design requirements, they do have some value if only from the standpoint of "prodding" the engineer to consider aspects that he had overlooked. A memorandum of control-display recommendations may not cause the designer to buy the particular configuration recommended, but it will cause him to think more—if only momentarily—about controls and displays. For example, when maintainability inputs were presented to subjects in study III, some of the engineers reported that they had completely forgotten about maintainability; the new input then required them to amplify their design. Consequently, even when it appears as if a particular H input has missed its mark because it does not produce an immediate effect in terms of a design change, it serves a useful purpose in reminding the engineer of factors he may otherwise ignore.

Subjects' attitudes were quite illuminating as these reflected their motives for accepting or rejecting inputs. The acceptability of an input is determined to a great extent by its source. An input may emanate from a higher level (e.g., customer, company management), from a level parallel to the designer (e.g., an engineering group recognized by the designer as having a technical capability equal to his own), or from what he considers to be a lower level (e.g., from a group which is not part of engineering or which does not have status equal to the designer's). An input stemming from a higher level is almost always accepted even when there is initial resistance; this is related to the designer's perception of the input as imposing a design requirement. Lateral inputs (i.e., from a level parallel to the designer's) are reviewed and accepted if they fit into the designer's already existent concept or if they more effectively satisfy system criteria. Inputs from a lower level (and H groups are often in this attitudinal category) are often rejected or accepted only after much resistance.

The engineer is very critical of anything which he considers to be technically incorrect. Vague inputs (i.e., phrased in generalities) are almost never acceptable. The engineer requires that an input be specific to his problem and spelled out in detail as well as being practical; consequently, the H specialist may have to demonstrate the practicality of his suggestions. Finally, the input has to tell the engineer something he has not thought of before. All of this can be summed up by saying that, if the specialist is accepted by the engineer as technically competent, what he has to say about the engineer's design will be accepted not unreservedly but with good grace.

Is design any less effective when engineers fail to make use of H

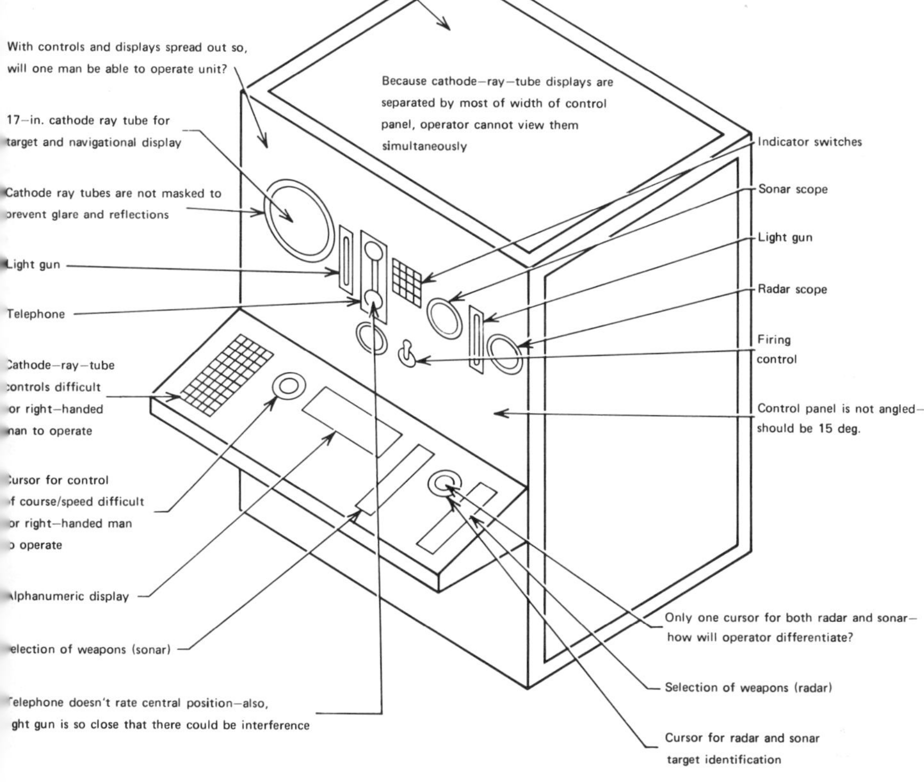

Figure 7-4 Typical human engineering discrepancies (Study I).

inputs? To answer this question we must differentiate between the design *concept* established initially and the design *detail* which implements that concept; the latter is an elaboration of the former.

The primary evidence comes from studies I and II which dealt with consoles because it is extremely difficult, in analyzing a complete subsystem design, to determine deficiencies that are the result of ignoring human factors. The reason is that human engineering design criteria do not evaluate design *concepts* but rather refer to design details such as the placement of controls and displays. Hence only the problems presented in the first two studies, dealing with control panel/console layout details, can answer the question we have posed.

In these two studies there were discrepancies which should have been, and we think would have been, avoided by proper consideration of H principles. Figure 7-4 illustrates a number of these discrepancies. (The illustration is a composite, of course, of drawings by several engineers. We do not suggest that any one engineer would be so careless.) Here the engineer's experience, the application of his rules of thumb, and his placing himself in the operator's role were insufficient to avoid gross errors.

The results of the subsequent studies also indicated very strongly that some engineers have an inadequate grasp of the manning required by their systems. In study III the subject producing the most automated system (two computers) needed three times as many personnel as did the subject who developed a semi-automated system and substantially more than the subjects who developed more manual systems. In study IV a small minority of subjects were unable to estimate the number of personnel they would need.

Utility of Human Engineering Standards

The human engineering standard (as we see it described in MIL-STD 1472 and similar documents) is a criterion or measure with which the adequacy of a particular design feature can be compared (from a human engineering standpoint). Since measurement is at the core of any discipline, we considered it important to determine whether human engineering standards could be utilized meaningfully by engineers.

The utility of the standard employed was defined in terms of two criteria: did the standard impose mandatory requirements on (did it constrain) design; could the information presented in the standard be used to assist design solution? These criteria were adopted because the efficient standard performs two functions: to *control* and to *guide* design, the latter by providing necessary information to the engineer.

A number of design problems similar to those described earlier were presented, and the subject indicated which of 22 randomly selected items from MIL-STD 803A-1, 1964, would restrict his solution of those problems. To determine the assistance human engineering criteria could supply, the subjects were asked to solve several design problems with the aid of the standard.

The largest number of MIL-STD items was considered to be completely nondesign restrictive. Only four of the 22 criteria were considered by all engineers as being mandatory. Of the remaining criteria all subjects considered two to be nonmandatory, and the others were accepted by some engineers and rejected by others. If the items selected were representative of the total population of human engineering criteria, this means that many (not all, but most) design engineers would consider *all* such criteria to have no significance for their designs.

As for the information provided by MIL-STD 803A-1 (which is really a small scale human engineering handbook and the predecessor of MIL-

STD 1472), it would be best to quote from the paragraphs describing this part of the experiment:

"In investigating the application of 803 to representative design problems, the difference between what the designer says and what he does was further delineated. Despite the designer's indication that 803 was at least fairly important to design, when he was asked to apply it to specific problems he reported that it was (a) too general, that it did not offer enough specifics; when the designer did consult it, he should be provided with an answer, not with vague generalities; (b) it was written at too low a level (e.g., "written so monkeys can do the job," and "this would be for the inexperienced man, an experienced man would know what to do,") and (c) it was too narrow; that is, it did not attempt to relate or even indicate the relationship that exists between the various elements that go to make up the system. By this the subjects meant that contextual relationships (e.g., as between number of fasteners, stress placed on the structure, and weight) were not sufficiently included.

"Regarding the generality of the standard, we found that our . . . designers approached 803 with specific questions in mind and wanted specific answers, not platitudes. They refer to the standard, when they do, as a source book for *an answer* to their problem. The implications of this should be obvious; if the standard is to be useful or used it must contain answers and these answers must be specific and oriented toward the problems that designers experience." (Meister and Sullivan, 1967).

It appears then that military human engineering standards (and what was written above about MIL-STD 803A-1 applies also to MIL-STD 1472 (DOD, 1969) do not really impose restrictive requirements on design, contrary to the impressions of certain governmental representatives. Because of their looseness, human factors standards do not control design although they may supply some useful information, about as much, we suppose, as that supplied by a human engineering checklist.

Industrial experience suggests that, if contractors wish to avoid implementing military human engineering standards, they have little difficulty doing so. Hence we cannot, as has sometimes been suggested, assign the responsibility for ensuring the inclusion of human factors in design to these standards. If it is argued that the human engineering standard is intended to be largely informational, it fails on that basis also because it provides the engineer (and the specialist also) with little information and less guidance.

Adequacy of Human Engineering Guides

The question that initiated the series of studies reported in this chapter was whether or not typical human engineering guides (e.g., Woodson and Conover, 1964; Morgan et al., 1963) supply the information engineers need, and, if not, how to improve them.

About all we can say of the engineer's response to human engineering handbooks which purportedly are developed to assist him in the human factors aspects of his design is that he largely ignores them. In study I none of our subjects possessed a copy of the two most popular guides, referred to above, and none had ever referred to them. In fact, only a few of the subjects (the two or three who were design managers) were aware that these documents existed. When at the conclusion of study I the investigators attempted to give a copy of Morgan free to the design group that had supplied the subjects for the study, the group supervisor refused on the basis that the book would be of no use. In study II one or two of the subjects actually had a copy of Woodson. Our subjects for studies III and IV were completely unaware of the two guides cited. Of course, it must be said of the company organization in which the latter subjects work that it possesses no human engineering group which can act as propagandists for H data. The companies in which subjects for studies I and II worked did, however, have active H groups.

An interesting subexperiment which casts some light on possible reasons for this negative response to human factors information was conducted as part of studies I and II. In this experiment 14 common items of information (e.g., specifications for the design of seats) were extracted from Morgan, et al., 1963, and modified so that the same item could be presented in each of the following formats: verbal (i.e., paragraph format); pictorial (i.e., sketch drawing); graphic curve; and tabular. Care was taken to ensure that the identical information was available in each format. Each of the 14 information items was presented in paired or triad comparison fashion in random order, and the subject indicated his preference for the format in which he wished the information to be given to him.

The results of the experiment are quite marked. Subjects preferred that human engineering material be presented in pictorial and graphic form most often, tabular next most often, and verbally least of all. The percentage of choices made is shown below:

Format	*First Place Choices*	
	Study I	Study II
Graphic	75%	78%
Pictorial	70%	77%
Tabular	35%	37%
Verbal	20%	15%

Because of the emphatic way in which verbal material was rejected and because of the largely verbal manner in which most human factors data are presented in handbooks, we would expect to find that our engineers rejected human factors information.

The overwhelming reaction was that the material was too verbal, insufficiently illustrated, overly qualitative, entirely too general, and lacking in equipment-orientation and design-relevancy. This comment in essence summarizes the general attitude of design engineers to many formal H̲ inputs.

The utility of human engineering handbooks could be observed objectively as well as by securing subjective responses to questions concerning their contents. In studies I and II copies of Woodson and Conover and Morgan et al. were available for use by engineers as part of the experimental situation. In study I they were almost completely ignored, as was the human engineering standard (U.S. Navy, 1963) which was called out as required in the design SOW. In study II Woodson and Morgan were available not only for general reference use but a specific set of subtests was developed in which engineers were required to make use of these guides to solve certain design problems. Subjects were asked to indicate (a) what sections of either guide were applicable to the solution of the problem; and (b) in what ways the guides were particularly helpful or unhelpful. An example of the type of problem presented is the following: assume you are responsible for designing a cockpit display that is intended to alert the pilot to an imminent lack of oxygen. You wish to select a display that will produce the fastest pilot reaction time. Your problem is to decide whether you should use a visual or an auditory display. What should the characteristics of the preferred display be?

In one of these problems nine of the 10 subjects answered purely on the basis of the experience they had had without attempting to search for appropriate information in the guides. (This also reflects the experiential bias in design solutions noted earlier.) In a second problem five subjects did look for and find the answer in Woodson and Conover; the others could not find the section containing the correct solution but indicated they would extrapolate from other (somewhat applicable) anthropometric data. The responses elicited a decided preference for information presented pictorially but also indicated a preference for a greater amount of factual data than can be supplied in pictorial form.

In some cases designers had information-retrieval (semantic) difficulties in using handbooks. These related to the terms with which the handbooks were indexed. Thus, if the designer conceptualized his problem as dealing with indicator flash rates and if he did not find the precise term in the index, he was quite frustrated.

The most recent work performed by the author suggests that it is possible to develop human engineering guides that are more compatible with expressed engineering desires (e.g., quantitative, graphic, less verbal, more design-relevant). A human engineering guide for design of visual displays (Meister and Sullivan, 1969) was developed in consultation with, and tested by exposure to, display engineers. The results of the testing were fairly positive, suggesting that it is possible to make significant improvements in the typical human engineering handbook. This does not ensure, of course, that the engineer will, in view of his general attitude toward human factors, make more use of such improved guides; but it makes it more likely that he will, and, if he does, it will give him more relevant information.

SUMMARY

It appears then that, if engineers do not utilize human factors inputs, the fault lies as much with the specialist as it does with the engineer. Much of the data supplied by the specialist during system development is late, design-irrelevant, and formatted inappropriately. Given the relative inflexibility of the engineer's attitudes, the only reasonable attack to the problem is for the specialist to improve the quality of his inputs so that the engineer *must* recognize their utility. To do this, however, the discipline as a whole must first make its research more meaningful. The way in which human factors research is secured is the subject of Chapter Eight.

REFERENCES

Coryell, A. E., 1967. The Design Process, *Machine Design,* November 9, 154–161.

Department of Defense, MIL-STD 1472, 1969. *Human Engineering Design Criteria for Military Systems, Equipment and Facilities.*

Marples, D. L., 1961. The Decisions of Engineering Design, *IRE Trans. Engineering Management,* 60–71, June.

Meister, D., and D. E. Farr, 1966. *The Utilization of Human Factors Information by Designers,* The Bunker-Ramo Corporation, 16 September, DDC Report No. AD 642057.

Meister, D., and D. J. Sullivan, 1967. *A Further Study of the Use of Human Factors Information by Designers,* The Bunker-Ramo Corporation, 16 March, DDC Report No. AD 651076.

Meister, D., D. J. Sullivan, and W. B. Askren, 1968. *The Impact of Manpower Requirements and Personnel Resources Data on System Design,* Report AMRL-TR-68-44, Aerospace Medical Research Laboratories, Wright-Patterson AFB, Ohio, September.

Meister, D., D. J. Sullivan, D. L. Finley, and W. B. Askren, 1969. *The Effect of Amount and Timing of Human Resources Data on Subsystem Design,* Report AFHRL-TR-69-22, Air Force Human Resources Laboratory, Wright-Patterson AFB, Ohio, October.

Meister, D., and D. J. Sullivan, 1969. *Guide to Human Engineering Design of Visual Displays,* Final Report, Contract N00014-68-C-0278, Bunker-Ramo Corporation, Canoga Park, California, September, AD 693 237.

Mitroff, Ian I., 1967. *A Study of Simulation-Aided Engineering Design,* PhD. Dissertation, University of California, Berkeley.

Morgan, C. T., et al., 1963. *Human Engineering Guide to Equipment Design.* New York: McGraw-Hill.

Rabideau, G. F., et al., 1961. *A Guide to the Use of Function and Task Analysis as a Weapon System Development Tool,* Report NB-60-161, Northrup Corp., Hawthorne, California, January.

Ramstrom, D., and E. Rhenman, 1965. A Method of Describing the Development of an Engineering Project, *IRE Trans. Engineering Management,* Vol. EM-12, September, pp. 79–86.

U.S. Air Force, MIL-STD-803A-1, 1964. *Human Engineering Design Criteria for Aerospace Systems and Equipment,* Part 1, Aerospace System Ground Equipment.

U.S. Navy, NAVWEPS OD 18413A, 1963. *Human Factors Design Standards for the Fleet Ballistic Missile Weapon System.*

Van Cott, H. P., and J. W. Altman, 1956. *Procedures for Including Human Engineering Factors in the Development of Weapon Systems,* Report 56-488, Wright Air Development Division, Wright-Patterson AFB, Ohio.

Woodson, W. E., and D. W. Conover, 1964. *Human Engineering Guide for Equipment Designers.* Berkeley, California: University of California Press (2nd ed.).

APPENDIX A: STATEMENT OF WORK FOR DESIGN PRODUCT TEST I

The following has been, for obvious reasons, considerably abbreviated from the original.

Applicable Specifications

The applicable sections of the following shall form part of this design requirement:

1-1 Federal Test Method Std. No. 151a, Method 811.1, Salt Spray Test, May 6, 1959.

1-2 MIL-STD-167 (Ships) Military Standard, Mechanical Vibration of Shipboard Equipment, 20 December 1954.

General

The WACS is designed to be operated with the MATRIX ballistic solution computer. On the basis of its radar, sonar, and navigational inputs, WACS shall initiate appropriate firing commands which shall cause the MATRIX computer to develop appropriate ballistic solutions for the various shipboard weapons.

Requirements

3-1 Performance

3-1.1 *Functional characteristics.* The Weapon Assignment and Control Station (WACS) shall have the following capabilities:

1. To initiate and direct radar and sonar searches during Naval operations. The control station shall be able to monitor radar plots of aircraft and surface vessels and sonar tracking data from the radar and sonar tracking stations. These monitoring functions shall be capable of being exercised simultaneously or separately.
2. To control and display own ship's course and speed during attack phases.
3. To select weapons appropriate to the threat involved; to determine against which targets to fire these; to monitor the readiness of weapons to fire; to initiate firing commands; and to display the availability of weapons during firing sequences.
4. To communicate with all other Combat Information Center (CIC) stations, the bridge, and weapons firing stations and batteries by means of telephone and dynamic displays; and to communicate through the CIC communications station to Combat Air Patrol (CAP) and other fleet units.

3-1.1.1 Primary performance characteristics

3-1.1.1.1 Modes of operation. WACS shall provide the following control and display modes, which should not be assumed to be inclusive of all necessary controls and displays:

1. *Radar monitor on/off* shall be accomplished by activation of the appropriate control(s).
2. *Sonar monitor on/off* shall be accomplished by activation of the appropriate control(s).

3. *Communication selection* (bridge, sonar and radar stations, CIC communications station, weapons firing stations) from WACS shall be accomplished by dial telephone.
4. *Identification of targets* shall be accomplished by marking the target with a light gun, joystick/cursor, or similar device that automatically assigns a coded designation of the target on display presentations.
5. *Availability status of weapons* shall be displayed by means of appropriate indicators. This information shall be continuously provided from weapons firing stations.
6. *Firing readiness of weapons* shall be displayed by means of appropriate indicators. This information shall be continuously provided from weapons firing stations.
7. *Selection of weapons* to be fired shall be accomplished by depressing controls that activate appropriate displays at weapons firing stations.
8. *Initiate firing sequence* shall be accomplished by depressing controls that activate appropriate displays at weapons firing stations.
9. *Control over own ship's maneuvering* shall be assumed by depressing a control that activates the appropriate displays at the bridge and simultaneously energizes WACS steering control.
10. *Relinquish own ship's control* shall be accomplished by depressing control(s) which activate the appropriate display(s) at the bridge and simultaneously release WACS control of steering.
11. *Own ship's course control* from the WACS station shall be accomplished by movement of an appropriate joystick or other rotary-type control.
12. *Own ship's speed control* shall be accomplished by movement of an appropriate joystick or other rotary-type control.
13. A separate *navigational display* shall display own ship's course and position relative to targets, which shall also be displayed on this display. When own ship's course is not controlled from WACS, ship position information from the bridge shall be displayed on the navigation display.

14. *Own ship's speed* shall be displayed under all conditions by an alphanumeric display.
15. Bias, gain, and other *adjustment controls* for adjustment of radar and sonar monitors shall also be provided.
16. Provision shall be made for both *automatic and manual* control of weapons. It is assumed that the manual mode shall be used until a favorable attack situation is displayed, at which time automatic control shall be given to the ballistic solution computer. Manual control may be regained at any time.

3-1.2 Operability

3-1.2.1 Reliability. The WACS equipment shall demonstrate a minimum of 300 hours MTBF with a 0.95 confidence level.

3-1.2.2 Maintainability. Provisions shall be made for displaying major WACS out-of-tolerance conditions to the WACS operator(s). Provisions shall be made for ready access to the internal components of WACS.

3-1.2.2.1 Maintenance. An MTTR of 30 minutes duration from diagnosis to verification of equipment function is required. The design of the WACS equipment shall be such that diagnosis and replacement of any malfunctioning component shall be accomplished in an average time of 30 minutes, with no repair requiring more than 45 minutes. The equipment shall be functionally designed so that repair can be accomplished with minimum downtime, personnel, personnel skills, and resources.

3-1.2.4 Environmental. The equipment shall be capable of meeting the requirements of this specification under the following conditions or natural combinations of conditions. . . . .

3-1.2.6 Human engineering. This equipment shall be designed to applicable sections of NAVWEPS OD 18413A, "Human Factors Design Standards for the Fleet Ballistic Missile Weapon System."

3-1.2.6.1 Design criteria.

1. WACS shall have a console configuration. While it is desirable that the equipment should be operated by a single individual, *the contractor should consider the possibility of multi-man operation.*
2. Because WACS will be located aboard frigates of the 5600 ton class, with consequent space restrictions, its dimensions shall be the minimum consistent with performance of its operational requirements. The equipment shall not exceed 1500 lbs with an external envelope of 6′ x 6′ x 6′.
3. WACS shall be operated under ambient lighting of 30 to 50 foot-candles (reduced illumination).

APPENDIX B: DESIGN INPUT TEST ITEMS PRESENTED IN STUDIES I AND II

Study I

Item *Description*

1-0 Conceptual analysis of specifications for DPT I and II. Eliminated before testing as redundant.

2-1 Subject is presented with list of equipment components (e.g., cables, connectors) and asked to anticipate possible maintenance problems. Rationale: designer's concept of potential problems may be related to the kinds of design alternatives he employs.

2-2 Subject is presented with list of equipment factors (e.g., control-display relationships) and asked to indicate potential design problems which might arise in relation to these factors. Rationale: same as 2-1.

2-3 Subject is asked to indicate steps he would take to incorporate human engineering principles in design of missile control panel. Rationale: to determine whether designer can act as his own human engineer.

2-4 Subject is asked to indicate steps he would take to ensure that shipboard navigation equipment is maintainable. Rationale: same as 2-3.

3-1 Subject is asked to rank alternative design solutions to a high voltage hazard problem. Rationale: to determine how equipment characteristics would be influenced by design requirements.

3-2 Subject is asked to list all possible means of eliminating cross connect problems. Rationale: same as 3-1.

4-0 Subject is asked to list and rank all the factors involved in design of portable battlefield TV camera set. Also, to rank a series of information items in terms of their importance to design of this equipment. Rationale: to determine design criteria and information priorities.

5-0 Subject is presented with three detail design drawings and asked for criteria he would use in evaluating their operability and maintainability. Rationale: evaluative criteria should indicate significant design principles.

6-0 Subject is presented with 14 paired or triad comparisons of the same information in four alternative formats: graphic, pictorial, tabular, and verbal. Rationale: to determine the engineer's preferred format for information presentation.

7-0 Subject is required to indicate in list of 15 design situations whether safety, maintainability, and operator considerations are important. Rationale: hypothesis is that the more situations in which the designer considers human factors to be important, the larger the number of applicable implications he will deduce from design requirements. (Eliminated as failing to differentiate among the three factors.)

8-0 Subject is asked to indicate on a five-point scale the actual availability of various types of human factors information assumed to be necessary for effective design. Rationale: to determine whether information requirements specified in human engineering handbooks are actually realistic.

9-0 Subject is asked to rank the order in which characteristic design activities are performed. Rationale: the order in which design activities are performed may have implications for the utility of various human factors informational inputs.

10-0 Subject is asked to indicate agreement/disagreement with 42 items dealing with the role of the operator in design, attitude toward human engineering, human factors information, etc. Rationale: to plumb engineer's attitudes toward human factors.

11-0 Subject is asked to rank nine criteria which enter into design. Rationale: to determine relative priorities of design criteria.

12-0 Subject is asked to indicate relative interest in 25 topics generally found in human engineering handbooks. (Eliminated since subjects had never read human engineering handbooks.)

Study II

The following DIT items were administered in addition to those presented in Study I:

1. Triad comparison of 12 design criteria. Rationale: to determine relative priority of design criteria.
2. Subject is asked to rank the importance to design of a ground equipment of 15 specification items randomly selected from AFSCM 80-3 and MIL-STD 803A-1. Rationale: to determine the significance to design of military human engineering standards.
3. Subject is asked to check 22 randomly selected items taken from MIL-STD 803A-1 in terms of their design constraint character. Rationale: to determine how mandatory for designers military human engineering standards are.
4. Subject is asked to make a triad comparison of 10 criteria to be used if the engineer were writing a human engineering handbook for his company. Rationale: to determine characteristics which engineers wish to see in human engineering handbooks.
5. Subject is asked to determine which requirements in MIL-STD 803A-1 would help him design the cover for a missile autopilot. Rationale: to determine how useful to the designer military human engineering standards are.
6. Subject has the problem of designing a cockpit display which will alert the pilot to impending lack of oxygen. He is asked to refer to MIL-STD 803A-1; to Morgan, 1963; and to Woodson, 1964, for information which will help him solve the problem. Rationale: to determine how useful military human engineering standards and popular human engineering handbooks are for design information.
7. Same as (6), except that the problem deals with the ambient lighting in an air traffic control center.
8. Subject has to solve an anthropometric problem dealing with minimum dimensions of a space pilot; he is required to use the sources cited in (6) and for the same reasons.

9. Subject is asked to review a draft chapter of an on-going revision of the Joint Services Human Engineering Guide and to evaluate its usefulness. Rationale: to determine the adequacy of popular human engineering guides.

APPENDIX C: PERSONNEL INPUTS PROVIDED IN STUDIES III AND IV

Item	*Definition*
1. Lists of personnel tasks	Tasks defined in terms of personnel functions and equipment acted upon.
2. Personnel/equipment flow diagrams	Diagrams illustrating the sequencing and interrelationships among tasks.
3. Personnel/equipment analyses	Description of equipment characteristics required by tasks or effect of equipment characteristics on task performance.
4. Task analysis, including	
(a) task structure	task descriptions in terms of function and equipment operated or maintained (see Item [1]).
(b) task criticality	consequences of task being performed incorrectly or not at all.
(c) team performance	number of personnel required to perform the task.
(d) probability of successful task completion	quantitative estimate of probability that the task will be completed successfully by personnel (the converse, error probability, also is provided).
(e) task location	approximate physical area (e.g., transporter, launch pad) in which the task must be performed.
(f) task duration	estimate of the time required to perform a task.
(g) difficulty index	estimated difficulty of task defined in terms of error probability and response time.

5. Time-line analysis, including task frequency	distribution over time, including overlaps, of individual task durations.
6. QQPRI data, including	
(a) number of personnel	quantity of personnel required to perform subsystem operations, defined initially in terms of maximum number to be utilized, later in terms of actual number needed.
(b) skill type	characteristics of the job to be performed in terms of demands upon personnel.
(c) skill level	Air Force skill levels required by the task, defined in terms of error probability, response time, and amount of assistance required.
(d) proficiency	skill characteristics which personnel should possess to perform the job satisfactorily.
(e) task error-likelihood	type of error which may occur during task performance.
(f) personnel availability	definitions of AFSC type possessing necessary qualifications to perform the job, together with the probability of such personnel being available for the job.
7. Training requirements, including	
(a) anticipated training time	time needed to train to given level of proficiency.
(b) required aptitude	job skills which training should provide.

VIII
THE PROCUREMENT OF HUMAN FACTORS RESEARCH

In these days of shrinking system development projects it is not uncommon to find much of the specialist's time and energy being directed at procuring and performing government-sponsored research studies. Kraft, 1969, indicates that 10–15 percent of the specialist's time is spent in this effort. The purpose of this chapter is to describe how that research is secured. There are two reasons for this: not only may the reader be engaged in this work but also the manner in which H research is performed has a direct impact on its adequacy and hence on its capability to support system development.

HOW HUMAN FACTORS RESEARCH IS PERFORMED

In our sense of the term "human factors research" we exclude research performed as part of, and to implement, on-going system development projects; that is, we do not consider exploratory, resolution and verification testing (described in Chapter Six). Almost all remaining human factors research is funded by the government through contracts specifically aimed at answering questions of interest to the government. Sometimes that research will attack general methodological questions and will therefore assist H specialists as a whole. At other times the research is highly specific to the governmental agency funding the research, in which case its utility to the H discipline as a whole is somewhat restricted.

Other than research contracted for by the government and performed by contractors, the remainder (a very small part) is divided between (a) that performed within the governmental agency by its own H specialists (so-called "in-house research") and (b) research funded by and performed for the contractor's own purposes (this last being almost miniscule in amount).

Because the government "buys" human factors research (as it presumably "buys" engineering, physical, or biological research), the procurement process is not only interesting as a reflection of our technological customs, but also critical because it determines to a large extent the results of that research.

Government-funded human factors research may be *solicited* or *unsolicited.* Solicited research is competitive; the procuring agency advertises its desire for research to answer particular questions, and any research organization which desires may apply. In unsolicited research, a research organization (often an individual within that organization) will propose to the governmental agency that a particular study be performed. If the agency likes the idea, it issues a contract to the proposer. In unsolicited research only the proposing and the procuring agency are involved.

In general most governmental agencies prefer solicited research. It gives them the option of initiating the research question and phrasing it in their own terms. In addition, governmental policy emphasizes the need for competition in letting contracts; otherwise the public might suspect (as some do) collusion (favored treatment) between the procuring agency and the selected contractor. Some governmental agencies do procure human factors research solely on the basis of unsolicited proposals, but it is our impression that these agencies are few in number.

Between the conception of a solicited research competition and the award of a contract a series of steps ensues:

1. Human factors research in most governmental agencies is highly personalized; that is, within the broad confines of the research area assigned to the agency and to the H̲ specialist within the agency, the latter has considerable latitude in the specific studies to be performed within that area. Three examples of a general research directive are (a) to determine the human engineering characteristics responsible for effective display performance; (b) to study man-computer relationships; (c) to improve the adequacy with which human factors inputs are provided to engineers.

Within (c) any or all of the following studies may be pursued:

i. A handbook to be distributed to engineers describing the human factors inputs available to them and how to use these inputs.

ii. A survey among H̲ specialists (or engineers) to discover how effective H̲ inputs are at the present time and when they are used.

iii. Analytical studies of the input methodology (e.g., a study to revise the task analysis format).

iv. A literature review of the data on a particular methodology (e.g., principles of designing keyboards).

v. A survey among pilots of cockpit layout deficiencies found in operational aircraft.

vi. Experimental studies of principles of laying out control panels.

Naturally this does not exhaust the specific research possibilities. There are, of course, qualifying factors that restrict the choice of study; the selection of a study is, after all, not a game of chance. Such factors as the personal interest of the government specialist or a predisposition to experimental investigations or the reverse play a part in determining the type of study which will be funded.

Once the governmental specialist has decided upon the outline of the specific research area he wishes to fund, he must sell the research concept to the head of the agency. The author cannot describe the problems within the governmental agency that arise in order to secure approval for the research concept, but undoubtedly many excellent ideas are still-born.

2. Usually (although not invariably) the agency's desire to procure research is advertised in the *Commerce Business Daily* (CBD) under "Sources Sought." The CBD is a daily publication of the U.S. Department of Commerce, in form much like the classified section of a newspaper, which contains advertisements of contracts to be let by the government for various services, among them research.

A sample notice might read as follows:

"Human Performance Laboratory:

"Organizations possessing appropriate qualifications are invited to submit these to perform A STUDY OF THE FACTORS AFFECTING PERSONNEL PERFORMANCE IN COMPUTER SYSTEMS. Applicants should possess the following characteristics: (1) personnel skilled in computer design and programming; human engineers; training specialists; (2) appropriate experience in computer design and human factors experimentation. . . ."

Since the agency is required by government regulation to evaluate all proposals received and since the evaluation process is fairly complex, it is to the agency's advantage to advertise its desires in the CBD. In this way it alerts those organizations which are best qualified to bid and at the same time it can establish a register of qualified contractors on the basis of responses to the CBD notice. Presumably only qualified contractors will be selected to receive an RFP when it is issued.

Since the CBD notice is often phrased in very general terms (see above), it may be difficult for the respondent to answer the question,

"What kind of study does the procuring agency really want?" However, at the CBD stage, this question is less important than the appropriate "mix" of experience and personnel requested. Although the CBD notice suggests that it would be useful for the respondent to submit ideas for solving the problem (in preliminary form only, of course), no one, certainly not the CBD initiator, expects the problem to be solved in the CBD response.

In the first place, considering the abstractness with which the CBD advertisement is phrased, the respondent may have difficulty merely interpreting the subject of the request, much less solving the problem. In the second place, the CBD response must usually be submitted in 10 days, which does not offer much opportunity for deep thought. In any event, if the respondent had an immediate solution to the problem (not very likely, but still within the realm of possibility), he would be advised by his management to save it until he has received the RFP. (Although it is entirely without foundation, contractors are sometimes suspicious that, if original ideas for research are proposed by the contractor in responding to the CBD, they will be swept up by the agency and translated into the requirements for a competitive proposal.)

There is, of course, no guarantee that a CBD response will place the contractor on the list to receive the RFP. There are two reasons for this: (a) the contractor's response may be unsatisfactory, meaning that he lacks appropriate experience or personnel; (b) the research concept which led to the issuance of the CBD notice may be canceled before an RFP is issued. One of the "overhead" burdens that the contractor bears is the cost of responding to CBD advertisements that fail to bear fruit in an RFP or at least in an RFP sent to the respondent.

3. The potential contractor notes the advertisement in the CBD and responds. A CBD notice of impending research will usually be responded to if the contractor possesses a reasonable amount of competence in the general area sought. After all, if the contractor does not respond, he stands little chance of being asked to reply to the RFP. Moreover, since the CBD response is not a proposal but merely an advertisement of qualifications, the time taken to write the response is comparatively short (an experienced respondent should be able to write an appropriate reply in a day or two); sometimes the contractor keeps statements of qualification on file for use in responding to CBDs.

4. The agency receives these responses and evaluates them. Again the author is unable to specify the rationale for selecting one contractor over another as being qualified to receive the RFP, but it seems reasonable that the ultimate determinant (short of a brilliant idea for solution of the research problem) is the respondent's past experience and his per-

sonnel (the latter being another form of experience). Thus, if the CBD asks for contractors experienced in developing crew simulators, for example, a contractor who has had specific crew simulator experience is much more likely to be qualified than one who does not have that experience or who has experience only in a related area.

5. If the contractor is one of the favored few who receives the RFP, he is faced with a decision: to propose or not to propose. The decision is not automatically yes, even though the contractor may have solicited the RFP through his CBD response. The actual statement of the research required is usually sufficiently different from that described in the CBD that the requested study requires fresh consideration.

The proposal decision is based upon such factors as the probability of winning the competition (should be high or at least as good as anyone else's), the chances of the contractor's competitors (should not be overwhelming), whether or not the research to be performed is an area which the company has been interested in and has followed over the years (desirable, but not essential), the cost of proposing (should be low or at least not excessive), the availability of time and personnel to write the proposal (absolutely essential), and, if the contract is won, personnel to conduct the study (desirable but not essential, since he can always hire people).

If the decision to propose on the research is positive, the contractor gathers a team together (the number of people dependent on the scope of the proposal) to write the proposal. The proposal itself will contain the following sections:

I. Summary. The proposal in summary form for management personnel who may be too busy to read it in detail.

II. Introduction.

(a) purpose of the study;
(b) why the research area is important to study;
(c) previous work performed in this area by others and by oneself (a sort of literature review to indicate expertise);
(d) major problems in the research area that need resolution. (These are probably the questions the customer wants answered.)

(Much of this introductory material may appear obvious to anyone experienced in the subject area, but the assumption must be made that the customer wants to know if the contractor understands the background of the problem and is not just "shooting from the hip.")

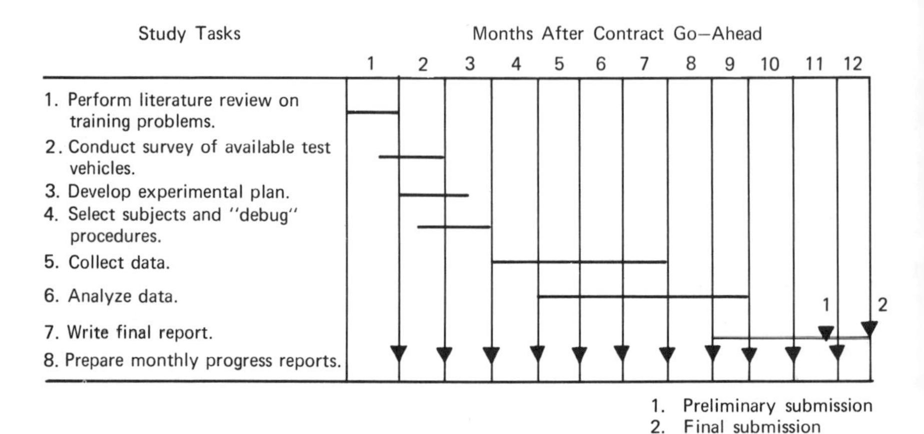

Figure 8-1 Proposed schedule for study performance.

III. Study Approach. What the contractor proposes to do to answer the research question and to solve the problem areas described in the previous section. This is the heart of the proposal and requires the most effort. The rationale for, and the steps involved in, performing each major task are explained and the outputs anticipated from the task are described.

IV. Statement of Work. This summarizes and lists sequentially, usually in one or two lines, each task to be performed by the contractor in the study.

V. Work Schedule. This consists of a chart listing tasks by calendar duration (see Figure 8-1). The number of hours (man-months, 160 hours to the month) to be spent on the project by each member of the research team (by name) is listed together with the number of trips to be taken in performing the study. If the contractor proposes to do the required work in less than the number of hours recommended by the agency, this is the place to explain why.

VI. Relevant Experience. This section and all further sections are what is called "boiler plate" since essentially it is the same material used in all proposals, with minor revisions to tailor the material specifically to the individual proposal. It describes (in glowing terms, of course)

(a) the company;
(b) the company's unique experience in performing related studies, each such study being described in one or two para-

graphs, with emphasis on its relevance to the proposed work.

(All this is said in such a fashion that it must be obvious to the customer that only the contractor proposing could adequately perform the job.)

VII. Personnel. This section contains the following subsections:

(a) a brief description of the outstanding characteristics of the key personnel who perform the study;
(b) detailed resumes (biographical sketches) of these personnel, including education, experience, and previous publications.

VIII. References. This assumes that the proposal writer has referred to the published literature in the previous sections.

The time made available by the customer to put a proposal together is rarely longer than a month, and at least a week of this time is generally occupied in meeting to decide whether to propose and in exercises to determine the lowest possible price to be submitted with (but not as part of) the technical proposal. A substantial segment of this time is also devoted to typing and reproducing the proposal document.

This time limitation means that unless the contractor has received, through *sub rosa* channels, word that the RFP will be shortly issued and has consequently done some preliminary thinking about it (or unless the subject matter is an area in which he has great experience and confidence), it is likely that the amount of creative thought entering into the proposal will be sharply limited.

6. The proposal is then sent to the procuring agency and the contractor waits to receive—if he has won the competition—a phone call from the agency's Contract Officer that negotiations for a firm price are to begin. This is almost always an indication that the contractor has won the contract. If the competition has been lost, a form letter will usually (but not invariably) be received stating that the company's proposal was unacceptable for reasons that may be difficult for the contractor to understand, much less accept; and that company X has won the contract. A debriefing (explanation of the reasons for losing the competition) can be requested, but it rarely gives the losing contractor much satisfaction.

7. In contrast to the solicited procedure, an unsolicited proposal is far more informal. Since the initiator of this type of proposal is the contractor, he must find a governmental agency which may be interested in his research. It is necessary for him to be aware of the special research

interests each agency has and to become as intimate as he can be with agency personnel. This will probably require periodic visits by the contractor to the agency offices, during which he discusses the nature of the research he is interested in. It is good policy to discover the particular biases and attitudes of the potential customer and to tailor his research to these. Occasionally the visiting contractor may have no very specific research concept he is selling; he is merely "fishing" to discover the agency's interests. When these have been tapped and a line of research interest ascertained, the visits are "selling" operations designed to persuade the customer to "buy" a particular study concept.

The buying-selling terminology with its connotations of the market place, as opposed to traditional concepts of how scientific research is performed, is perhaps unfortunate (certainly the author feels this), but it is used because it is highly descriptive of what goes on. Both the solicited and unsolicited proposals represent an effort to persuade a customer to purchase a particular research product in competition with other research products.

Although anyone may submit an unsolicited proposal at any time, the wise contractor is unlikely to do so without first receiving some encouragement from the customer. To do so would be to waste both his and the customer's time and effort (and burden the latter, since governmental regulations require that all formal proposals—as opposed to informal "inquiries of interest"—be evaluated rigorously). On the other hand, if the potential customer indicates his interest in an unsolicited proposal, he may even suggest how much money is available for the study and describe the wording that should be used in the proposal.

From this description of human factors research contracting, it is possible to extract certain common features which influence the usefulness of that research.

RESEARCH SUBJECT MATTER

One of the consequences of the fact that the government is the primary if not only customer for human factors research is that the government largely determines the subject matter of that research. This is not to say that the H specialist who works outside of government has no influence on that subject matter because obviously the governmental H specialist reads the literature, attends symposia, and comes in daily contact with others outside his agency. Nevertheless, the one who most significantly determines the nature of human factors research is the customer representative.

For marketing reasons it is not unusual, for example, for a contractor with an idea for a study to tailor that concept to what he thinks the customer will most readily accept. For example, if the customer is interested in computer models, the question for the contractor becomes, "How can we modify a study, otherwise unrelated to modeling, to include a computer model?"

The author does not wish to assign too cynical and self-serving a character to the research contractor and the governmental representative. Indeed, it is possible to argue that the tactics that have been described are inherent in any contract procurement process and are necessary if any research is to be performed.

Whatever the justification, the consequence of the customer's research domination is that much (certainly not all, but a good deal) of the research being performed has little relevance to the major system development problems of the discipline. But why should this be so?

One of the reasons for this unfortunate state of affairs is that the government's H personnel often (again not all, but many) have little recent knowledge of system development problems even though they may at one time have worked on industrial system development projects. It is conceivable that any kind of prolonged governmental activity tends to isolate its personnel from contact with extragovernmental problems.

Most human factors research is performed to answer some question that the governmental agency wishes answered. If the questions asked do not relate to real-world problems, then the answers supplied by the research will not have general usefulness. To determine what is relevant is a highly subjective matter. The only criterion available to decide the point is to ask of each piece of research supplied: will the research contribute to the solution of some system development problem (however obscure)? The author believes that much human factors research will not meet this test. However, anyone is free to apply the test himself and draw other conclusions.

In part the inadequacy of much human factors research results because the selection of a research subject is a highly idiosyncratic process (within broad research charters, of course). It may in fact be impossible for it to be anything else since someone or some group must eventually take responsibility for a decision about where to expend human factors research money. The selection of a research study is particularly difficult in the case of behavioral research, however, because so many of the personnel who can influence research directions have little or no acquaintance with the problems faced by the industrial human factors specialist. Paradoxically, much human factors research is performed by special consultant research firms who, because of their concentration on research, may have little knowledge of system development problems.

One of the characteristics of the governmental research procurement process is that it tends to ask for solutions to highly complex problems which, upon examination, cannot be answered without first performing much preliminary work. Nevertheless, the research proposed assumes that that preliminary work has been completed even when it has not. There are two possible reasons for this: governmental agencies do face complex behavioral problems for which they need answers; and they may in fact believe that the methodology already exists to solve these problems.

It is instructive to review a hypothetical example of an RFP requirement for solicited research in terms of what its requirements really imply.

"The study will describe the pilot training to be required in the 1980–1990 time frame. It will consist of three phases. First, identify and catalogue common pilot training requirements in terms of the detailed tasks to be performed and the skills required to perform these tasks. Second, develop alternative methods of training pilots, applying the most advanced learning and training concepts (e.g., computer-assisted instruction (CAI), television, adaptive training, etc.). Third, develop quantitative measures of the effectiveness of pilot instruction and proficiency and incorporate these into a cost-effectiveness model of training effectiveness."

If we were to list the implicit requirements of this research, they would look like the following:

Research Requirement	*Availability of Data or Method*
Perform task analysis.	Method available but very crude.
Identify common tasks.	Criteria for specifying commonality among tasks are presently not available but certainly within the state of the art.
Define/identify skills.	Methodology is available to identify gross motor and simple psychomotor skills but is largely lacking for higher order cognitive processes (see Finley et al., 1970).
Develop alternative methods of training.	Within the state of the art; however, the effectiveness of the alternative methods is not known.
Determine the applicability of CAI, TV, adaptive training, etc., to alternative training methods.	Very little is known about the adequacy of advanced training technology; some speculation is possible, but little more.

Develop measures of pilot instruction effectiveness/proficiency.	Although considerable research on this has been performed (see Smode, 1966), the data are confused and contradictory. The large number of potential pilot proficiency measures do not intercorrelate highly. Pilot proficiency is a significant research area in and of itself. The development of measures of instructional effectiveness is within the state of the art (e.g., transfer of training) except that the opportunity to utilize these measures in real-world situations is usually lacking.
Develop cost-effectiveness models of training effectiveness.	Although we can model the *cost elements* of training, it is almost impossible presently to model related effectiveness aspects since behavioral measures of effectiveness are very difficult to quantize.

It is apparent therefore that it is difficult to perform the desired investigation unless more basic research results are available.

Research problems tend to telescope upon each other. The problems for which immediate solutions are requested require more fundamental solutions which in turn require more basic research. Suppose, for example, that the H specialist is asked to *predict* the operational performance of personnel performing maintenance tasks. This requires that he have in hand an effective method of describing these maintenance tasks; this in turn means that he must know the behavioral factors responsible for effective maintenance performance. None of these is presently available. Assuming, however, that he knew what these behavioral factors in maintenance were and could define maintenance tasks, it would still be necessary to possess a store of data describing success in performing these maintenance tasks. This data store is not yet available although a preliminary effort can be found in Irwin et al., 1964. Indeed, unless we know the behavioral dimensions which characterize maintenance tasks, it is doubtful that we can even begin to develop such a data store. And so it goes.

If the investigator proceeds to study problems such as predicting pilot training or maintenance performance without the needed prior research, the answers he receives are likely to be (a) superficial or (b) of dubious validity. The potential contractor bidding on such an RFP may be fully aware of the dubious nature of the study he is proposing. However, since his livelihood depends on winning the contract, he submerges his doubts (or at least does not overtly express them) and "goes along with the game."

It would be unfair to say that governmental agencies are uninterested in the fundamental research needed to provide valid answers to complex behavioral questions. The governmental customer would reply to this comment that a solution to problems—any solution, even a partial one—is needed *now,* not after more basic research is performed. Nevertheless, it is questionable whether the partial answers which much human factors research provides are actually satisfying either to the agency which funds these researches or to the specialists who perform them.

RESEARCH DURATION

Ironically, governmental funding policies, which are geared to the fiscal year, in part determine the quality of the research performed because they often determine the duration of that research. Because the government runs on a 12 months funding cycle (starting July 1), most human factors research (as well as other) projects are allotted a year (or 14 months) to do the job. Very occasionally more or less time will be allotted, but it is likely to be less rather than more.

There are significant implications of this policy. One consequence is that, although the questions posed by the government often require a more prolonged examination, the annual limitation prevents this and instead forces the kind of research solution which can be achieved in one year. Another consequence is that agencies may be reluctant to attack more fundamental problems because these require more than annual funding to be productive.

It would be unfair to say that there have been no continuing studies of a basic, methodological nature. An example of a line of research which has been continued from year to year is the work by Siegel and his co-workers on digital simulation models for predicting operator performance (Siegel and Wolf, 1969).

Nor is it true that there are no research questions which cannot be answered in a year; these, however, are likely to be minor ones; for example, it is perfectly possible within a year (or less) to survey pilot opinions on the human engineering characteristics of new aircraft, but such a study, while important to the agency which commissions it, is of somewhat limited utility to the discipline as a whole.

Unfortunately, the governmental agency is as helpless about the situation as the H community as a whole. The agency too is constrained by fiscal funding policies. Although we may be able to estimate the required duration of engineering development projects fairly well (and even then not very well, see Peck and Scherer, 1962), it must be nerve-racking to make such estimates for human factors research projects, considering the complexity of most behavioral variables.

The consequence of the yearly limit on human factors research is to encourage overly facile, superficial answers. It also encourages manpower tradeoffs by both the customer and the contractor. It is assumed (without any real evidence) that, if a study actually requires two men working for 24 months, then 4 researchers working in 12 months or 8 in 6 months can do as much. If the research tasks to be performed are in fact independent and concurrent, this solution will work. However, many research tasks are sequential and dependent; this means that performing them concurrently ignores some of the necessary interrelationships.

PRICING CONSIDERATIONS

From all that has been said it is apparent that in many if not most solicited, competitive proposals pricing—the cost to the customer—plays a significant (although perhaps not overriding) part in the determination of who gets to do what research.

The research proposal has two components: *technical* and *financial.* Under the competitive concept of research, that proposal which is technically adequate *and* is lowest in price wins the competition. Proposals which are otherwise adequate technically but higher in cost lose out. This determination is made by separating the proposal into its technical and pricing packages when it goes to the government specialist—who sees only the technical package when he evaluates adequacy. Only when the technically adequate respondents have been identified are the cost packages examined, with the result that the lowest satisfactory bidder theoretically wins.

We would think that under these circumstances financial considerations would not influence technical ones; but they do. Not necessarily at the customer's agency, but definitely in the potential contractor's offices. Any research program, like any design program, can be performed in a number of ways (at least two: superficially or in depth) with varying amounts of manpower expended. In an effort to shave the price below what the competition can offer, the contractor may modify his technical proposal to provide only that minimum of manpower (the primary cost in most research) which will permit him to do the job. This can be done in two ways: by reducing the number of hours estimated to perform the job; or by substituting less experienced (hence less well paid) personnel for those who would otherwise do the work. This does the customer an injustice, because he is not getting the best possible research, but it is inevitable as long as financial considerations are critical.

In a few cases the contractor may be so eager to secure a particular contract that he "buys" it (i.e., he bids an artificially low price to do the research), intending to absorb part of the cost himself in return for the opportunity to work on the problem. This is rarely altruism. Although the opportunity to research a stimulating problem is a strong inducement for the research scientist, in most cases, however, the contractor anticipates that a follow-on contract (possibly involving the development of equipment) will result from the initial research effort and is willing to gamble on the improved chances he has as the first stage researcher. Under these circumstances the customer stands a chance of being disappointed in the initial research product; business being what it is, one rarely gets adequate research for nothing.

Theoretically the technically adequate bidder with the lowest price wins the competition. We say "theoretically" because in some cases (how many it is impossible to say) the lowest of the adequate bidders will *not* win the competition.

It is a popular concept among research contractors—the validity of the concept obviously cannot be determined—that often the customer has a favored contractor or at least leans toward one or two; that in a number of competitions the customer desires the contract to go to the favorite, but regulations prevent this; and that, in an effort to defeat the "rules of the game" which require competitive bidding, the customer creates a competition which is apparently "open" although he already has a significant bias to a particular contractor.

Whether the foregoing is true or not, the cost criterion in research selection has unfortunate implications not only for the customer but also for the contractor. Since the contractor wishes to win the competition, he tends to shave his price as much as he can. He does this, and at the same time he feels impelled to promise the customer as many research outputs as he can. Should the contractor win the contract, he may find himself committed to an unrealistic level of performance. From a research viewpoint it is unfortunate because, in an effort to meet the enlarged requirements, the contractor may spread his effort thin, thus making his total research effort shallow. Some contractors assume that once the contract is negotiated, they may be able to secure a dispensation from some of the research tasks that have been promised; and in a few cases this dispensation is granted.

If the effect of the marketplace criterion is that the potential contractor has to promise more than he can realistically perform, for the customer the consequence is that he is led to greater expectations than can be fulfilled. Hence the customer often feels cheated and disappointed.

Granted a limited supply of money for human factors research, it is not difficult to see why the marketplace criterion developed. Those who have been trained in a scientific research framework, however, in which money is—or should be—only a secondary factor in research performance have difficulty feeling at ease in the marketplace and wonder if some better way of procuring research cannot be discovered.

CUSTOMER DESIRES

The problem is compounded by the difficulty the potential contractor often has in determining exactly what the customer wishes from the research. Perhaps because the governmental management structure demands it, the RFP is often phrased in highly abstract terms which are difficult to interpret. Responding to such RFP's often becomes a sort of guessing game for the contractor, a highly intellectual one, of course, but a game nevertheless.

Often the guessing continues after the contract has been awarded, and it takes several conferences with the government's contract monitor to ascertain what the latter's goal is. The customer may, for example, ask for the development of a computer model of decision-making behavior when he really wants the contractor to write a justification for the agency to secure the services of a computer. He may specify that he wants to secure data on the relationship between maintenance behavior and design variables when he actually wants the contractor to develop a maintainability model.

In many cases it is not the customer's fault but that of the obscurantist way in which the RFP is often written. Sometimes of course the customer is not completely aware of what he really wants, and the entire study becomes an effort to help the customer elucidate what his needs really are.

THE "CORRECT" ANSWER

If this last is the case, the customer is again likely to be disappointed because the contractor may not be able to lift the veil of obscurity. From a research standpoint the answers secured may be less than optimal because much time will have been spent in adopting alternative strategies in an effort to find the answer the customer really desires.

If the title of this subsection sounds a little ambiguous, it is because it is difficult to characterize this concept. Basically it expresses the idea that, when the customer "purchases" research to answer a question or solve a problem, there is an inclination on his part—expressed subtly—

to insist that the research answers supplied be positive. It is possible of course that such an inclination does not actually exist but that the contractor imputes this motivation to the customer. If so, this may in part be because the research contractor is always aware that the customer is evaluating his performance against the customer's own standards.

Let us assume that the customer solicits research to investigate the relationship between computer-assisted instruction (CAI) and learning effectiveness. There are several possible questions we can ask about the relationship, such as what characteristics of CAI most improve learning; what principles of learning should be emphasized in CAI, at what stage is CAI most effective, etc.; but answers to these questions all assume that CAI improves learning effectiveness.

Since there would be little point to spending money to investigate the relationship between CAI and learning unless the relationship were positive, it is inevitable that a researcher returning to the customer with an answer which negates or downgrades CAI would be received coldly (even though there is some evidence for this, see Oettinger, 1969 and Meister and Sullivan, 1970). Such an answer is admittedly non-cost-effective, and it may cast doubt on the wisdom of a governmental agency that provides funds to investigate CAI. The customer may feel that perhaps the study was not clever enough or that perhaps the data were not interpreted properly; otherwise it would have achieved the desired answer.

It is inevitable that a researcher should gather the impression that positive answers are better than negative ones, however understandable it is that in scientific investigations negative answers occur more frequently than positive ones.

The researcher's impression of the desirability of positive answers is amplified by the fact that some agencies fund research to prove a point they would like to have proven. For example, if the agency wishes to purchase CAI devices but must first convince its management of the wisdom of doing so, what better strategy than to commission a study demonstrating that CAI is a tremendous aid to learning?

There is a great temptation, therefore, if not pressure—and what we are referring to here is self-initiated-pressure—for the research community to be responsive to these impressions and for research conclusions to be distorted subtly to produce positive answers.

It is not that research is falsified but merely that the conclusions drawn from the data may be interpreted in such a way as to cast the basic assumptions or hypotheses which initiated the study in the best possible light. If there is indeed a research bias stemming from financial considerations, it appears in an anxiety to achieve desirable results.

One of the consequences of this point of view is that it produces a marketplace response to the RFP. If the research contractor knows the customer's point of view (and, without knowing the customer in advance, responding to an RFP is like playing roulette), the former tends to channel his research concept in a manner that will agree with the latter's framework. Again, this is not to suggest falsification or distortion, merely a sinuous bending of the research concept to the prevailing wind.

One possible consequence of all of this is that research performed in this framework will tend to perpetuate as fact somewhat dubious assumptions and hypotheses.

RECOMMENDATIONS

Because of the marketplace framework which is imposed on human factors research, both the governmental agency and the contractor bidding on the research are locked into a situation which neither considers satisfactory. Until research funds become unlimited (which is hardly to be expected), some selection among competing research proposals must be made, and cost will obviously be a significant factor. It is difficult to conceive of any arrangement that does not require that financial decisions play a part.

Even accepting the financial constraints, however, improvements can be made. Two changes might prove beneficial:

1. All human factors research sponsored by the government might be coordinated by a single governmental agency, which would allow the salient aspects of that research to become more visible and which would tend to avoid research overlap at the lower echelons that carry this research out.

One of the major difficulties in developing a systematic human factors research program is the number of individual governmental organizations, each of which sponsors research, sometimes overlapping, sometimes not; for example, training research is supported by the Air Force in their *Human Resources Laboratory* and by the Navy in the *Personnel Research and Development Branch* of the Bureau of Personnel, the Personnel and Training Research Branch of the *Office of Naval Research,* and by the *Naval Training Device Center.* The *Engineering Psychology Branch* of the Office of Naval Research and the *Human Engineering Division* of the Aerospace Medical Research Laboratory both support human engineering research, as does (but to a much more limited extent) the Army's *Human Engineering Laboratory* at Aberdeen, Maryland. Other agencies involved in various aspects of human factors research include the

Behavioral Sciences Research Laboratory (Army), NASA in its various centers, and several Naval laboratories. And this is by no means a complete list.

Unless each of these organizations has such individual research requirements that they cannot be subordinated to an over-all Department of Defense structure, a central coordinating agency could permit a more realistic appraisal of what has been done and what needs to be done.

2. Such a central agency could conceivably develop a set of human factors research priorities that might eliminate much of the idiosyncratic nature of the research sponsored by the individual organizations. (Of course, there is always the danger that the managers of this central agency would allow their own personal interests to bias research allocations; but this might be tempered by appointing a consultative group of H specialists to advise the agency—among whom should be industry representatives.)

It would be foolish to believe that all research is equally important, although it is an article of scientific faith that all research concepts should be heard. The author believes that it should be possible (although with some controversy) to secure a consensus among H specialists from government, industry, and research contractors about *areas* of greatest importance. These areas should be broad enough so that there would still be room for individual research interests within them and so that specific opportunities to present the unusual concept would be preserved. The specification of those research areas would at least indicate the general direction in which human factors research would be performed. Periodically (say every two or three years) research progress achieved could be reviewed and changes in approach made, if warranted.

REFERENCES

Finley, D. L., et al., 1970. *Human Performance Prediction in Man–Machine Systems: I. A. Technical Review.* CR-1614, National Aeronautics and Space Administration, Ames Research Center, August.

Irwin, I. A., et al., 1964. *Human Reliability in the Performance of Maintenance,* Report LRP 317/TDR-63-218, Aerojet General Corporation, Sacramento, California, May.

Kraft, J. A., 1969. *Human Factors and Biotechnology—A Status Survey for 1968–1969,* Report LMSC-687154, Lockheed Missiles & Space Company, Sunnyvale, California, April.

Meister, D., and D. J. Sullivan, 1970. *Future Undergraduate Pilot Training System Study: Investigation of the State-of-the-Art in Instructional Technology,* Bunker-Ramo Corporation, Westlake Village, California, June.

Oettinger, A. G., 1969. *Run, Computer, Run: The Mythology of Educational Innovation.* Cambridge, Massachusetts: Harvard University Press.

Peck, M. J., and F. M. Scherer, 1962. *The Weapons Acquisition Process: An Economic Analysis,* Division of Research, Graduate School of Business Administration, Harvard University, Cambridge, Massachusetts.

Siegel, A. I., and J. J. Wolf, 1969. *Man–Machine Simulation Models.* New York: Wiley.

Smode, A. F., et al., 1966. *An Assessment of Research Relevant to Pilot Training,* Report AMRL-TR-66-196, Aerospace Medical Research Laboratories, Wright-Patterson AFB, Ohio, November.

IX

H GROUP ORGANIZATION

In Chapters Five and Six we discussed the functions the H group can perform during system development. How efficiently the group performs depends to no small extent on its organization. The purpose of this chapter is to describe how the group should be organized.

In describing this organization we assume that the company in which the group functions is of reasonable size (e.g., has about 500 employees, of whom at least 50 are engineering professionals) and that the number of specialists is large enough to warrant organization as a group. Just how large that number should be is indefinite, but a reasonable estimate is that the group should have at least 5 specialists before the problem of its organization becomes pressing.

Organization is a rather global term which includes many factors (some of them rather intangible). Specifically this chapter will describe the following:

1. The H group's organizational relationships with other engineering departments.
2. Nature and selection of its personnel.
3. Size of the group.
4. The directives under which it works.
5. The facilities it enjoys.
6. Its financial support.
7. Its method of indoctrinating engineering.
8. Factors influencing the effectiveness of the group's performance.

This chapter is directed to the engineering manager who has to create such a group or who wishes to improve the efficiency of one already in existence. It will also interest the specialist who is concerned about the administrative (nontechnical) factors which may influence his work.

ORGANIZATIONAL RELATIONSHIPS

Our recommendations for H group organizational structure are based on certain assumptions:

1. That the major role of the H group (whatever else it does) is to assist in system development; and therefore whatever facilitates that assistance improves its efficiency.
2. That direct contact between the H specialist and the engineer renders that assistance more effective.
3. That the organization likely to be most effective is one which places the specialist and the engineer in most direct, immediate contact.

These assumptions are based both on logic and experience. Direct contact should logically facilitate communication; and this improves the likelihood that the H input will receive more favorable consideration from the designer. The experience of some H groups also tells us that, when the specialist is organizationally isolated from the rest of engineering, his personal contacts with other engineers may also contract; as a consequence he may become passive, awaiting invitations for his services that may not materialize.

At the same time the effective H organization enables the specialist to retain a certain independence of action; it does not make him excessively dependent on the good will of the engineer. The specialist should be free to criticize engineering design without fear of repercussions. (This is not to suggest that specialists are more critics than performers; if anything, most specialists are overly sensitive to the susceptibilities of their engineering peers.)

There is no basic contradiction between an organization which intimately relates the specialist to engineering and one which simultaneously ensures his freedom to act. Nevertheless, as between the two factors, the first is more important because it is a prerequisite for any meaningful activity on the part of the H group.

A number of organizational alternatives exist (without, however, exhausting the possible minor variations in these alternatives):

The Staff Organization

The group may be staff to a manager who functions at a relatively high level in engineering. In that case (illustrated by Figure 9-1) the group is part of engineering but not at a working engineer level.

The specialist in a staff position may have no direct responsibility for day to day design problems. In the organization illustrated in Figure

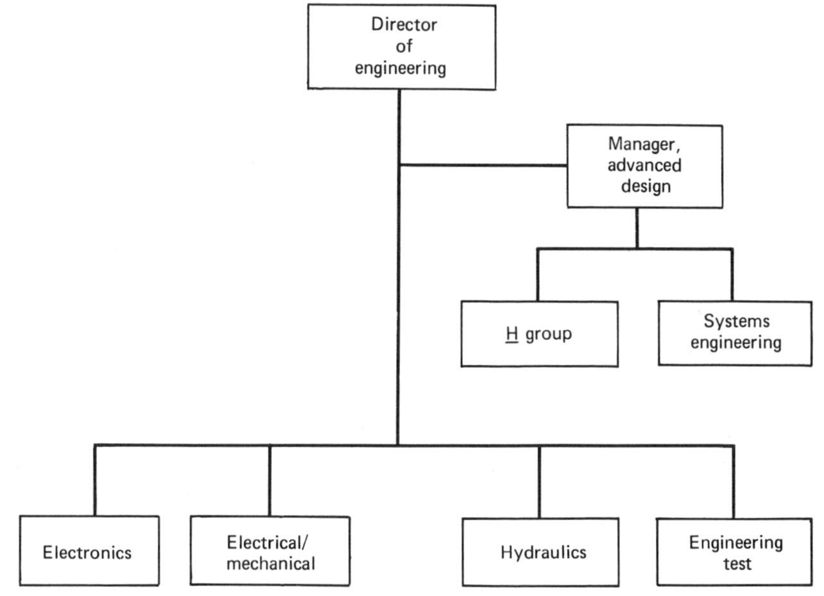

Figure 9-1 The H group as staff to engineering.

9-1 the specialist takes his orders from the Manager of Advanced Design, who is often concerned primarily with studies of systems in their conceptual ("paper") stage or systems to be proposed by the company to a potential customer. This arrangement does have the advantage of ensuring that the specialist is in on initial, fundamental decisions; however, its disadvantage is that many of these advanced designs may never be implemented so that the H activities required by later detail design stages cannot be pursued.

It is also possible in this type of organization to locate the group as staff to the Vice President (or Manager) of Engineering (or some manager with a direct responsibility for an on-going system project). The specialist in this position avoids the restrictions involved in working solely on advanced design problems, but he still suffers the disadvantage that problems encountered at a high staff position are apt to be divorced from daily on-going development situations. The specialist may have only planning or monitoring functions, or he may be exposed only to problems of exceptional severity (naturally only those problems involving obvious human factors aspects); he may never encounter H problems capable of solution at engineering working levels.

We cannot reasonably object to the specialist being such a troubleshooter, but his inputs would necessarily be confined to only a relatively few situations. His opportunity to contribute to on-going design decisions, thus avoiding problems in the first place, would be reduced. Moreover, it is conceivable that, without a working level H organization, engineers would have difficulty recognizing the existence of incipient H problems. The author's prejudices are obvious; he does not believe in staff positions unless they are supplemented by, and serve to direct, a working-level organization.

A high level staff position for H does have one advantage, however. The military customer (or at least those H specialists representing the customer) is often concerned about the apparent low status of the H group, as portrayed in their location on the organizational chart. They prefer that the group report to higher managerial levels as a means of implementing its recommendations and influencing design. When H recommendations have been ignored in the past, the customer has often felt that this was because the group was "too low on the totem pole." From a proposal standpoint, therefore, it is advantageous to show H in a relatively influential position. Kraft, 1969, indicates that most H groups in his sample reported a fairly high reporting level.

The Specialist in a Nonengineering Organization

A second structure, one that may almost be called "segregated," is described in Figure 9-2. In this structure the H group is one of a number of speciality groups located outside of engineering (e.g., design assurance, integrated logistics) although with lines of communication to engineering at the top level, as shown by the dashed line. Although the group's responsibilities may require that it work directly on system development projects and intimately with working level engineers (see the dotted line which represents the interaction between H personnel and the working design groups), its personnel are not part of the engineering organization. The inherent suspicions that many engineers hold about nonengineers may be intensified if the specialist is in a different department. Recall also from Chapter Seven that the acceptability of a design input is conditioned by, among other factors, the status of the group making the input; that status is handicapped when the group is outside the direct stream of engineering activity.

The engineer may also feel less obligation to consider the H recommendation because it is provided by the member of an outside organization. The specialist's reports do not automatically go to engineering management but must first be filtered through his own departmental management (e.g., the Director for Product Assurance).

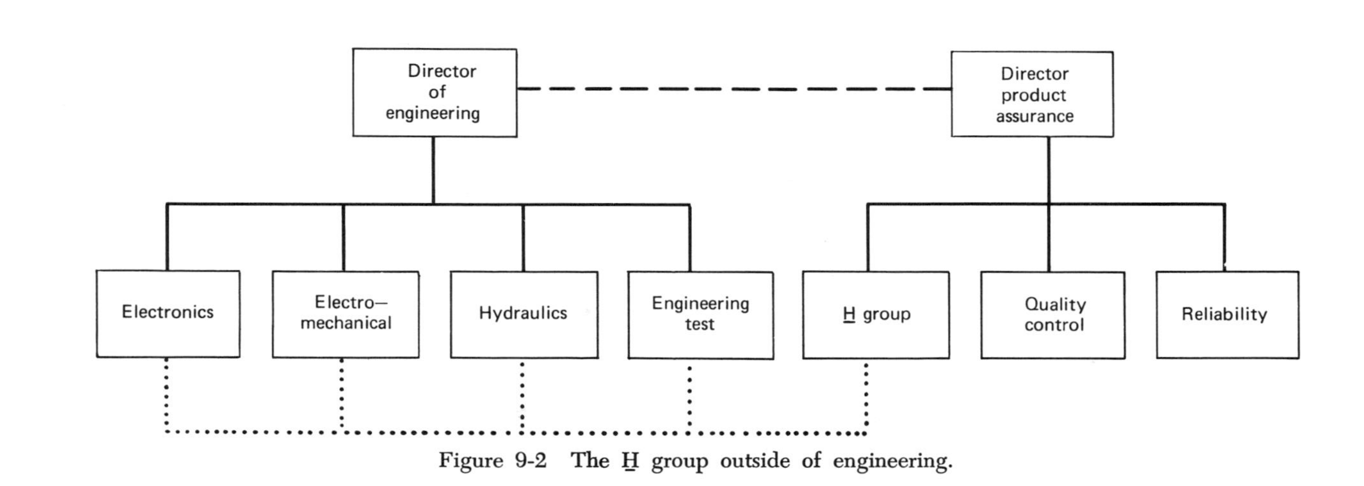

Figure 9-2 The H group outside of engineering.

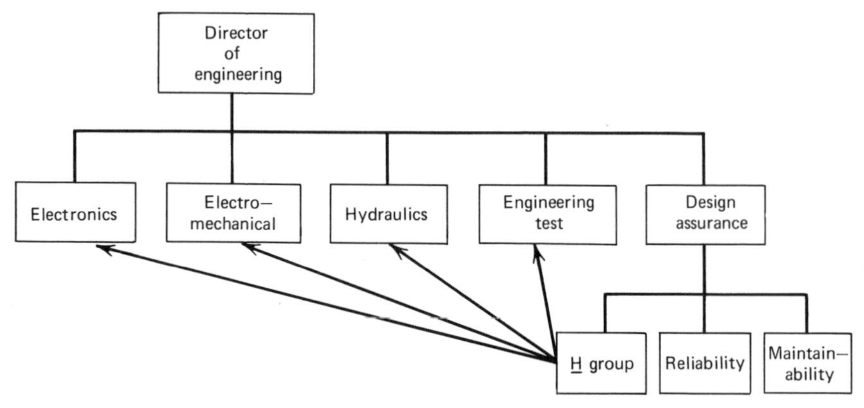

Figure 9-3 H as a speciality group within engineering.

The Speciality Group Organization

Figure 9-3 illustrates another organizational alternative. In this the H group is part of engineering at a level roughly parallel with that of other engineering groups and hence at a working level; it is one of the speciality groups, such as Reliability, which have a specific mission to contribute to, and monitor, engineering's system development. The H group often forms part of a Reliability or Quality Assurance organization within engineering. Although H personnel report organizationally to their own group, they use it largely as a base of operations, fanning out from that group daily to service the individual design and test groups.

The possibilities for effective H work on the various aspects of system development are good for this third type of organization, but also depend on the initiative of the individual specialist. Although the group is part of engineering, it is not part of any specific design group. Like other speciality groups, therefore, the call for its services depends on the willingness of the individual engineering group to make use of those services. This willingness depends to a great extent on the directive under which the H group functions and which will be discussed in detail later.

As a consequence the H group can find itself isolated from design if its personnel are not sufficiently aggressive. In addition, just because the H group is working on the line, so to speak, it may be some levels removed from top management; this can be a disadvantage if it cannot report at a sufficiently high level in order to exercise some pressure, if necessary, on the working engineering level.

The Design Group Organization

In a fourth organizational structure (Figure 9-4) the individual specialist is attached to or becomes part of the individual engineering design group. If an H group exists as an independent entity it serves as a sort of headquarters group, the specialist being detached to work directly within the design group, often maintaining his desk in the design group and taking his orders from the head of the design group. The specialist may report back to the H group at intervals for various administrative and consulting services which the design group cannot provide (for example, advice on H techniques to be applied).

This structure is distinctly different from the one illustrated in Figure 9-3, in which the specialist never formally joins the design group, however intimately he works with it.

When there is no formal H group as such (e.g., when the number of specialists does not warrant it), the individual specialist may be administratively attached to a speciality group like Reliability; or he may become formally a member of the individual design group, in which case the problem of H group organization does not arise.

The advantage of the Figure 9-4 situation, in which the specialist is attached to or becomes part of the design group, is that here he has the most direct contact with the engineer. He is intimately aware of design as it proceeds. This arrangement may also have disadvantages:

1. To make use of the specialist's services, the design group will probably find something for him to do, but, since he is responsive only

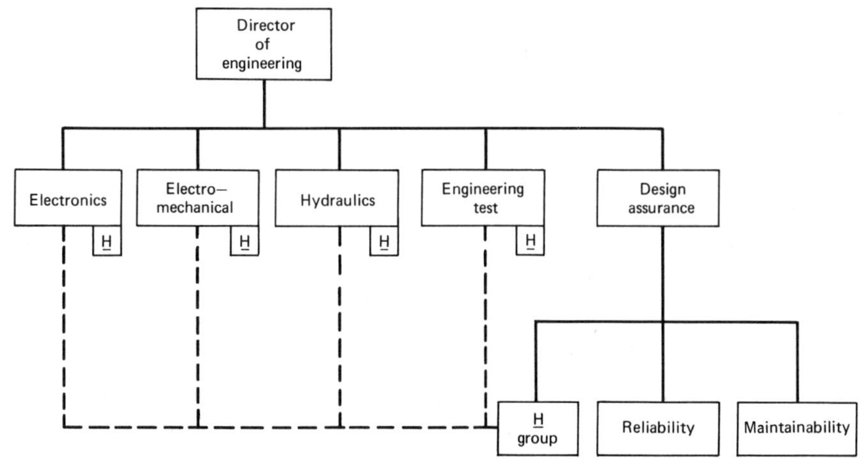

Figure 9-4 The H group as part of design.

to the desires of the design group chief, that "something" may make incomplete use of the services for which the specialist is trained. In other words, when direct control over H activities is exercised by a design group which does not know precisely what the specialist should do, it makes less than maximally efficient use of his services. He may, for example, be excluded from basic design decisions on the grounds that H is not concerned with these decisions.

2. In this arrangement, even if the specialist is occupied in meaningful H activities and even if in theory he can be as independent in his judgments as any other engineer, in practice, since he is directly responsible to the design group, he may be overly influenced by the judgment of the engineers with whom he is working. Like anyone else, he will be swayed by majority opinion.

3. If there is no H group, the specialist working within engineering by himself may lack anyone with an H background with whom to consult or on whom he can rely for assistance or consultation. This may be satisfactory for highly experienced and qualified personnel but may present difficulties for the less skilled.

Which organizational structure is most effective for H activities? It is impossible to provide an unequivocal answer to this question, since much depends on the engineering climate within the company; but, all things considered, we tend to prefer an independent H group within engineering as the best compromise. Our recommendation assumes that the H group will have the support of engineering management in the form of controlling directives and that the specialists within the group are sufficiently aggressive to maintain the interaction with designers needed to make their work effective.

Should either of these assumptions prove unrealistic, it is desirable that the specialist be attached to or be a formal part of the design group with which he is to work. An organizational structure in which the specialist is not formally part of the engineering department should be avoided, if at all possible. The staff position is satisfactory only as long as the H activities to be performed do not require the specialist's ongoing assistance to system development.

SELECTION OF H PERSONNEL

It may appear presumptuous to suggest who should or should not be selected as an H specialist. If any standards are set, presumably they should be set by some organization, like the Human Factors Society, which represents a consensus of specialists. Unfortunately, neither that organization nor anyone else has seen fit to provide standards. Nevertheless, an effort should be made to specify at least the minimal requirements for performing certain types of H activities. What these requirements should be poses a problem for anyone faced with the task of hiring new specialists.

Although it may appear somewhat undemocratic in an aggressively egalitarian society to set standards, to the extent that the specialist lays claim to a certain expertise, it is necessary to specify the qualities (and the amount of these qualities) that make him an expert. Lacking that specification, anyone, qualified or not, may claim to be a specialist. If he is unqualified, this opens the way to many inefficiencies in specialist performance, some of which already exist in the discipline. Besides, there is the purely pragmatic problem: faced with an applicant for a position, how do we determine that he is qualified?

Those are the pragmatic reasons. Over and above these, the H discipline as a whole and as a matter of self-regulation should set standards of training, experience, and performance so that it can encourage personnel to achieve and maintain those standards. This is especially necessary in H because it integrates a number of disciplines, such as psychology, physiology, and engineering; its personnel come from a variety of backgrounds with different ways of working. This makes it even more difficult to say what combination of training, experience, and skill entitles a person to call himself a specialist.

It is particularly important to set standards if the specialist considers himself part of a scientific discipline. It is of little consequence if anyone who wishes to calls himself a shoe salesman, because the quality of his performance as a salesman affects only himself and his employer's profits. This is not true of the specialist-scientist, for two reasons:

1. Since part of his activity (e.g., testing) involves the gathering of data which become part of a common data pool, his performance or lack of performance feeds back and influences the knowledge achieved by the H discipline as a whole.
2. That may be thought of little significance to anyone beyond the specialist, but the specialist also attempts to influence the development of man-machine systems, some of which are of great importance to the country; the inefficient performance of these systems may affect the safety of the system personnel, as we saw exemplified in Table 2-2 of Chapter Two.

With this rationale concluded, we can proceed to the factors that influence whether or not an applicant will be an effective specialist. These factors are

a. the training the applicant has had;
b. the type and amount of his experience;
c. his capabilities and interests;
d. the type of H position for which he is applying.

The requirements specified below are considered minimal requirements. As pointed out in Chapter One, the ideal specialist is a superman, possessing advanced degrees in psychology, engineering, medicine, physiology, statistics, and many years of experience. Since finding such a combination of training and experience is well nigh impossible, it is necessary to approach the problem from thc standpoint of minimal qualifications.

Table 9-1 Educational Background of H Personnel *

Psychological Training	% of Total	Engineering Training	% of Total	Other Training	% of Total
Experimental	16.2	Electrical	10.0	Physics	4.5
General	9.6	Mechanical	7.9	Education	4.2
Industrial	4.8	Industrial	2.3	Bio-chemistry	3.4
Engineering	2.1	Aeronautical	1.6	Mathematics	3.4
Education	1.2	General	1.0	Biology	3.0
Clinical	1.0	Civil	0.8	Physiology	1.9
Physiological	0.9	Chemical	0.8	Industrial design	1.8
Social	0.8	Nuclear	0.4	Business	1.7
Other	0.4	Bio-engineering	0.2	Medicine	1.1
		Other	0.2	Operations research	0.9
				Anthropology	0.8
				Chemistry	0.8
				Microbiology	0.8
				Sociology	0.5
				Computer technology	0.4
				Genetics	0.2
				Clerks/technicians	8.4
Total	37.0		25.2		37.8
					100.0

* Based on 3804 replies to questionnaires. Includes all types of organizations.

In specifying certain standards the author makes one assumption which may invalidate his judgments for those who do not accept that assumption. We consider that H is primarily and most importantly a behavioral discipline, however much it overlaps with engineering and other disciplines. In fact the type of individual who in most cases will apply for a specialist position will be a behavioralist or have a behavioral orientation, even if he is not a psychologist.

This is not to suggest that engineers should be excluded from specialist work. Indeed, as we see from Table 9-1 (taken from Kraft, 1969), engineers form a sizeable subset of the total H population. However, the engineers who work in the H discipline as specialists must possess a behavioral orientation if they are to be more than superficially effective.

The largest single group within H consists of psychologists of one sort or another. This is also the group which is most preferred by all hiring organizations (see Table 9-2, also taken from Kraft, 1969). Perhaps the reason for this is that management realizes that only a behavioral orientation (with which the psychologist is most familiar) will permit effective performance of H tasks. We would prefer to believe this. Perhaps it is because the supervisors of most H groups, being psychologists themselves, prefer to hire people in that discipline.

Table 9-2 Industry Preferences for Future Specialist Hires

Academic Specialty	Psychology			Engineering			Other		
Percent preferred		64.3%			27.4%			8.3%	
Degree level	AB–BS	MA–MS	Ph.D.	AB–BS	MA–MS	Ph.D.	AB–BS	MA–MS	Ph.D.
preferred	11.1%	70.4%	18.5%	60.8%	30.4%	8.8%	28.6%	42.8%	28.6%
Years of experience preferred	2.5	4.4	4.1	2.9	4.5	3.6	2.8	2.9	3.8

It may be possible to employ someone who lacks any extensive psychological training. This may be done by routinizing his work by, for example, allowing him to make design recommendations or evaluations in accordance with standard checklists. It may then be a comparatively simple process to train an unskilled individual to perform the mechanics of H analysis and evaluation.

However, underlying the checklist (and all the other basic H tools) is, as has been pointed out, a number of very subtle behavioral problems (e.g., the nature of the dimensions which make up the equipment configuration, their relationship to operator performance, etc.) all of which should be taken into account in performing H analyses and evaluations. It is possible to perform H activities on either of two levels: a routine surface in which we are unaware of the behavioral processes underlying the activity and one which probes below the surface of the design judgment. Although it is possible to perform H tasks superficially (at the level of filling out checklists, for example), the results are usually superficial also.

Part of the problem is the lack of a highly precise H methodology. As long as the cues on the basis of which the specialist makes a judgment (that one man-machine configuration is better than another or that this or that design feature is or is not acceptable from a human factors standpoint) are essentially nonquantitative and difficult to specify, the specialist will require a more than superficial capability to do his job. If we can look forward to the day when H principles are highly quantitative and operationally precise, then much H activity may be performed by paraspecialists (i.e., technicians with minimal training and experience) just as a laboratory technician, nurse, or orderly can function in a hospital under a physician's supervision. We are, however, a long way from this.

Nonetheless, there is an increasing tendency to routinize the specialist's craft. As engineering becomes more familiar with H activities and processes, the latter lose their mystery, and there is a tendency to expect that H procedures can be so routinized that personnel with minimal background will be able to perform them. One of the major factors in creating this attitude has been the spate of governmental procedures for performing in step-by-step fashion what are actually highly subtle processes. For example, AFSCM 375-5 (USAF, 1964), AFSCM 310-1D/AFLCM 310-1D (USAF, 1966), and AFSCM 80-3 (USAF, 1963), which have governed the H aspects of Air Force system development for approximately 10 years, imply that very complex activities such as function allocation, task analysis, and human engineering design and evaluation can be performed more or less "by the numbers." We have seen in Chapter Three that this is not the case.

There is also a tendency—perhaps because Americans are great "do it yourself" exponents—to believe that almost any technique can be performed by almost anyone if its procedures can be written down. As a result, we see, for example, an increasing number of human engineering handbooks and guides to permit the design engineer to be his own H specialist. All of the above tendencies have contributed to an increase within H of personnel with only minimal academic and experience requirements.

As the demand increases and exceeds the available supply of qualified H personnel, less well trained personnel enter the discipline to satisfy the demand. Whereas 15–20 years ago it was difficult for someone who lacked the MA degree in psychology to secure even a beginning position, a much larger percentage of specialists now have only the bachelor's degree plus a minimal amount of experience.

In addition to a practical working knowledge of H methods (which is assumed in the levels of experience described in Table 9-2), what are the attributes we would like to see in the specialist? The two most important qualities are (a) an attitude of strong interest in and understanding of equipment functioning (i.e., a "feeling" for equipment in general, akin to that felt by the automobile enthusiast for his vehicle); (b) an appreciation of the complex of behavioral variables that interact with that equipment functioning.

The first is more practical, the second more theoretical. How can we tell when the applicant has the first quality? One index of an interest in equipment is whether the specialist has ever built or repaired equipment on his own, or likes to "tinker" with equipment. The second quality can in part be ascertained by asking the specialist to discuss the behavioral variables in some system development project he has worked on previously. Ultimately of course the determination of the requisite qualities must be highly subjective.

A less important but still highly desirable quality is some familiarity with the behavioral research relevant to H and the ability to understand the capabilities and limitations of that research in terms of system development problems. It need hardly be added that the specialist should believe that the solution of man-machine system problems is vitally important; in other words, he should feel that being a specialist is more than "just another job."

Given the general attributes described above, what specific capabilities should the engineering manager or H supervisor look for in the applicant?

As far as experience is concerned, no one who has had less than two years' experience in system development should be considered as fully qualified. This does not mean that we should not employ someone with less than two years' experience; but he should be considered as being in apprentice status (i.e., a *junior* specialist). This applies regardless of the amount of academic training he has had (up to and including the Ph.D.) since no academic training fully prepares the specialist to solve the kinds of man-machine problems he will encounter in system development. Academic training may provide an essential background for system development work, but further on-the-job training is always required.

We assume that the *minimum* academic degree for the psychologist-applicant is the BA/BS, plus 12 to 20 *graduate* credits in experimental psychology. Six psychological credits could be exchanged for a corresponding number of engineering credits. If the applicant is an engineer by training, he should have his BS degree plus 12 *undergraduate* credits in experimental psychology. None of these psychological credits should be exchangeable for additional engineering credits.

If we seem to be harder on the engineer than on the psychologist, in the sense that we do not require engineering training of the psychologist, it is because we assume that psychology is more critical to H than is engineering. We also make no distinction between types of academic engineering training (i.e., between electrical/electronics, mechanical, civil engineering, etc.). The academic background is, of course, in addition to the required two years' relevant experience.

The most acceptable combinations of academic degrees to a prospective employer should therefore be (but not necessarily in the order listed) the following:

1. Psychology (BA or higher degree), plus graduate psychological work (or equivalent graduate work in physiology, anthropology, mathematics, etc.).
2. Psychology (BA or higher degree), plus some number of undergraduate engineering courses.
3. Engineering (BS), plus some undergraduate psychological course work.
4. Anthropology/physiology/pre-medical, mathematics, operations research, etc. (BA or higher degree), plus undergraduate psychological course work.

The combination of psychological and engineering background is considered particularly desirable by the prospective employer, principally because it may help the specialist structure his recommendations to the engineer in design-relevant terms.

What has been discussed so far is the *minimum* for accepting a specialist as fully qualified in his profession. The nature of the work the specialist must perform will determine any additional training, experience, and personal qualifications required. At the risk of oversimplifying the range of H functions which may be performed, we have attempted to develop a scale (described in Table 9-3), the major divisions of which demand different qualifications. (Naturally such a scale must be quite crude and will not include, except in terms of broad generalities, all the specific activities required of any individual specialist job.) Moreover, the scale is to a certain extent only an abstraction because it is unlikely that H activities will be so neatly ordered as we have indicated.

The activities described in the scale range from simple to complex. The activities are these involved in system development only, and include only that research which is incidental to system development. In addition, it assumes no significant supervisory responsibilities; at each level the amount of applicable experience required would be substantially increased if supervision is assumed.

Table 9-3 Range of H Activities

Least Complex

1. Control-display analysis only (i.e., design/evaluation of control panels and consoles during detail design).
2. Work station analysis, including
   (a) the design and evaluation of all equipment features influencing or influenced by operator capabilities;
   (b) development and evaluation of operator procedures;
   (c) surveillance of engineering design.

   (All the above during detail design only.)
3. Performs same functions as in (2), but
   (a) throughout system development, with special emphasis on basic design decisions early in development;
   (b) man-machine system analytic functions, including functional and task-equipment analysis.
4. Performs all preceding functions, and in addition
   (a) performs mockup development and testing;
   (b) test and evaluation of the prototype system, including statistical analysis of test data.
5. Performs all preceding functions, and in addition
   (a) assists in writing proposals for new systems; and/or
   (b) performs H-relevant research.

Most Complex

Certainly the scale at best is ordinal; the intervals between scale points are *definitely* not equal. At the risk of repetition, it must be emphasized that the scale does not deal with all possible H activities, merely with those that are most common. Despite its inadequacies, however, it will serve as a starting point for considering required specialist capabilities.

First Scale Level

The first point on the scale is design and evaluation of control panels and consoles during detail design. This function demands least of the specialist. Not that control panel design/evaluation is unimportant (quite the contrary); however, the number of ways in which we can lay out a control panel or console are severely limited; and the principles which can be applied to the problem are relatively crude (see Meister and Farr, 1966). In the present state of our H knowledge the control panel design/evaluation solution is largely intuitive; we know so little about underlying equipment/personnel relationships that we cannot be very sophisticated in performing this function. Assuming a substantial increase in knowledge relating equipment characteristics to operator performance, the control panel/console design task may become more, rather than less, complex because we will know more about (and so can influence) the factors which influence operator performance. Finally, it is our impression that this level of H effort—as long as this is *all* that is asked of the man—is usually assigned to junior personnel.

In any event, for this type of work (again assuming that this is all that is performed by the specialist) we want a junior man, meaning a bachelor's degree in psychology or engineering, plus minimum graduate training, plus *one year's* experience. The specialist should have a strong feeling for the hardware characteristics of control/display components because that is what he will be largely dealing with.

Second Scale Level

There are of course many things which the specialist can do besides designing and evaluating control panels. Still staying within the detail design period, in addition to control panel/console design and evaluation the specialist can be concerned with the characteristics of the equipment which influence its maintainability, the placement of equipment within a facility (workplace layout), environmental factors (e.g., lighting, noise), the development and evaluation of procedures for equipment operation, etc. It is assumed that these H responsibilities include surveillance of all the engineering design related to the above.

To expand the specialist's range of responsibilities in this way we need a more experienced man than the junior specialist who concentrates on control panels. Possibly he needs a few more graduate course credits (if the man is a psychologist), but the primary need here is for greatly expanded experience. The man performing this range of H activities should have at least *four years* relevant experience.

Third Scale Level

Up to this point our hypothetical specialist has been confined largely to detail design, where the H problems and the decisions to be made are largely (although not exclusively) molecular. Let us suppose that we wish our specialist (at the third level of the scale) to perform all the functions so far described, but now throughout system development (except for test and evaluation) with special emphasis on functional/task-equipment analysis required in the very early design stage. Now we ask him to perform the functions described in Chapter Five (e.g., function allocation, the analysis and comparison of alternative man-machine configurations, the analysis of tasks to be performed, and predictions of operator performance).

The man who will perform these functions must obviously be more capable and experienced than the previous two. This is because effective work in the early design stage is highly creative; the decisions to be made involve far more uncertainty, and such rules as exist are far less structured than those of detail design.

Our preference here would be for somewhat more academic training, not because the academic background is necessarily more relevant to work on early design, but because an advanced degree often serves as an excellent selection device. We assume that the ability to secure an MA degree in experimental psychology ensures a minimal amount of creative ability. This should be combined with at least *five years* relevant experience. Three additional years of experience can be substituted for the MA. The applicant should manifest a strong feeling, not so much for equipment functioning although that too must be present in some degree, but for system *interrelationships* (a more abstract level of functioning).

Fourth Scale Level

Level four on the scale is different from level three, not so much in amount of skill as in terms of requiring a different type of background. We consider that the ability to perform human factors tests and evaluations requires a somewhat different breed of specialist than the man who concentrates on design alone. This is of course an oversimplification because design work ordinarily requires evaluation of drawings, mockup tests, etc. And certainly it would be ridiculous to segregate these evaluation tasks from the task of recommending design changes resulting from the evaluations.

However, if we assume that the test/evaluation activity is fairly heavy, will involve exploratory studies to answer the question of operator capability, will require development and operation of sophisticated mockup or simulator equipment, will extend into quasi-operational testing of the prototype system, then we want a specialist whose training and experience has emphasized a knowledge of experimental procedure, the conduct of research, the design and construction of test equipment, and the statistical analysis of performance data. Since this activity closely parallels the activities performed in experimental psychology, to perform this work in a subordinate capacity we want an MA in psychology plus *five years* relevant experience. Additional academic training in experimental psychology plus additional relevant experience would be required for increased responsibility in these functions, to the extent that someone who assumes supervisory responsibilities for major system tests (thus controlling the activities of a number of more junior people) must definitely have a Ph.D. in experimental psychology and *eight* years experience, most of it in testing.

Fifth Scale Level

The final point on the scale (level five) includes all the previous ones but specifically involves two further responsibilities: the writing of the human factors section of the system development proposal and performance of H̲ research. If the human factors section of the proposal is relatively minor, presenting only such relatively simple material as a description of control panel design/evaluation activities and a justification of a particular control panel/console layout, then a junior man with a capability for written communication can handle the job. If, on the other hand, the human factors role in the proposed system development is a major one and if the H̲ section of the proposal analyzes and suggests an approach to human engineering all aspects of the new system, then the man writing this section should be a senior man because he is in effect planning the entire H̲ effort. This latter specialist should have an MA degree and *five years* experience; but the essential thing is the ability to conceptualize and describe the varied facets of the H̲ program and its interrelationships.

H̲-related research is assumed to extend beyond the test and evaluation activities described for the fourth scale level. As Kraft (1969) points out, two-thirds of all industrial H̲ groups are expected to help support themselves by performing contract research which is completely distinct from system development activities. Personnel performing such research at a junior (beginning) level should have a minimum of an MA in psychology (or a related behavioral discipline) plus *three years* of research experience.

Of necessity this brief description of specialist requirements is arbitrary (especially with regard to years of experience) and oversimplified. In particular we have failed to consider special training and experience requirements implied by special types of man-machine systems. For example, the development of life support equipment requires knowledge of human physiology, as well as relevant equipment development experience; H̲ assistance in the development of visual displays requires knowledge of visual parameters and measures; H̲ inputs to computer-driven systems of any character are tremendously facilitated by some knowledge of computer concepts. We could go on and on pointing out similar specific knowledge requirements. The standards we have suggested assume that the specialist has as part of his training and experience the specific knowledges required by the systems he will help develop.

In comparison with the hiring preferences noted by Kraft, 1969, and described in Table 9-2, our standards may appear slightly higher, at least for greater levels of responsibility. This conflicts with the tendency, noted previously, for an increased demand for specialists to bring in less well trained and experienced personnel (who, in these times of constricted budgets, are generally cheaper than their seniors). These standards conflict also with the tendency of some engineering departments that do not have an H staff but that are suddenly presented with the need for one to select an engineer who may have had some apparently relevant experience and training and to designate him as an H specialist, thus fulfilling a contractual requirement. On large scale system development projects like Atlas, Titan, Minuteman, Apollo, and C5A, the need for personnel, at least in junior-level positions, may be so great that some are designated as specialists who would not meet the standards we have specified. The absence of formally recognized personnel standards created by some organization like the Human Factors Society, whose opinion commands respect, accentuates the tendency to accept less trained personnel as specialists.

Not only are these standards high, but they have minimized the engineer's role in H activities. It is often quite useful for an engineer to function *as an engineer* within the H organization, acting as a consultant on engineering problems to the specialists, but the engineer who wishes to perform H functions during system development must have an irreducible minimum of behavioral training and experience.

How can the H supervisor determine whether an applicant is properly qualified on the basis of experience? A choice is always a gamble. However, the applicant should be questioned about his work on previous system development projects. He should mention or otherwise demonstrate familiarity with some of the engineering details of these systems and indicate some of the H inputs he made to these systems. It is also useful to ask him to describe how he interacted with engineers on these projects. Published reports (although not necessarily in journals) are an indication that he is at least semi-literate. Because of the necessity for direct verbal contact with designers, some verbal fluency is necessary. For a senior man, a discussion of the problems involved in applying H concepts to design would be illuminating. In addition, the senior specialist should indicate awareness of the issues underlying H activities, some of which were described in Chapter Four.

The author would be remiss in concluding this discussion of specialist qualifications if he did not mention a problem which occasionally occurs. Despite the fact that many H groups solicit outside research contracts, H activities in system development are largely nonresearch oriented. The academically oriented psychologist with an advanced degree who lacks any real experience in the field may seek a specialist position with the mistaken notion that the work is essentially research (which is of course precisely what he wants to do). If he is hired and is flexible enough, he will adjust to the constraints of the system development situation; but in some cases he may not adjust, and the usefulness of an academic researcher in a system development context is at best minimal (as well as potentially a source of distress to the researcher). It would be a kindness therefore for the interviewer to alert the applicant specialist that his opportunities to perform basic (or even applied) research may be few and far between.

SIZE OF THE H GROUP

Obviously we cannot specify an absolute number of specialists as that number generally required. Kraft, 1969, in his survey of 390 industrial, consulting, governmental, and teaching organizations found that the average industrial H group numbers 20 people; this figure appears to be independent of the total number of personnel employed by the company.

One hypothesis that can be advanced—if we remember that it is only a hypothesis—is that the relatively stable size of the H group despite wide variations in company employment may be related to the number of major design groups in the company (e.g., electronics, electromechanical, hydraulic/pneumatic, mechanical test). In other words, as long as engineering is generally organized among companies in the same way, the size of the H group may be some function of the number of design groups (e.g., 3-4 specialists for each design group).

The size of any individual H group is of course determined by a number of factors, of which the most important is the amount of work required of it; this in turn determines the number of people needed to perform that work. Naturally, that work effort is influenced by the number and size of the system development projects that the company has under contract. A rule of thumb which is often heard is that the cost of H activity on an average sized system development contract runs to about 1 to 2 percent of the gross cost of the total contract; when H activities are enlarged because of the criticality of human factors in a system, it may rise to 6–8 percent. From these admittedly ideal values, it is possible to deduce the financial resources capable of supporting an H staff.

Another major determinant of group size is the range of responsibilities assigned to H as part of the system development project; these responsibilities vary depending on the character of the project and its size (a large project demanding more responsibilities than a small one). The willingness of the customer to purchase H activity is also an influential factor.

Beyond all these factors, however, what the H group will be asked to do depends on what engineering management thinks it *should* be doing. If H responsibility is limited to control panel design and evaluation, the number of specialists required may be small; if H responsibility is expanded to encompass all man-machine aspects of the system, throughout system development, the required number will be significantly larger.

Kraft (1969) indicates that the size of H groups has about doubled over the past 10 years. The increasing size of these groups may in part have been produced by increasing demands for formal documentation of analytic functions. The imposition of very formal documentation requirements, such as those of the Air Forces 375-5 Personnel Subsystem program (which requires detailed task analyses, time line analyses, QQPRI, etc.) tends to produce larger groups of personnel because documentation demands large numbers of personnel. The Personnel Subsystem group on the Atlas project, for example, had between 100 and 200 people.

As the size of the group increases, its organizational structure must change. It is difficult for a group (as distinct from a few individuals) to be attached directly to a design group; this means that it must be set up as a separate organization. Thus there may be a slight tendency to lose contact with other engineering groups as the size of the H staff increases beyond some threshold point.

H GROUP DIRECTIVES

H group effectiveness depends to a large extent on the way in which its relationships with other engineering groups are defined by directive. Ultimately the effectiveness of that relationship is determined by informal factors such as mutual appreciation of capabilities and an attitude of cooperativeness. However, these informal factors are encouraged by, and in some cases cannot be developed without, formal published directives.

The measure of H group effectiveness is often the ratio of what the group *could* do compared with what it is required, or asked, or permitted, to do by engineering. That ratio is often inordinately small, meaning that the group is doing far less than it could. The cause of this inefficiency often rests with the engineering management which has not adequately accepted and established the H group's role. Unless that role is quite explicitly defined in directives, it will be difficult for many H groups to secure sanction for the things they feel they must do. The paradox is that, given that the H group has certain goals and that the company is supporting the group financially, the company is effectively depreciating its investment (in the vernacular, wasting its money) if it does not support the group with the executive directives which any group requires.

(The engineer-reader may object to the previous statement. Many equipment engineering groups do not have, nor do they need, a formal directive describing responsibilities. If so, it is because the engineering group's role has been defined by custom and practice. However, the H group's role is still in the process of definition, which means that it needs formal directives strenuously enforced.)

Performing H Functions Without Directives

The case for directives can be more forcefully made if we contrast two very different situations. In the first the H group has no formally described responsibilities, although an informal understanding exists between management and the group that they are to perform certain functions such as control panel design and evaluation during detail design. (It will be recalled that this is the popular engineering concept of what H specialists can and should do.) Despite or perhaps because of this understanding, no formal procedures have been specified to implement the function. For example, there is no obligation on the part of design groups to call specialists in on the design of the control panels or even to notify them that their design is proceeding. (Hence the need for the many specialist visits to the individual design groups to check on design progress.) If there are formal design reviews in which control panels are involved, the H group is neither notified nor invited. (If this situation appears ridiculous or far fetched, the reader should be assured that it is not uncommon.)

In this situation H specialists are in the position of salesmen soliciting design groups to make use of their services. It is necessary for the specialist to visit the individual design group, to explain his purpose, to ask to see design drawings, and to perform his evaluations hurriedly, on the run as it were, because there is no requirement for an H design review. If he does not do these things, all too often there is no call for his assistance; so, if he lacks the aggressiveness to push his way in where he may feel he is not wanted, he may wait at his desk for the summons that does not come.

The H group in this situation may even lack a defined reporting channel. By this is meant that there may be no provision on a periodic or emergency basis for the group to make formal recommendations based on its findings to design groups.

The H group should be concerned about two different types of reports:

1. *Progress reports.* If the H group reports organizationally at a low engineering echelon, it may lack visibility because its reports are progressively edited and summarized as group reports are transformed into section into departmental reports at increasingly higher level. This is annoying, but not too serious, except that for reasons that hardly need repeating, H activities need as much visibility as they can get. In addition, engineering management may not know what it is getting for its H dollars. This is more serious, because an engineering management which is hardly aware of its H activities is more likely to dispense with them if budget becomes tight.

2. *Technical reports.* The inability to insure that technical findings are seriously considered by responsible design engineers is much more serious and ties in with the organizational structure. It is usually possible to send a memorandum to anyone in engineering, but the difficulty is that it may have no force behind it unless a directive requires that the memorandum be considered seriously. The H group may, for example, determine that a serious deficiency exists in some aspect of system design. It may send a memorandum to this effect to the head of the responsible design group (the level above the individual design engineer with whom the specialist has interacted). However, lacking a directive requiring that some substantive answer be given to the memorandum, no action may result.

What are the H group's alternative courses of action? The problem could of course be described in a progress report, but we cannot describe a problem in detail in a progress report. A memorandum might be sent to the Manager of Engineering, but, unless there is a directive which prescribes a reporting channel that high, this procedure means bypassing normal channels. For lack of defined reporting channels, then, nothing may be done about an important H recommendation. (We emphasize the importance of the H recommendation; obviously we would not go to such lengths for routine recommendations.)

Given all the preceding disabilities, there is a great temptation for specialists to sit at their desks and wait to be "invited to the waltz." Some specialists may do this. Others, more aggressive and energetic, will earnestly pursue direct contact with individual design engineers, with greater or lesser effectiveness. They visit the various engineering groups and ask to see any control panel drawings that have been made recently. Even if it is after the fact and the drawings have been "frozen," they evaluate the designs, write critical reports, and suggest design modifications which both they and the design groups know will not be implemented. However, they consider that submitting such reports represents at least a moral victory and may even serve to indoctrinate the recipient. Indeed, when it is difficult for them to discern concrete changes which the H group has been able to secure, they hypothesize that engineering indoctrination may be the most effective way in which they can function, and they concentrate on creating tutorial aids like checklists and brochures which they force on the design groups.

This is a situation for which lack of strong management directives is at least partially responsible. Contrast this situation with the one following, in which the role of the H group is clearly and concretely defined.

Performing H Functions with Directives

The H group's responsibility for specific areas of man-machine design is recognized by engineering design groups. For example, certain types of drawings in which man-machine aspects play an important role must be signed off by the specialist; or, if this is not the case, a formal review of those designs must be held in which specialists play an important part (although certainly not the sole part) and in whose findings they must concur.

Responsible design groups are required to consult periodically with specialists on designated man-machine design areas. If the signoff referred to in the previous paragraph exists, natural self-interest on the part of engineers will lead them to consult with the specialist. A formal report describing H design recommendations must be sent to the design group with copies to engineering management. Periodic progress reports are sent unedited to the level of Manager of Engineering, thus providing maximum visibility.

This favorable contrasting situation does not mean that specialists can afford to be complacent and to wait at their desks for engineering calls. They will still have to visit design groups regularly to advance their services. However, we assume, and logic supports the assumption, that engineering receptivity to the specialist's recommendations will be substantially greater than in the previous situation, particularly if the specialist performs his functions competently. The philosophy that the establishment of requirements may eventually produce a change in engineering attitude is the same philosophy applied to the civil rights movement. Although it may not be possible to change a man's attitude directly, it is possible to affect the way he responds, which may then change his attitudes.

What the H Directive Should Contain

Let us now consider in detail what the H directive should contain:

1. *A definition of the H task,* but in concrete, not abstract, terms. First, what it should not be. A task definition which specifies, "The function of the H group is to assist in optimizing the man-machine relationship," is true, but it is also twaddle because it is impossible to determine from the definition what the group is supposed to do. Depending on how the individual engineer or manager chooses to interpret the definition, the task could involve the full range of H activities; or none of these. Abstract definitions leave the specialist at the mercy of individual attitudes toward H.

In contrast, the following, which is much more specific, is also much more useful: "The H group has the following responsibilities: To

(a) assist engineering groups in the design and evaluation of all equipment which includes a man-machine interface;
(b) assist in the development and evaluation of operating and maintenance procedures for new equipment;
(c) develop plans and procedures for human engineering design and testing of prototype equipment."

This list obviously does not include all the H functions which could/should be performed; it merely attempts to illustrate a level of required detail. Obviously specific responsibilities depend on many factors, not least of which is the nature of the company's business and the systems being developed. The important thing is the relative specificity of the task definition. Note that the statements above are still broad enough to encompass a wide range of more specific responsibilities. For example, the development of plans and procedures can be interpreted to give authority to the H group to participate in the writing of the H part of system development proposals, to develop checklists for distribution to engineering groups, etc.

2. The task definition should be supplemented by a *definition of procedures,* by which the task will be implemented. Again the following is only illustrative:

"H responsibilities will be performed in accordance with the following procedures:

1. All engineering groups responsible for the design of control panels and consoles or other equipment which is intended to be directly operated by personnel will keep the H group informed of the design schedule for the subject equipment.

2. It is expected that H̲ specialists will assist in the design of such equipment by participating in relevant planning meetings.

3. All drawings of equipment requiring personnel operation of controls and displays or other components (e.g., cabling) will be reviewed by a representative of the H̲ group prior to design "freeze." The responsible H̲ group representative will initial the drawing to indicate its acceptability from an H̲ standpoint. The level of drawing detail subject to H̲ review will be that required to ensure compliance with applicable H̲ requirements.

4. Should any equipment design described in (3) above not satisfy H̲ requirements, the H̲ group will submit a report of its findings and recommendations with supporting data to the head of the cognizant design group. The cognizant design group will reply in writing within one (1) week, and both reports will become part of the documentation supporting the design of the equipment.

5. A representative of the H̲ group will serve as a member of the Engineering Design Review Board with responsibility for chairing meetings called specifically to review man-machine interface designs."

Additional procedures specifically describing other H̲ functions (see Chapter Three) could be called out, but the illustration above should be sufficient to indicate the flavor of what is required.

3. *Reporting channels* should also be specified; for example:

"The head of the H̲ group reports to the head of the Product Assurance Organization. Monthly progress reports describing H̲ activities will be submitted to the latter and will be incorporated in reports to the Chief Engineer. Technical reports of H̲ design recommendations for major equipment units will be submitted as required to the head of the Engineering Design Review Board."

The sample directive above is of course only an ideal. In practice few engineering directives are of such specificity or accord such rights and responsibilities *in writing*, and few if any engineering groups possess such explicit charters. Few engineering groups need such a charter since their working relationships are recognized and rooted in a tradition which the H̲ organization does not have. It is, however, a fact, which most managers recognize, that engineering departments would work much more efficiently if group interrelationships were spelled out and enforced; it is a common complaint that the individual design and test groups often fail to communicate adequately with each other; this tends to retard system development. Nor is this simply a function of company size; lack of communication is characteristic of medium and small engineering departments, although perhaps not to the same extent.

From a cost-effectiveness standpoint the engineering manager protects his investment in the H group by specifying the rules under which it is to work. Such directives are also useful in persuading the customer of the H group's ability to perform its system development functions, and in particular to get adequate consideration for the H viewpoint.

Even, however, when such a directive is provided, the specialist cannot afford to remain passive about his job; he must be energetic in following up the directive or, like any other charter, it will fall into disuse. Nor does a directive diminish the need for making H design inputs meaningful and effective; if these inputs do not have substantive value to engineers, the directive, one way or another, will be subverted. Assuming, however, that H inputs are design-relevant, the availability of a directive will enhance the value of those inputs.

The directives described in this section are company-developed and regulate internal company activities. H specifications and standards which are customer developed and applied, such as MIL-H-46855 (Dept. of Defense, 1968) and MIL-STD 1472 (Dept. of Defense, 1969), and which also have a significant effect on the specialist's activities are described in the Appendix to this chapter.

FACILITIES

The facilities the H group will require are primarily: a laboratory for conducting studies and for testing subjects; space and tools for constructing and storing mockups; and laboratory instruments. Facilities also include, of course, reference books and journals.

Logically the H group should need extensive facilities (beyond the customary office furniture and space) only if its function includes research and research-related activities like the development of mockups. If the H function involves no research or evaluation responsibilities (except at the drawing design stage), then the need for these facilities does not exist.

Apparently, however, a large number of H groups do perform research functions because Kraft, 1969, reports that the 43 industrial organizations which provided facility information in his survey invested an average of over $600,000 in these facilities. Of course, this represents only about 25% of the companies responding; this suggests that H research activities are not overly important in industry. Of those companies reporting H research activity, 80% of the research effort was directed at applied problems, presumably supporting system development. Understandably, consultant, nonprofit, and government organizations (particularly the latter) support research much more extensively than does industry.

The statement that H research facilities should be provided only if there is a need for research should be qualified. Facilities should be provided for mockup development and subject testing as part of routine design evaluation. Mockup development is an essential part of the H̲ design evaluation methodology. If the mockup evaluation requires the testing of subjects, as it should, then space for subject testing should be provided. How much space? An absolute minimum would be 400 square feet (20 by 20 feet), but double or triple this area would not be excessive.

The simplest type of mockup consisting of control panels and consoles, with minimal reproduction of equipment functioning, can be developed from cardboard, styrofoam, or plywood with conventional indicators and controls; the facilities needed for its construction consist primarily of carpenter tools and artist's supplies (e.g., saws, hammers, etc.). More elaborate functional mockups can be created more efficiently in experimental factory shops, which contain equipment such as drill presses, lathes, etc. A number of H̲ groups perform research on true simulators, but in most cases simulators are developed for over-all engineering purposes, and H̲ personnel merely make use of them. Consequently, space for lodging and utilizing the simulator would not ordinarily be an H̲ responsibility.

True laboratory facilities should be developed much more circumspectly since they are quite expensive. Only the largest companies can afford to develop and support a laboratory (and the personnel required to make use of a laboratory). In the context of system development the need for research and the laboratory facilities that go with it must be thoroughly justified. The specialist who views H̲ as primarily an opportunity to perform research may agitate for a laboratory even when there is little or no need for it. However, if the H̲ group proposes with company support to compete effectively for governmental research contracts, minimal laboratory facilities will probably be a prerequisite.

Whether or not the group is provided with a laboratory, some standard laboratory instruments should be provided to support mockup tests and testing in the later phases of system development This equipment includes (but should not be limited to) sound level and light measurement meters, timers (e.g., stop watches, reaction timers), oscillographs and polygraphs, an inexpensive still camera, and tape recorders. Sound and light level meters are used to evaluate the noise and illumination characteristics of facilities in which new equipment is to be placed; timers are used to record operator reaction time; oscillographs and polygraphs (e.g., pen recorders) to record errors and response time; tape recorders and cameras to record verbal and physical responses in test situations. A calculating machine should also be available for performing statistical analyses.

The nature of the instrumentation needed by the H group also depends on the type of equipment that the company is producing. If the company is designing visual displays or visual display consoles, apparatus for presenting visual stimuli will be needed. If the company product emphasizes auditory stimuli, specialized audio equipment will be quite important.

Since the specialist often becomes a consultant on special controls and displays, the company will find it a useful and not overly expensive investment to support a collection of the new control and display devices which annually hit the market. The specialist can mount these on special display boards and demonstrate their operation to engineering groups which can make use of them. A file of manufacturers' catalogues of controls and displays should also be started. Information on new devices can then be disseminated to design groups in periodic news bulletins. The control-display facility will supplement the capability of the engineering parts group and make available to engineers more specialized information on controls and displays than the usual parts group can supply. From an indoctrinational standpoint, information on control-display hardware helps to create a positive relationship between the specialist and the engineer.

FINANCIAL SUPPORT

The H group is in a special position with regard to funding. As with other speciality disciplines, such as Reliability, a new system development project does not automatically mean that the group will be employed on that project. Although it is, for example, impossible to develop an electronics system without electronics engineers, it is possible to do so (although perhaps less effectively) without H specialists.

The extent to which the H group is funded depends upon the willingness of the customer and the company management to require its services on the project. They cannot complain if the H effort is minimal because funding is minimal. When the customer explicitly calls out H functions in the contract statement of work, the group will in most cases be funded for these functions. This also occurs when an H specification is applied to the contract; however, the probability of H funding, and the amount of money supplied, is less when only the specification is applied to the contract (without a call-out of functions) since many companies feel that they can satisfy the over-all specification requirement by using design engineers as specialists or by providing only minimal H activity (e.g., control panel evaluation). If neither of the above occurs, the company management may decide after the contract is negotiated that H should "get a piece of the action" (participate in system development) as an assist to other engineering functions.

Some company managements feel that H should be supported completely by direct contract funds for system development although they may supply some small amount of overhead funds for company-sponsored research, to support administrative supervision or to pay for assistance on proposals. In other words, the H group is considered to be a company-support group only. However, to adopt this point of view (which has a certain logic behind it), the management must support the H group in playing its legitimate role in system development. This requires, among other things, that the company should include H functions in proposals for new system development. It also requires that the company not dispense with the H activity when the development budget becomes tight or at least not to cut H more than any other engineering group is cut.

Other managements expect that 60–70% of the funds to support the H group will come from direct system development projects, with the remainder derived from contract research. For this 30–40% the H group serves essentially as a research organization, seeking out governmental contracts on its own and competing with other organizations for available research contracts. When this occurs, however, the company must supply the necessary research facilities (e.g., laboratory space, instruments).

Very occasionally the H group is asked to support itself primarily through contract research. The 60–40 or 70–30 ratio noted earlier of system development to contract research is reversed. Under these circumstances the company has the services of a human factors organization on its system development projects but pays very little for this because the H group largely supports itself while making itself available for any system development projects on which it may be required. This arrangement imposes severe difficulties on the H group, because in effect it must compete with research organizations having relatively low overhead rates (administrative and housekeeping costs) while paying the high price in overhead with which a company devoted to system development is often burdened. Fortunately, this type of situation is not very common.

Wherever the money comes from and however much it is, the H group must control its own specific, allocated budget and cannot be dependent on contingent funds doled out by engineering whenever design groups decide to use H services. If it lacks control of its own budget, the group will never know how much money it can afford to expend in bringing new people into the group and supporting those already "on board." This contingency arrangement makes everyone in the H group nervous because it is impossible to plan ahead under the circumstances. This is particularly true if the H group secures its funds on allocation from individual design groups rather than from the engineering department as a whole. Design groups will be reluctant to make use of H services if the money for the services must come out of funds they consider were allotted for their own use.

ENGINEERING INDOCTRINATION

One of the central themes of this book is the necessity for developing positive relationships between the engineer and the specialist. For reasons which have been explained earlier (see Chapter Seven), this is difficult; the typical engineer is unaware of and uninterested in H, and when he is aware of it at all, views it sceptically.

It is therefore essential for H to influence engineering attitudes, at the very least to make engineers aware of H and its purposes, uses, methods, and benefits. Even if it is not possible to induce a significant change in engineering attitudes, it is at least necessary to make engineers aware of H, for the engineer is the most immediate customer H has.

That others be aware of his operations is important to the specialist because it partially determines how effective he will be. We cannot market a product unless the customer (in this case the engineer) is aware of it. Lack of awareness poses a special problem when the H group is first formed. However, even when the group has been functioning for several years (and not necessarily in one of the colossi of industry like IBM), it is not uncommon to find that a substantial proportion of the company's engineering population is unaware of the H group or, even if it is, has only the vaguest idea of what the group does.

It follows therefore that the specialist must consider engineering indoctrination as one of his more important functions, although an ancillary one. He may wish to move on to more substantive activities, but he must consider that without some form of propagandistic effort the present status of H would never have been achieved. Moreover the indoctrination program must be a continuing one, repeated at intervals, although with well-established groups the magnitude and frequency of the indoctrination can be reduced. Fortunately, the indoctrination effort need not be presented in any blatant "pitchman" style in order to be effective.

Indoctrination has several goals:

1. To call to the individual engineer's attention the existence of the H̲ group.
2. To inform him of the purpose of H̲ activity.
3. To describe H̲ functions and responsibilities, particularly as they influence engineering design.
4. To provide the engineer with an outline of general H̲ principles such as the need to consider operator capabilities and limitations in design.

Two types of indoctrination can be distinguished: *formal* and *informal.* Formal indoctrination is more common when the H̲ group begins its work; it is employed only at intervals and involves groups of engineers rather than individuals. Formal indoctrination consists of short (about 30 minute) meetings with small groups of engineers (anywhere from 10 to 20) and can be conducted in simple lecture style, but as informally as possible, emphasizing audience participation.

An outline of the topics covered in an initial presentation might include

(a) definition of human factors;
(b) how H̲ developed (historical background);
(c) the need for H̲ as evidenced by human engineering discrepancies, the frequency of operator errors, and their effect on system performance;
(d) how the specialist interacts with the design engineer;
(e) H̲ functions, that is, what the specialist can and, equally important, what he cannot do (the latter to reassure the engineer that the specialist is not trying to assume his design responsibilities);
(f) evidence for the usefulness of H̲ efforts (culled from the literature or personal experience);
(g) desired working arrangements with the engineer (e.g., provisions for sitting in on design and reviewing drawings).

Later lectures should deal with more specific topics and should emphasize the transmission of useful information which may be of value to the engineer, for example, new control-display devices just out on the market, principles of man-machine design (e.g., control panel layout, maintainability provisions), and brief summaries of any new design-relevant human factors research.

Formal indoctrination also includes such tutorial devices as developing special human engineering checklists to be distributed to engineers, monthly brochures or news bulletins containing items of human factors information which may be of interest to engineers, and posters emphasizing human factors.

Are these indoctrination efforts worth the time and the money? It is doubtful that they convert any engineers to whole-hearted support of human factors, but at least they familiarize engineering with H group responsibilities and functions, and they reinforce, even if only verbally, positive attitudes toward H. From this standpoint lectures are more valuable than written brochures or memoranda because they expose the specialist to the engineer as an individual.

Informal indoctrination is what the specialist does every working day. His contacts with engineers, the conversations he has with them, the questions he answers, his work outputs: all serve to create an image in the engineer's mind which may make the latter more receptive to H. Since the specialist is often most effective as a verbal information source, the more efficiently he answers the engineer's questions and provides information, the more effectively he serves the indoctrination function. The same is true of his written contacts with the engineer. To the extent that his memoranda are simple, cogent, and design-relevant, they will have indoctrinational value.

A word of caution. Although formal engineering indoctrination is important, it should not take precedence over effective H performance. The author has participated in meetings of H specialists in which the great cry was to "sell" the engineer, as if the H gospel had been finally codified and all that was necessary to do now was to convert the heathen. Despite the commercial influences acting upon him, the specialist must not forget that H is not like selling used cars, in which persuasion is a strong motive force. Before we can effectively "sell" the engineer on H, the discipline must demonstrate its capability for solving man-machine problems. Some specialists (e.g., Leuba and Wallace, 1965) feel that the capability exists, if only the engineer can be persuaded to make use of it; others like the author feel that the capability exists only partially and that the effort to persuade the engineer should not divert attention from the pressing need to improve H methodology.

H GROUP EFFECTIVENESS

Unless we can measure the amount of system performance efficiency or improvement resulting from a specific design input, it is necessary to define its effectiveness in purely pragmatic terms.

An input is always a recommendation for one alternative over another even though the alternatives may be implicit rather than explicit: to select one design over another, to incorporate one design characteristic in equipment over a number of possible other characteristics, or to modify design in one way rather than another. (From that standpoint design can be conceptualized, despite its obvious creative elements, as a series of branching decisions.)

Obviously, if a course of action recommended by an input is rejected by the engineer, it cannot be put into effect and its utility, whatever its inherent qualities may be, is zero. Like the first law of politics—to get elected—the first prerequisite for any input is to be accepted. After the system is fabricated and tested operationally, it may be possible to determine objectively how effective that system was and, by inference, how effective the inputs that made up the system were; but this determination during system development is extremely difficult because of the perturbations that any single design decision sets up in the developmental process. In any event, since system design is a product of, and a compromise among, many inputs, how is one to disentangle them, to determine that one individual input was good or bad? It appears then to be impossible for the specialist to measure the value of his contributions other than by their acceptance by engineering. (This may account also for the emphasis on persuading and indoctrinating the engineer.)

The primary factors responsible for the utilization of human factors inputs in system development are (a) input *timing,* (b) the amount and type of *information presented,* (c) *cost* of the recommendation, (d) input *impact,* (e) its *visibility,* and (f) the *perceived value* of the input. (There are other factors too, of course, such as the technical adequacy and appropriateness of the input, which are not discussed here; we assume for purposes of this discussion that the inputs being considered are both technically adequate and appropriate. The factors discussed in this section are essentially attitudinal.)

The *timing* of the input refers to the phase in system development at which the input is made relative to the "freezing" of a particular aspect of hardware design. All other things being equal, the earlier in development that a human factors input is made, the more likely it is to be incorporated in design. However, the input timing must be appropriate to the particular stage of design activity that has elicited the input. A design recommendation may not be acceptable if the over-all engineering team is not prepared to consider such a recommendation at that time.

Amount and type of information is, of course, crucial, regardless of timing. The author's previous research suggests the great importance of detailed information, phrased in concrete design terms, with the behavioral and design logic clearly enunciated. Regardless of any other factors, general design recommendations (e.g., "make it simpler") will not be accepted. Properly phrased information is therefore essential to input acceptability but will not, of course, make a human engineering design recommendation acceptable unless further conditions are satisfied.

In system development the monetary *cost* of a design recommendation largely determines its impact on hardware. This cost includes the time required for engineers to make the changes necessitated by an input in already existent drawings and the cost of producing hardware with the recommended design feature as opposed to the cost of producing the same hardware without the design feature. Management typically views cost in terms of hardware, even when greater hidden costs (e.g., training/maintainability) exist.

Early in system development, when hardware has not yet been developed, the cost of a human engineering recommendation is very slight since few changes to existent drawings or hardware are required. Although the specific shape of the cost curve is unknown, that cost increases significantly as development progresses and engineering comes closer and closer to the point of no return—the time when design is stabilized ("frozen"). The closer one gets to this point the greater the cost of a design recommendation.

The reason for this is that every modification produces increasing perturbations down the line; the sheer cost of revising drawings for even a minor change in locating a control on a panel is exorbitant because of the volume of related documentation which often must be changed.

The greater the cost, then, the more difficult it is to secure engineering approval of a design recommendation. If the recommendation can be made in terms of what engineers consider a nonhardware factor (e.g., a change in procedures, training, personnel), it is much more likely to be accepted than a recommendation which must be implemented in hardware.

Impact can also be defined in terms of the extent of the influence exercised on over-all design by an input. The impact of a minor human engineering recommendation (e.g., the choice of a control position) is slight because it affects only the control panel and even that perhaps only slightly. The impact of a human factors recommendation dealing with a basic design concept (e.g., automatized or manual operation of a system) is great because many consequences flow from that recommendation—if it is accepted.

The impact of an input is greatest early in system development. Because basic design decisions are made early in system development human factors recommendations made early in development should be more fundamental and hence have greater impact than those made later.

Visibility refers to the perceived prominence of the input, the degree to which it stands out in the engineer's consciousness. Visibility is related to both timing and impact.

The early human factors contribution is likely to be less visible to engineering for several reasons: in early development many inputs are competing with each other, and concepts are often quickly accepted or rejected; the interaction of these inputs is such that they tend to be modified by each other, and the final accepted design only partially resembles any individual input. The degree of ownership of that accepted design is difficult to ascertain. Moreover, many design engineers are not concerned with the visibility of their individual inputs and think of the final design product as a group output; they tend to downgrade the excessive prominence of any single input and are pleased when human factors inputs fit smoothly in with other engineering contributions.

One factor which, however, tends to make early human factors inputs more visible is that they deal with basic design concepts which are themselves more visible because they have potentially greater impact.

The human factors input in later development, despite its concentration on molecular design characteristics, is also likely to be quite visible when it requires a change in already accepted design features. For that reason its visibility is likely to be negatively tinged.

The *perceived value* of an input to an engineer reflects the latter's concept of its worth. It is positively related to the impact of the input and negatively related to any changes the input may require in already existent design. The engineer is more impressed by major design inputs than he is by minor ones. He also prefers a smooth-running developmental process and dislikes perturbations, particularly in the later phases. Because the later human factors input deals almost always with minor equipment characteristics, its perceived value in the eyes of engineering is likely to be comparatively slight. Moreover, because the later input usually requires some change in already existent design, the late input is devalued.

Acceptability therefore interacts complexly with timing, cost, and impact. We can pose the following rules:

1. The lower its cost, the more acceptable the input.
2. The earlier the input, the more acceptable it will be. However,
3. The greater the impact of the input, the less favorably engineers will receive it.

The greater the cost, the less acceptable the input will be. For that reason inputs made later in development, when design is nearing the hardware stage, are less acceptable because the input will probably require changes which perturb the developmental process, changes which, however slight in themselves, are costly to procure. Moreover, recommendations for changes made late in development are by their very nature almost always critical of previous decisions already incorporated into drawings. Consequently, such late recommendations face a double barrier: not only do they implicitly censure previous engineering work but, because they are inherently more costly than early inputs, they demand a greater amount of proof to be convincing.

As was pointed out previously, inputs made early in development are likely to be less costly since many drawings have not as yet been finalized or hardware programmed. From that standpoint such inputs are more likely to be accepted. On the other hand, early inputs deal with or influence major organizing design concepts rather than individual design characterisics, and consequently they are likely to have more impact if accepted. Human factors inputs which have major impact on the system are less likely to be accepted, for several reasons. One is that design engineers usually conceive of the specialist's role as related only to molecular human engineering factors. They also demand greater proof that more fundamental human factors recommendations are meaningful than they do for minor human engineering inputs.

Obviously, there are a number of contradictory elements in the engineer's attitudes that influence his acceptance of the H input; for example, the early H input may not be very visible because it blends with other engineering inputs; on the other hand, because it is early, it involves decisions on operator factors fundamental to system design, and this gives it greater visibility than it might have otherwise. The engineer is more impressed by major decision inputs than by minor ones but may be more disturbed by these major inputs than he would be by minor ones. Thus the engineer is genuinely uncertain about his attitudes toward human factors. It is difficult for people trained almost exclusively in physical hardware concepts to believe that behavioral factors have or should have any influence on the products of those physical concepts. On the other hand, they are urged not only by the H practitioner but also by the customer and sometimes by their own management to include these behavioral factors in their design. If the engineer's attitudes toward H appear from the outside to be somewhat schizophrenic, they may be a true reflection of the attitudes of technological society in general, which has yet to come to terms with human behavior.

The major conclusions we can draw from the factors influencing H effectiveness is that the earlier the H group can make its contributions to design, the more likelihood there will be that the contribution will have impact, visibility, and perceived value, the less it will cost, and the more acceptable it will be. The specialist must be prepared, however, to encounter genuinely ambivalent behavior on the part of the engineer toward these early inputs.

APPENDIX: H SPECIFICATIONS AND STANDARDS

Introduction

This Appendix describes MIL-H-46855 (Military Specification, Human Engineering Requirements for Military Systems, Equipment and Facilities) and MIL-STD 1472 (Military Standard, Human Engineering Design Criteria for Military Systems, Equipment and Facilities), which govern H functions in the development of military systems and specify the human engineering characteristics those systems should have when finally produced.

These documents were recently (1968) put into effect by the Department of Defense to integrate a number of specifications and standards which were issued by the individual military services and applied only to their systems. MIL-H-46855 supersedes MIL-H-22174, 27894, 46819, and 24148; MIL-STD 1472 supersedes MIL-STDs 1248 and 803A-1, -2, and -3. Because of this supersession the two documents are the controlling H documents on all military systems being developed.

These documents, 46855 and 1472 (as we shall refer to them in the future for the sake of brevity), apply only to H and human engineering characteristics. A number of equipment specifications controlling the development and characteristics of electronics, electro-mechanical, and mechanical equipment may each contain a section imposing human engineering requirements on the development of these equipments. However, when a human engineering requirement is applied to a contract, these two documents are the most likely to be cited.

These specifications/standards can be used in various ways. The procuring agency may, if it wishes, ignore 46855 and 1472 completely and write into the contract SOW its own version of an H requirement (assuming such a requirement is appropriate). Alternatively it may assume that an H requirement is not needed and may not apply any H provisions to the contract at all. The SOW may refer to 46855 and 1472 *in toto* with a proviso that these sections which are appropriate to the system being developed be applied. Alternatively the SOW may list the specific paragraph items of the document to be applied.

(The author prefers that the applicable items of the specification/standard be specifically listed in the SOW since the contractor may ignore significant parts of these documents if he is given the opportunity.)

The degree to which we apply these documents depends on the same factors which govern the extent of the H group's responsibilities: the criticality of the human to the system; the size of the system; the money available for human factors activity; whether or not the new system is a redesign or well within the state of the engineering art.

The *functions* and *processes* to be performed by the H group are controlled by 46855. In other words, it specifies what the specialist will do and, to a lesser extent, how he should do it. The H *characteristics* which the system should have, if it is to be properly human engineered, are described by 1472.

The two documents obviously have an effect upon the three major role players in the system development process: the customer, the contractor (engineering management or design engineer), and the H group. The three parties take a somewhat different view of what they want from the documents and how to achieve their desires. The customer presumably wants the best human engineered equipment at the least cost. The contractor would prefer to be relatively unconstrained in his development of the system. The goal of the H group more closely corresponds to that of the customer because the group also wishes to develop the most effectively human engineered system. From this standpoint there is probably greater community of interest in interpreting the specification and standard between the H group and the customer than there is between the contractor and the customer.

From the customer's standpoint the purpose of the two documents is to control the design process and the product. From the H group's standpoint the documents function as *charters* to permit the group to perform its tasks and also as a guide to effective design. From the company's standpoint, and more particularly from the individual engineer's vantage point, the documents act as design constraints, to be avoided if possible; and, if this is not possible, to be tolerated as well as it can. Industry's philosophy is that it does not wish to be told how it should do its job but only what it should do.

From the standpoint of this book, both documents reflect the present conceptual and methodological status of H. Any specification or standard can require of specialists and engineers only those system inputs and outputs which are within the state of the disciplinary art. If, for example, 46855 requires that specialists analyze function flow diagrams to allocate functions, the methodology must exist to do this; or the requirement is pointless. Therefore 46855 and 1472 can serve as a means of measuring how far H has progressed and what methodological improvements are required.

This Appendix therefore discusses the two documents in terms of their

(a) impact on the H group;
(b) impact on engineering (i.e., the design engineer, the manager, the company as a whole);
(c) reflection of H status.

MIL-H-46855. This specification is divided into two parts: a set of general requirements and an Appendix which describes data requirements for naval systems. Although it is a reasonable inference from the Appendix that the Navy feels it has certain functional requirements which the other services do not, the Appendix generally follows the provisions of the specification proper. When Naval requirements are significantly different from those provisions which can be applied to other military systems, they are noted in the following discussion.

Scope and Nature of Work (3.1.1). The document specifies that the human engineering effort will include "active participation" in system analysis, engineering development, and test and evaluation. It is assumed that human engineering participation in system development will begin with initial system analysis and continue throughout engineering development and test and evaluation. The intent is to ensure that human engineering is an integral part of all system development phases. The reason for emphasizing this point is to counter the general engineering concept that the H role is largely confined to detailed design. The Naval Appendix indicates that the Appendix "may be invoked in its entirety or selectivity"; this presumably applies to the document as a whole. This is important because the use of the specification should be tailored to the kind of system one is developing. Every project does not require the application of all specification provisions.

Human Engineering Program Plan and Other Data (3.1.2). A Human Engineering Program Plan must be submitted concurrently with or as part of the system proposal. (This provision is generally applied only when the system to be developed is a major one or the human is critical to the system—in other words, when H activities would be extensive.)

In essence the plan is a description of how the H effort will be conducted through development. The plan is an attempt to force the contractor to think seriously about his H program at the earliest possible development time. It should be as concrete and detailed as possible although contractors generally attempt to make it rather general. It should indicate not only what H activities are to be performed but how (e.g., design reviews, sign-off, etc.). The necessity for developing such a plan is one reason for emphasizing the specialist's role in helping to prepare the proposal.

The Appendix specifies that the plan will include a milestone chart and PERT diagram of H activities, a description of the organizational structure, and the names and qualifications of key H personnel. These special provisions in the plan are included to exercise a degree of customer control over the H effort which the procuring agency has felt it has not been able to exercise in the past. Company management may have some objection to including a list of key H personnel in the plan because it implies that these people will be "pledged" to the project and cannot be used on other projects; hence it reduces the flexibility with which management can transfer personnel from one project to another.

Human Engineering in System Engineering Analysis (3.2.1). This consists of four major activities:

1. Definition and allocation of system functions.
2. Information flow and processing analysis.
3. Estimates of potential operator/maintainer processing capabilities.
4. Equipment identification.

All are analytic, that is, they represent how the specialist examines the system design problem and develops appropriate solutions. These activities were described in Chapter Three.

The Naval Appendix calls out similar activities but in somewhat greater detail. For example, it requires functional flow charts, detailed time-line analyses, a table of functional allocations, Operational Sequence Diagrams (OSD's), tables of information requirements, and a definition of control, display, and communications requirements.

Several comments must be made about the four activities:

1. If the procuring agency has done the initial system analysis required in the concept formulation (predesign) stage of activity (see AFSC 375-5), activities 1, 2, and 3 will have been largely performed, at least to some gross level of detail. The SOW supplied to the contractor by the procuring agency often contains at least a preliminary function allocation and identification of roles to be performed by equipment and personnel.

2. Even if the contractor's SOW has not allocated functions and identified personnel roles and equipment (even at a gross level), the typical speed with which design proceeds make it necessary for the specialist who hopes to have any influence on basic design concepts to perform these analyses very quickly. Since the equipment designer

proceeds almost immediately after contract go-ahead to the design of hardware (see Chapter Seven), detailed H requirements (e.g., flow charting, OSD's, etc.) may make it difficult for the specialist to make his inputs with the necessary celerity.

3. Although we might assume that the first three activities are performed sequentially, they are actually performed concurrently and iteratively. For example, to allocate functions we must perform some sort of information flow analysis and also determine whether personnel can perform required tasks on the system. In other words, the activities described are actually different ways of analyzing the same system development problem.

We consider now each of the activities in greater detail.

Defining and Allocating System Functions (3.2.1.1)

"Human Engineering principles and criteria shall be applied to specify man-equipment performance requirements for system operation, maintenance and control functions and to allocate system functions to (1) automatic operation/maintenance, (2) manual operation/maintenance, or (3) some combination thereof."

The human engineering principles and criteria to be applied are unspecified, which is a reflection of the fact that no substantive methodology of function definition and allocation presently exists. This is not to say that such a methodology does not exist in theory (see Rabideau et al., 1961), but at best it is rather general and leaves much to subjective judgment.

Information Flow and Processing Analysis (3.2.1.1.1)

"Analyses shall be performed to determine basic information flow and processing required to accomplish the system objective and shall include decisions and operations without reference to any specific machine implementation or level of human involvement."

It is difficult, if not impossible, to conceptualize decisions and operations without framing them in terms of some man-machine configuration. Even though we may not specify it, a decision or operation automatically assumes an equipment which is involved in the decision or which is to be operated. The statement is an attempt to divorce artificially the consideration of system functions from their man-machine context.

Estimates of Potential Operator/Maintainer Processing Capabilities (3.2.1.1.2)

"Plausible human roles . . . in the system shall be identified. Estimates of processing capability in terms of load, accuracy, rate and time delay shall be prepared for each . . . function. These estimates shall be used initially in determining allocation of functions and shall later be refined at appropriate times for use in definition of operator/maintainer information requirements, and control, display and communication requirements."

The requirement for these estimates specifies that they shall be made "in terms of load, accuracy, rate and time delay." The implication is that these estimates are quantitative, utilizing predictive error and time data. Unfortunately, as has been pointed out in previous chapters, the store of such data is very slight. Even if the predictive data to perform such an analysis did exist, their application to "definition of operator/maintainer information requirements and control, display and communications requirements" would be extremely tenuous since again a standardized methodology for doing so is lacking. The estimation requirement as applied to maintenance operations is even more difficult to accomplish than that applied to equipment operation because most maintenance behaviors are more complex than operating behaviors.

Equipment Identification (3.2.1.2)

"Human engineering principles and criteria shall be applied to identify and select equipment to be operated/maintained/controlled by man. The recommended design configuration shall reflect human engineering inputs, expressed in quantified or 'best estimate' quantified terms . . ."

This section again requires quantification although an adequate data base to support the requirement does not exist.

In all fairness to the developers of this specification, it must be noted that they would say that the items cited above are not designed to indicate *how* H functions are to be performed but *what* these functions should be. It is impossible, however, to differentiate between *what* and *how* because the *how*, at least in part, defines the *what* of a method. If a method required to perform an H function is missing, effectively that function does not exist.

For the H group the sections cited previously represent a charter to enable it to work in early system design. The specialist cannot take these specification statements too seriously, however, unless the procuring agency also requires that specified design outputs must stem from performance of the required functions and unless the agency also requires that those outputs be considered in making design decisions.

If no formal outputs of the analyses are required, it is probable that the contractor will feel no urgency to perform the analyses or at least that he will perform them only superficially. Even if formal outputs are required, the procuring agency must insist that they be used as the basis for design decisions. Otherwise the company will concentrate its effort solely on the production of "paper."

The contractor is unlikely to consider these requirements seriously unless documentation is specified in the contract, and the customer monitors its application. The documentation serves the purpose of forcing the contractor to attend at least minimally to analytic requirements. Without documentation the H group is unlikely to be able to implement the requirements unless the group is especially aggressive. Without documentation requirements the H activities to be performed under the specified sections cited could range from very extensive to merely token, more probably the latter.

The Navy Appendix is therefore on firmer ground because it does require specific documentation. The Air Force has something called a Contract Data Requirement List (CDRL) which specifies the documentation required and for which it is prepared to pay. Since the company makes money on the documentation, not on the performance of the analyses which underlie the documentation, it will undoubtedly emphasize documentation in any contract negotiations involving H functions.

In developing its SOW for a new project the H group should therefore emphasize the documentation outputs it will supply and use these as a lever to ensure engineering cooperation in performance of the system analyses. The same caution applies to activities performed in other phases of system development.

Another reason for emphasizing documentation is that it will be difficult for the H group to price out its activities unless these are in terms of concrete outputs (i.e., documentation requirements).

Analysis of Tasks (3.2.1.3). Both gross and refined task analyses are called out as required.

"The analyses shall provide one of the bases for making design decisions, e.g., determining . . . whether system performance requirements can be met by combinations of anticipated equipment and personnel, and assuring that human performance requirements do not exceed human capabilities. The analysis shall also be used as basic information for developing preliminary manning levels, equipment procedures, skill, training and communication requirements . . ."

The refined analysis of critical tasks is supposed to identify: (a) information requirements; (b) information availability; (c) evaluation process; (d) decisions; (e) actions taken (by operators); (f) body movements; (g) workspace required and available; (h) location and condition of the work environment; (i) frequency and tolerances of action;

(j) the time base; (k) feedback; (l) tools and equipment; (m) number of personnel required and their job aids; (n) communications; (o) special hazards; and (p) operator interactions. These analyses must be kept current with the system analysis and applied to system/subsystem designs represented by design criteria documents, performance specifications, drawings, and data. The overlap between task analyses and the analyses required under 3.2.1.1. and 3.2.1.2 should be noted.

The Navy Appendix calls out similar requirements although in somewhat less detail.

What are we to say about this requirement? The intent is laudable, and we certainly wish to see all operator/technician tasks so analyzed. However, if we were to apply this requirement literally, it would overwhelm the H group with work unless the system were quite small or the H group were enlarged. It is obviously necessary to exercise some judgment about which tasks the analysis shall be applied to. Some tasks are quite routine, pose no special problem to the operator, and consequently need little or no analysis. As the specification states, the refined analysis, which is time-consuming and effortful, should be applied only to "critical tasks" (for which a definition is supplied in the *Notes* section of the specification). (It is suggested that an essential part of the Human Engineering plan supplied by the contractor under 3.1.2 should be a listing, not an analysis, of all the critical system tasks. Only these tasks should be intensively analyzed.)

Whatever tasks are analyzed, the specialist will face the problem of deciding upon the detail in that analysis, even for what the specification terms "gross analysis." This detail level, which is not identified in the specification and which must be mutually agreed upon by the contractor and the procuring agency, is often a major source of difficulty for the specialist. Adding to the difficulty is the absence of a universally agreed upon format in which the analysis will be reported.

Just as in the case of system analysis, almost certainly no formal task analysis will be performed unless the procuring agency specifies a documentation requirement and an analytic output which is funded.

The specification refers in a somewhat off-handed manner to the application of the task analysis outputs to manning, training, skill, and communication requirements. The methodology for applying task analysis to design and to these factors is largely logical but subjective. This, combined with the length of time it takes to perform a detailed task analysis, may make it difficult for the specialist to make use of these analyses to influence design, manning, etc. With regard to training and skill level specification, the specialist has some opportunity to influence the requirements for these, provided that the training group is also organizationally a part of the H organization.

Human Engineering in Equipment Detail Design (3.2.2)

"During the detail design of equipment the human engineering inputs made in complying with . . . systems analysis requirements . . . shall be converted into detail equipment design features. Design . . . shall meet the applicable criteria of MIL-STD 1472 . . . Human engineering personnel shall participate in design reviews of equipment and items to be operated or maintained by man."

Human engineering requirements during this period include the conduct of experiments, laboratory tests (see 3.2.2.1), the development of mockups (3.2.2.1.1), and dynamic simulation (3.2.2.1.2).

"Human engineering principles and criteria shall be applied . . . to equipment drawings . . . The approval of these drawings by the contractor shall verify that human engineering requirements are incorporated . . ." The design of work environments and facilities (3.2.2.3) which affect human performance shall include, where applicable, consideration of atmospheric and weather conditions, acceleration, noise, weightlessness, adequate space, walkways, provisions to minimize fatigue and stress, clothing and equipment handling provisions, protection from toxic conditions, illumination, and life support and crew safety requirements.

The Navy Appendix presents similar requirements, although in less detail, with the addition of the following critical point: "Engineering detail drawings shall bear the signature approval of the responsible human factors engineer" (30.3.10). This requirement, which is something that most specialists have been fighting for for years, is not included in the body of the specification. Most engineering managements still refuse this requirement because they view it as an infringement on their design prerogatives.

Again, requirements for human engineering in detail design serve mostly as a charter for the specialist to perform necessary activities. Attention should be paid particularly to the provision that he "shall participate in design reviews of equipment and items to be operated and maintained by man" (3.2.2). Although methods of performing human engineering in detail design are not specified, they are well within the capability of the specialist. The provisions for dynamic simulation are unlikely to be exercised unless the system under development is very large and operator performance is highly critical to its operational success. It is also unlikely that all or even most of the factors to be considered in work environment and facilities design would in fact be considered in any single development project, but system design is not complete unless it includes the applicable work environment and facilities factors.

There is also a provision that "Performance design specifications, prepared by the contractor, shall reflect applicable human engineering criteria of MIL-STD 1472 . . ." (3.2.2.4), but the wording of the statement makes it largely a design goal.

Particular attention should be paid to the wording of 3.2.2.2 (Equipment Detail Design Drawings) which specifies that "the approval of these drawings by the contractor shall verify that human engineering requirements are incorporated therein and that the design complies with applicable criteria of MIL-STD 1472. . . ." It is the author's opinion that this statement has no binding force and that many drawings are approved by contractors with minimal or no human engineering review. Only design review and special sign-off provisions can ensure human engineering effectiveness.

Equipment Procedure Development (3.2.3)

"The contractor shall apply human engineering principles and criteria to the development of procedures . . . to assure that . . . human functions and tasks . . . are organized and sequenced for efficiency, safety and reliability and that these are appropriately reflected in the development of training and technical publications."

It is our impression that not much H activity is performed on procedure *development*, primarily because the procedure is usually derived more or less directly from equipment design. However, if we can believe the respondents to Kraft's 1969 survey (Kraft, 1969), 55.8% of the industrial H groups responding frequently helped to develop operator procedures. To the extent that the specialist participates actively in design decisions, he will have a corresponding influence on procedure development. It is more likely, however, that H influence on procedure development results from review of the completed procedure to ensure that personnel will be able to perform the procedure within specified requirements or that error-predisposing conditions are eliminated. The absence of quantitative data with which to predict the adequacy of personnel performance from procedures is a severe limitation on any formal, sophisticated procedural analysis. Hence the function described is primarily one of *review* and review that is highly subjective at that.

In view of the large number of procedures which are developed for even moderately sized systems, procedure review, like task analysis, should concentrate on critical procedures only. As in the case of detailed equipment drawings, contractor approval of a procedure does not necessarily mean that any human engineering review of the procedure has taken place. Before the customer accepts this statement, he had better determine exactly what the H staff did with the procedure.

Human Engineering in Test and Evaluation (3.2.4). The specification requires that engineering development tests

> "shall include tests demonstrating that man-equipment combinations can accomplish the intended operation, control, maintenance functions in accordance with system objectives . . . The objective of human engineering participation in major development tests shall be to verify compliance with the system requirements. . . ."

The system test plan will also include a human engineering design verification program.

When major development tests are performed (these are the quasi-operational tests performed with prototype hardware), human engineering participation will involve, where appropriate, the following elements: simulation of a mission or work cycle, tests in which human performance is critical, a representative sample of noncritical maintenance tasks, proposed job aids, utilization of personnel representative of the user population, collection of task performance data, and identification of discrepancies between required and obtained task performance. Equipment failures will be differentiated between those resulting from equipment factors alone and HIF (3.2.4.3). This differentiation has been discussed in previous chapters.

Although this program does not explicitly require a human engineering test plan (see Chapter Six) prior to performing test and evaluation, it is clear that no effective tests can be performed without such a plan. The Navy Appendix does however emphasize the development of such a human engineering plan for design verification.

Several points must be made about the test and evaluation section (3.2.4). Because of their training in experimental procedures, the methodology necessary to perform required tests is generally available to H specialists, although there are, as was pointed out in Chapter Four, technical problems in determining the criteria and measures to be used and in securing sufficient control over the test situation in order to secure sufficient data samples and unambiguous results.

The objective of H testing is to compare human performance with that required by the system. From this follows a more significant difficulty: that system performance requirements specified in the contractor's SOW usually do not include human performance requirements described in time and error terms. System requirements may be phrased in terms of reliability, in terms of a maximum maintenance duration, or in terms of a probability of mission completion; but these apply primarily to gross system or subsystem functions. To determine therefore whether personnel can perform at the detailed function or task level to satisfy these system requirements presents problems.

When procedures have time or error limits associated with them, it is relatively easy to determine whether these time/error criteria have been exceeded, hence whether personnel do or do not meet requirements. When these time/error criteria do not exist (which is unfortunately a great deal of the time), the specialist may be constrained to describing the kinds of errors that occur and the design/procedural factors which should be modified to eliminate or prevent these errors. As a consequence, much of the data resulting from tests turns out to be of the critical incident type (e.g., error incidents illustrating a design weakness which should be fixed). Here the difficulty is generated not by an H deficiency, but the reluctance or inability of the procuring agency to establish meaningful human performance requirements.

The customer's project manager can avoid this problem by requiring the contractor to develop quantitative human performance requirements which can then be imposed on the contract.

Cognizance and Coordination (3.2.5)

"The human engineering program shall be coordinated with maintainability, reliability and other related programs . . . (and) shall be conducted under the direct cognizance of human engineering specialists."

Methods of accomplishing the desired coordination are not specified and are largely up to the H group itself. In any event, the requirement is essentially only a goal and has whatever direct impact on the H group that engineering management and the H group's own inclination desire it to have. Obviously there are aspects of H work which interact with maintainability design (e.g., access spaces, coding of cables, maintenance control panels, etc.) and with reliability (the prediction of human performance as it affects over-all system reliability predictions), but the extent to which H specialists will interact with reliability/maintainability specialists depends on the interest each has in the other's work.

It is noteworthy also that the specification requires the human engineering program to be performed under the direct cognizance of human engineering specialists. As was pointed out earlier, there has been a tendency on the part of some contractors without an H staff to attempt to perform H activities with unqualified personnel or to subordinate the H group too directly to engineering. The requirement above is intended to prevent these abuses.

The Navy Appendix is concluded with a section on human engineering progress reports, although the specification proper does not contain a similar section.

The specification concludes with a set of notes describing the intended use of the specification and an explanation of terms, including "critical," over-all layout drawings, panel layout drawings, system and task analysis. The notes are of course not contractually binding and would not be called out in the contract.

SUMMARY

How can we sum up MIL-H-46855? Obviously it is necessary to have an H specification. Primarily it serves to guard the customer's interests, but it also serves as a charter for the H group. It is sufficiently general that a contractor who does not wish to comply or who wishes to comply only minimally can do so largely with impunity. Unless concrete H outputs are specified (and paid for) as contractually binding, 46855 has little impact.

From a more theoretical standpoint it tends to reflect some severe methodological shortcomings in the discipline, particularly in terms of analytic methods and quantitative data. In summary, the specification is broad enough so that the customer, the contractor, and the H group can make of it whatever they mutually want it to be.

MIL-STD 1472. The purpose of MIL-STD 1472 is to *control* the human engineering characteristics of an equipment by prescribing what those characteristics should consist of. Whereas 46855 describes process, 1472 describes content. Indirectly the standard also serves as a reference handbook for both the specialist and the design engineer, because it contains a great deal of information needed to design human engineering into equipment.

Because the standard discusses many types of equipment and many aspects of those equipments, it is quite lengthy. This makes it impossible to cite its major provisions in this section. Although it would be desirable to describe each statement in 1472, doing so would result in an Appendix almost as long as the rest of the book. We must therefore content ourselves with commenting on the standard's general characteristics.

1472 is divided into five sections, of which only the fifth is of any significance, the preceding ones being devoted to an introductory scope and purpose (section 1), referenced documents (section 2), definitions (section 3), and general requirements (section 4), the last describing the general goals of human engineering design.

The fifth section, which contains about 90% of the approximately 150 pages in the document, presents detailed requirements under the following headings: control-display integration; visual displays; auditory displays; controls; labeling; anthropometrics; ground workspace design requirements (e.g., console arrangement, ladders, etc.); environment; design for maintainability; design of equipment for remote handling (e.g., manipulators); small systems and equipment (the Army's contribution: backpacks, tracking and optical instruments); operational and maintenance ground vehicles (e.g., tanks and trucks), hazards and safety; man transportability (more Army: weights to be lifted); aerospace vehicle compartment design requirements (anthropometry applied to cockpits).

As with any equipment standard, certain questions should be asked about 1472:

1. Does it contain those requirements needed for the customer and the contractor to exercise the desired control over design?

2. Does it contain that information which will enable the specialist and the engineer to do a better job of designing equipment to satisfy human engineering requirements?

The answer to the first question—does 1472 contain all the information needed to design human engineering into equipment—is yes. It would be difficult to find a relevant topic which has not been included.

There are, however, problems with making the standard effective in controlling design so that the answer to the question—does 1472 serve to control design from a human engineering standpoint—is a qualified no. That is to say, the standard will not be effective unless the customer (i.e., his project manager) maintains continuing surveillance over the contractor's activity.

There are several reasons for this:

1. The standard emphasizes molecular human engineering details but says little about major design concepts which are crucial to design, such as the effect of automation on human performance. The reason is that, as was pointed out previously, we know very little about basic design concepts like automation and about how they affect human performance.

As a result, the provisions dealing with general design concepts tend to be statements of principle, rather than fact. For example,

"4.6. *Simplicity of Design.* The equipment shall represent the simplest design consistent with functional requirements and expected service conditions. To the maximum extent possible, it shall be capable of operation, maintenance and repair by personnel with a minimum of training."

The result of this generality is that the provision cannot be used effectively either as a guide for the design engineer or as a means of exercising some control over his design. How, for example, do we decide whether or not the contractor was in violation of the simplicity provision?

2. Another problem is that many of the statements are phrased in purely qualitative terms, for example, "5.1.1.4. *Precision*—The precision required of control manipulation shall be consistent with the precision required by the display." Such a statement is, when analyzed logically, meaningless, because neither display nor control precision has been defined.

From the standpoint of the contractor, any human engineering standard which is described in qualitative terms *only* can be safely ignored unless it would otherwise be implemented as part of normal "good design." Any human engineering standard which has a qualifying phrase such as "generally," "where applicable," "may be," etc, can also be safely ignored because the qualifying phrase makes the implementation of the standard dependent on the engineer's definition of applicability.

Understandably, any statement which is intended to apply to a great variety of equipment types must be qualified to avoid its becoming overly rigid. The fact is, however, that such qualifiers sometimes become loopholes for an unscrupulous contractor to creep through.

Lest the reader suppose that the standard consists of nothing but generalities and abstractions, it must be emphasized that there are quantitative statements for many other human engineering aspects, such as "For critical functions indicators shall be located within 15 degrees of the operator's normal line of sight" (5.2.2.1.8). Indeed, there are many statements in 1472 of comparable specificity, but they are largely based on anthropometrics. Lack of specificity and nonquantitative statements in other areas of 1472 reflect the fact that the discipline lacks all the data it requires, even for well-researched aspects.

In summary, what can be said about 1472? From the government's standpoint it leaves much to be desired since, except for those provisions involving mostly anthropometric factors, it cannot be used to control fully the human engineering characteristics of design. From the contractor's standpoint it has little or no effect except on molecular design features. The H group, which would like to be able to use the standard as a club (among other things) to ensure adequate consideration of human engineering in design, finds it more of a willow wand.

REFERENCES

Department of Defense, MIL-H-46855, 1968. *Military Specification, Human Engineering for Military Systems, Equipment, and Facilities.*

Department of Defense, MIL-STD 1472, 1969. *Military Standard, Human Engineering Design Criteria for Military Systems, Equipment, and Facilities.*

Kraft, J. A., 1969. *Human Factors and Biotechnology—A Status Survey for 1968–1969,* Report LMSC-687154, Lockheed Missiles & Space Company, Sunnyvale, California, April.

Leuba, H. R., and S. R. Wallace, 1965. *Human Factors and System Effectiveness,* paper presented at the 9th Annual Meeting, Human Factors Society and EIA System Effectiveness Conference, both in October.

Meister, D., and D. E. Farr, 1966. *The Methodology of Control Panel Design,* Report AMRL-TR-66-28, Aerospace Medical Research Laboratories, Wright-Patterson AFB, Ohio, September.

Rabideau, G. F., et al., 1961. *A Guide to the Use of Function and Task Analysis as a Weapon System Development Tool,* Report NB-60-161, Northrup Corp., Hawthorne, California, January.

United States Air Force, 1963. AFSCM 80-3, *Handbook of Instructions for Aerospace Personnel Subsystem Designers (HIAPSD),* Air Force Systems Command, Andrews AFB, Maryland, edition of 15 January.

United States Air Force, 1964. AFSCM 375-5, *Systems Engineering Management Procedures,* Air Force Systems Command, Andrews AFB, Maryland, edition of 14 December.

United States Air Force, 1966. AFSCM 310-1D/AFLCM 310-1D, *Management of Contractor Data and Reports.* Air Force Systems Command, Andrews AFB, Maryland, edition of 3 January.

X

THE STATUS OF *H* THEORY, PRACTICE, AND RESEARCH

A ROUNDTABLE BY MAIL

The preceding chapters have directed the reader's attention primarily to details of H methodology and practice. It is just as important for the reader to assume a larger perspective. Therefore we will conclude our examination of the human factors discipline by attempting to summarize its present status and to anticipate where the discipline is heading during the decade of the 1970's.

Any effort to make such a summary statement is bound to be tentative and conjectural. Because it would be presumptuous for any one individual to perform this task, the author hit on the device of asking some extremely well-qualified H specialists to do this for him. Not only will the reader be exposed to other points of view, some of which differ markedly from those of the author, but these opinions highlight, as the previous chapters may not have, the strong and weak areas of human factors. It also permits the author to comment on the differing points of view and, quite frankly, to use them as a sounding board for his own ideas.

The material in this chapter is taken largely from an unpublished manuscript (Meister, 1970). Twelve researchers and specialists of recognized stature were asked to answer 10 questions dealing with what the author considered to be the major issues in Human Factors.[1] Because the

[1] In the development of these questions he had the assistance and encouragement of Dr. Julien M. Christensen, Director of the Human Engineering Division of the Aerospace Medical Research Laboratory, for which he is particularly indebted. Needless to say, the opinions expressed by the respondents are their own and do not represent the organizations with which they are affiliated. The author wishes to ex-

answers, when compiled, formed a rather lengthy manuscript, only the most pertinent parts of the answers were extracted for this chapter. However, the author does not believe that in editing this material he has in any way distorted the participants' responses. To this material he has added his own commentary.

Although an effort was made to select people from the four major areas in which human factors specialists work (government, industry, consulting, and teaching), those whose comments are included in this chapter cannot be considered a completely "representative" sample of the human factors discipline. Although the phrase lacks felicity and they would probably reject the appellation indignantly, they are all what has been termed "elder statesmen," having practiced their profession for many years; their contributions to human factors research, teaching, writing, and practice are familiar to many workers in the discipline.

Among them we count two in high-level governmental positions (Christensen and Regan); a third (Orlansky) works for a nonprofit organization which is closely tied to the government); three hold high rank in industry (Wissel, Kraft,[2] and Bauer); four (Van Cott, Fucigna, Fogel, and Woodson) perform major research and consulting functions; and two (Lyman and Weldon) are eminent academicians. Included are three past presidents of the Human Factors Society (Christensen, Kraft and Lyman); the co-author of probably the best known H handbook in the country (Woodson); another author of a basic H text (Fogel); the heads of two major governmental Human Factors laboratories (Christensen and Regan), etc., etc. In other words, if the H discipline contains an "establishment," these people represent that establishment. If any group of people can be said to reflect what H is at the present time (and what it may develop into in the future), it can be said of this group.

The titles and affiliations of each respondent are supplied at the end of this section, and more detailed biographical sketches are appended to the chapter.

Participants

Mr. Harold Bauer, formerly Head, Human Factors Engineering, McDonnell Douglas Corporation, Long Beach, California.

Dr. Julien M. Christensen, Director, Human Engineering Division, Aerospace Medical Research Laboratories, Wright-Patterson AFB, Ohio.

Dr. Lawrence J. Fogel, President, Decision Science, Inc., San Diego, California.

press his appreciation to all the respondents for the effort involved in responding to these questions.

[2] This was written before the untimely death of Dr. Kraft.

Mr. Joseph T. Fucigna, Executive Vice President, Dunlap & Associates, Inc., Darien, Connecticut.

Dr. Jack Kraft, formerly Assistant Division Manager, Biotechnology, Lockheed Missiles & Space Company, Sunnyvale, California.

Dr. John Lyman, Professor of Engineering and Psychology, Director, Biotechnology Laboratory, University of California at Los Angeles, California.

Dr. Jesse Orlansky, Institute for Defense Analyses, Arlington, Virginia.

Dr. James J. Regan, Chief Psychologist, Naval Training Device Center, Orlando, Florida.

Dr. Harold P. Van Cott, Director, Office of Communication Management and Development, American Psychological Association, Washington, D.C.

Dr. Roger J. Weldon, Professor of Systems Engineering, University of Arizona, Tucson, Arizona.

Dr. Joseph W. Wissel, Senior Staff Engineer, Human Engineering and Maintainability Staff, Lockheed Missiles & Space Company, Sunnyvale, California.

Mr. Wesley E. Woodson, President, Man Factors, Inc., San Diego, California.

QUESTIONS ON THE STATUS OF HUMAN FACTORS THEORY, PRACTICE, AND RESEARCH

Assuming that the term "human factors research" covers both general/basic and applied work, performed in the laboratory and in the operational environment, and including human engineering, training, personnel, etc., aspects:

1. How effective has government sponsorship of human factors research been? Has the government gotten its money's worth? Has government support of this research helped or hindered the solution of human factors problems?
2. Can you list one or more areas of research by subject matter or topic which you would say deserved the highest priority in human factors research today? (Do not list your current research project.) Why do you consider these should have top priority?
3. How successful has human factors research been in supplying the data human engineers working on defense projects need? Can you give any reasons for your point of view?
4. What theoretical or methodological deficiencies can you see in human factors research as it is performed today and why? Are the

methodological tools human factors research employs sufficiently precise or sophisticated or valid to perform the needed research? to supply us with the requisite data? If not, what do these tools lack?

5. Is human factors research performed in the laboratory being adequately applied to actual system development and/or operational problems? If not, why not? What are the problems in applying that research?
6. Is human factors research today attacking correct, that is, meaningful questions? What would you consider to be meaningful questions, and why?
7. Are we making proper use of the real world, that is, operational environment as a source of problems to solve? hypotheses to test? as a data source? If not, why aren't we, and what should be done?
8. Engineering involves three-dimensional physical parameters, human factors deals with multidimensional behavioral parameters. What kind of research should be performed to develop equations (mathematical or otherwise) which will translate the behavioral into their physical equivalents? Can you supply an example of research which was highly successful in providing these translation equations?
9. What is the influence of their psychological training on the performance of the people who perform human factors research? in terms of applying that research to actual engineering development problems? Is the influence of their training positive or negative for achieving human factors research goals—whatever these are?
10. Should human factors be considered solely as a technology which supports engineering technology, or should it have any goals beyond this? Specifically, should the goal of human factors research be to answer specific questions raised by engineering problems; or should it be to develop principles describing how personnel perform in a man-machine context?

The Answers

Question 1. How effective has government sponsorship of Human Factors research been? Has the government gotten its money's worth? Has government support of this research helped or hindered the solution of human factors problems?

Hardly anyone would disagree that without government sponsorship the H discipline would not exist, at least in its present form. Certainly nondefense industry has never made any extensive use of human factors technology. The dependence on government, however, raises the question whether the discipline is viable in a completely commercial environment.

Orlansky indicates that

"the government has been virtually the sole sponsor of all human factors research in the United States since World War II. Some exceptions to this remark may be found in research on vehicle control and display supported by automobile companies, a few doctoral dissertations, and some efforts of industrial designers. The government has supported both basic and applied research but not systematically towards the development of a broad base of knowledge. Rather, this support has reflected the dominant military programs of the moment."

This is backed up by *Fucigna*, "Without the government, I do not believe the Human Factors profession would be where it is today," and *Lyman* agrees, "Government sponsorship of human factors research has been the only thing that has made the field possible. Industry has been singularly slow in picking up human factors areas. . . ."

Kraft points out what happens when government support is restricted:

". . . Consider what has been happening during the past several years when USAF human factors research funding has been curtailed and NASA funding is being drastically reduced. Within industry, the effect of such action means the surplusing of the specialized researcher for whom in-house money cannot be secured and who, by virtue of his research orientation, is unable or unwilling to perform the mundane jobs of providing human factors support for company projects or conducting the "quick and dirty" types of applied investigations which frequently have to be done. Where do these people go when they leave industry? If government agencies are not hiring and if research based academic programs are becoming moribund, these people may leave the field completely."

While government support has been indispensable, there are some reservations about the worth of the resultant research. Although *Van Cott* believes there has been no more waste in Human Factors than in other system areas, the application of human factors research to system design has been less than optimal:

"One measure of the impact of research on system performance might be obtained by an examination of the research literature of the human factors field. I tried to obtain such a measure by examining Volumes 7 to 9 (1965–67) of the *Human Factors Journal,* a representative time sample of research in the field. I sorted the articles in this sample into four categories: (a) professional affairs, (b) new knowledge or methods not in a form suitable for direct application to systems, (c) theory, also not directly applicable to system design, and (d) information or principles directly applicable to one or a family of systems. The percent of articles falling into each category were: (a) 25%, (b) 37%, (c) 13% and (d) 25%. Crude as this approach is, one could conclude that only 25% of the research in our field has any chance of ever being applied to the design of improved man-machine systems.

"The next question to ask is, 'Of what is potentially applicable, how much was actually put into practice vs. how much was simply a research by-product that surfaced into the literature, but was not realized in human factors design?' The answer is likely to be discouraging: probably less than half or about 12% of the information and principles that the discipline produces is likely to see the light of day in terms of actual implemented applications.

"Looking at the problem in this "operational light," one could conclude that the government hasn't gotten its money's worth in terms of the dollar-for-dollar effectiveness of its sponsorship of research that really paid off in improved systems."

Woodson tends to agree and supplies several reasons:

"I frankly don't think government sponsorship of human factors research has been cost-effective. Several reasons for this are: (a) there is no over-all planning and coordination to establish objectives against needs and prevent overlapping and omissions, (b) there is too much of the small package purchase approach in order to skimp on cost and to keep a study within a 12-month time constraint, (c) there are too many people responsible for disbursement of research funds who either are not experienced or are biased by their limited interest and experience, and (d) too often human factors research proposals are judged and/or approved, not by human factors personnel in parceling out the budget, but by physicists, engineers, biologists, physicians and the like."

On the other hand *Christensen* disagrees:

"Government sponsorship of human factors research generally has been quite effective. This is a relatively new field and in any pioneering field one is bound to enter a few cul de sacs. Positive contributions have been made not only in areas such as workplace design and layout, including the design of individual instruments and controls and control systems, but also at the advanced development or conceptual stage. For example, probably very few people know that the basic idea for the lateral firing concept originated with Ralph Flexman when he was in AMRL on two weeks active duty.

"Lieutenant Colonel Simons, with the support of other in-house human engineers, was able to take this idea and advance it to the point where it has been adjudged the single most effective weapon in the Southeast Asian conflict. Government support has definitely helped; I doubt that there would be a national human factors program today were it not for government support."

Kraft feels that relatively speaking the government has gotten its money's worth: "The government and the research community has gained many more times than they lost by virtue of government spending for research." However, the gain may be related to support of particularly promising H̲ specialists:

"In some cases, the gain may only be in that qualified people received government support for their efforts. These people probably would not have been employed by industry in the absence of such support and quite likely would not have entered or remained in the human factors field. In other cases, the gain from government research has been in the area of facility development. Laboratories have been established which probably would not have come into being and these laboratories subsequently produced good research."

As to suggestions concerning means of improving the effectiveness of government support of human factors research, *Regan* has an interesting concept with which the author tends to agree:

"As to the further matter of *effective* support—i.e. could the human factors R&D money have been better spent? I'm sure it could have been. How? Principally by better statements of the problems and more effective review of the output from both government labs and industry. A widespread practice of forcing human factors efforts (if not at the outset at least within the past decade) to "pay their own way" would have been and still would be the key. By this I mean if, instead of calling on human factors activity solely to meet some general or particular specification, it were invoked because some measurable improvement in the product resulted, a better human factors program would evolve. For one thing, more competent people would have been required. The problem selection would have been more precise and appropriate. When particular R&D efforts failed to improve the system they would have been discontinued. Finally and importantly, once provided, the human factors R&D would be *used* because it would be relevant and the managers and users would clearly see in what way."

Although it is generally acknowledged that H would not exist without government support, the participants may not have considered the price that was paid—that is being paid—that the discipline, because of its dependence, becomes a prisoner of governmental attitude and interest, not only in its financial but more importantly in its research aspect. As was pointed out in Chapter Eight, the governmental agency influences (to some unspecified degree) the research topics and methods to be followed by the discipline. To its great credit, this benevolent sway has been in many, perhaps the overwhelming majority of instances, highly productive; but the condition of dependency always has the potential that it may operate to the disadvantage of the dependent discipline.

H dependence on government support, coupled with the indifference manifested by commercial industry, is vaguely frightening. Every discipline requires continuity of research effort if it is to advance [3] and that continuity is threatened by the cyclical perturbations and inadequacies of government financing. The situation would not be so threatening if the necessary continuity were maintained by specialists working in the shelter of universities. Although there are a few such scholars (Lyman comes to mind immediately), the work being performed by these people is largely addressed to problems of applied experimental psychology [4] rather than to the types of human factors problems described in Chapter Four. It seems to the author that for the discipline to survive as more than a temporary phenomenon of the 20th century it needs to develop a base of support in addition to that of the government, and, since commercial industry will apparently not supply that base, attention should perhaps be directed to the university. (Of course, the question arises whether a human factors discipline based largely on the university, and divorced from the system development problems that initiate research, would in fact remain H as we know it.) To make H into a distinctive and acceptable academic discipline, however, it must be able to convince other academic disciplines that its goals and methods extend beyond the support of engineering (see question 10 for a further discussion of this point).

***Question* 2.** Can you list one or more areas of research by subject matter or topic which you would say deserved the highest priority in human factors research today? (Do not list your current

[3] I assume that one of the goals of human factors is to study in a purely scientific sense the functioning of people in a technological environment.

[4] Applied experimental psychology studies the human as the machine impinges upon him. Human factors studies the interaction of the human with the machine and the effect of the human on the technological system as well as the effect of that system on the man. See the discussion also in Chapter Four.

research project.) Why do you consider these should have top priority?

The reader should compare the responses to this question with the human factors research program suggested in Chapter Four. Although some of the same research problems are noted as deserving priority, there is, as we might expect, much greater variability in the topics suggested by respondents; for example, *Orlansky* feels that

". . . primary support should be given to the development of command and control systems, including their supporting communications networks. This area deserves the highest priority because of its overwhelming importance for the effective utilization of our military forces and our political decision making at the highest levels. This area, so important for strategic and tactical reasons, is one in which the requisite basis for significant improvement exists at the present time. In general, R&D efforts should be directed towards the development of theory test and measurement in command and control systems rather than towards design."

Fucigna has a rather interesting suggestion:

"A fruitful area of research (would be) . . . what has and can the H profession contribute in *measurable terms?* Until this question is answered . . . we will continue to be confronted with an uphill fight in justifying our existence. Certainly members of the profession perform useful functions in the system development context—analyze jobs, design equipment and facilities, establish manning and training requirements, test and evaluate the man-machine interface. However, until we convince system engineers and management that members of the H profession have something special to contribute—convince them in terms they understand and can accept—they will continue to regard our efforts as superfluous and something each of them can do while performing their primary functions. Strong words and by no means applicable to all organizations or systems but a situation prevalent enough to warrant serious consideration by the profession as a whole.

"Such a situation suggests more specific areas of research. There is a need for a meaningful data base on human performance capabilities. There is a need to model and quantify the human contribution to system performance. What are the consequences in cost/effectiveness terms, of not adhering to accepted H practices?"

A more traditional area, decision making, is noted as particularly important by *Fogel, Christensen, Bauer,* and *Weldon. Van Cott's* list includes the following:

"(a) Development of a model of human performance which brings together separate models of human signal detection, decision-making, information processing, and motor and verbal response.

(b) Verification of accepted human factors principles that have been inadequately researched: e.g., principles for display grouping, principles for work-place layout.

(c) Research on the effects of changing motivational states on human performance."

Van Cott believes that

"the following technique development areas are weak and need further development:

"(a) Fast time simulation of human performance.

(b) Real time simulation of human performance in unusual environments.

(c) Standardized methods and instrumentation for human performance measurement.

(d) Prediction of field performance from laboratory simulations.

(e) Improved methods for storing, indexing, and retrieving human engineering data in support of design decisions."

Woodson's priorities include

"(a) dynamic anthropometry studies directed toward collecting data pertinent to the active operator and specialized populations, (b) development of sensitive, valid and reliable measures of human performance, which can be used in operational settings as well as the laboratory, (c) develop a catalog of task element performance time estimates, and (d) expand the effort in basic studies of the effects of stress on human behavior, especially combined effects.

"These are the areas where the same questions appear over and over, and there aren't any good answers . . . we keep extrapolating and experting."

Christensen emphasizes manufacturing processes:

"I am surprised that industry has not tried to exploit human factors information more in the area of manufacturing processes. Most of the improvements in manufacturing that have been made over the past 50 years have been made by industrial engineers or others in the engineering profession. The standard of living that we enjoy today attests to the success that they achieved. However, much more should be done in areas such as individual differences, fulfillment of the aspirations of the individual worker, development of better tools, development of more effective work stations, etc."

In addition,

"the entire broad area of non-defense products deserves more attention . . . More of what has been learned here should be applied to other complex systems such as the transportation system, to say nothing of the improvements that could be made in the everyday items that we use in our offices and in our homes. Fitts was among the first to envision a society in which the fulfillment of every individual is enhanced not only during his leisure time but also during his working hours. I think, further, that we are in danger of losing our individuality because of the restrictions that a modern engineered society imposes on us. This idea was expressed in an article entitled 'Individuals and Us' which appeared in the *Human Factors Journal,* February 1966."

Regan feels that "The key problem is quantifying human performance. This lack underlies almost all of the problems of assessing the consequences of human factors activity and of evaluating design and employment alternatives." *Kraft* has a shopping list of concepts which overlap in part the previous suggestions:

"1. Human error production: its causes and control.
2. Determining validity of man-in-the-loop simulation training to reduce expense and hazards of in-flight aircraft training.
3. Nonfatal or damaging riot and law enforcement control techniques.
4. Valid measurement of human performance in systems operations.
5. Prediction of human performance in new systems.
6. Application of computer and aerospace technology to teaching.
7. Highway accident prevention and mitigation of injury through application of human factors research to transportation and highway systems.
8. Application of human factors principles and techniques to consumer product development and to industrial work processes."

Bauer's list of six items includes a number which we have heard of before (information processing, decision making, instructional methodologies, command and control) but also several which might be considered to transcend traditional human factors research topics: civic action of emergent cultures and oceanographic ergonomics. *Wissel's* suggestions have the flavor of the capability research referred to in Chapter Four, although centered primarily on physical abilities:

"There are many answers about human capabilities that we do not have. How much can the average man lift, push, pull or manipulate in a variety of positions? How long can he do this? When can we expect him to exceed the normal and approach his physiological limits? What are the impairments in judgment under stress or emergency conditions? The highest priority in human factors research should be assigned to a complete description of human physical capabilities. A complete description of mental capabilities would be a close second."

Weldon's topics are closely related to traditional psychological preoccupations: information-retrieval; parameters in efficient training; process of internal selection and judgment.

We get the impression that what *Christensen* calls the traditional human factors emphasis on the input-output area (controls and displays) is yielding to a preoccupation with higher mental processes, motivation, stress, the need for inquiry into "large scale operations, macrosystems, involving such matters as social needs as well as individual needs for optimizing human beings" (*Lyman*). From that standpoint, if the research suggestions were to be followed up, human factors would become more, rather than less, psychologically oriented and certainly far more than an adjunct to the engineer or a means of assisting the development of new equipment.

At the same time the variance in the suggestions made is a bit disturbing. We find command/control, decision making, the necessity to justify human factors activity, modeling of human performance, verification of principles, motivation, simulation, learning/training, anthropometry, performance measures, information retrieval, cataloging of performance time estimates, manufacturing processes, highway research, and capabilities research.

All of these are important, of course, but are they equally important? Some of the items suggested (e.g., valid measurement) represent goals rather than research areas. The dispersion in research interests reflects different ways of viewing the same phenomena (which is not uncommon in behavioral science). It also suggests that human factors researchers have not developed a unified approach toward their discipline. Different research interests reflect differences in conceptualizing the importance of parameters affecting human performance. In other words, those who consider decision making as being of primary importance in relation to man-machine functioning will emphasize that over anthropometry or other parameters.

The question also arises whether H possesses the theoretical and methodological sophistication to deal with macrosystems and, in particular, social problems.

Question 3. How successful has human factors research been in supplying the data human engineers working on defense projects need? Can you give any reasons for your point of view?

There appears to be almost unanimous agreement that the application of human factors research to system development projects has been less than optimal, and in some cases rather poor. At best we can say, as does *Orlansky*, that "findings are generally relevant and useful for . . . projects, although not always timely and complete." *Christensen* feels that "while much additional information is needed, there is nevertheless an enormous amount available that is not being exploited." *Van Cott* feels that "The real problem is not in providing human engineering data to human engineers, but in providing it to design personnel in a form that they can and will use." This is backed up by *Bauer:* "Data are not usually in a form or format or concept that an engineer can either understand or utilize."

On the other hand, *Woodson* feels that human factors research has been

"not too successful in many cases. This is due to many different reasons such as: (a) not time enough to get valid answers by the time the questions arise, (b) not enough money or right equipment to do a good job of research, (c) people in management don't want valid answers, rather a publicity stunt to make it look good, (d) many researchers are stifled by their book methods and try to make their problem fit their method . . . with an inappropriate answer as the result, and (e) too many researchers themselves who don't want to tackle mundane questions . . . they can't be bothered with anything that won't put their names in the future psychology texts."

Lyman agrees that

"the success of human factors research in supplying data that human engineers need on defense projects has been somewhat spotty. Laboratory studies have been the principal source of data and, since the data needed is often situation-specific, the skill of the human factors engineer is as often as not more art than science. As the years have passed a larger and larger body of data has been built up, of course, and even a few principles have emerged. Unfortunately, much of the research that has been done for on-the-spot problem solutions has not been properly disseminated so that it can be used by others."

One of the difficulties is

"the inability of the research organization to respond quickly, say in one or two weeks. Engineering schedules in design and development activities do not allow the luxury of the three months to one year study. Perhaps it can be argued that the system is wrong and that problems should be anticipated in advance of design and development activity. This may be true, but in the realities of the present, it has not been established as (to) how this can be done. In some cases, absolute limits are being sought, e.g. how many aircraft can a controller keep track of; what is the absolute maximum load under the worst conditions. The answer often is "it depends" and the interpretation of the data is left to the design engineer." (*Wissel*)

"Much of the research that has been done has not been translated into terms that would admit of application. Again the problem is one of joining (translating) research finding and applications. Additionally, there is no *systematic procedure* for collecting from practitioners problems from which research issues can be derived." (*Regan*)

Kraft indicates that

"much of the needed data is not available in text or reference books, nor in technical reports because they are not addressed to the specific systems problems. Human factors research ofttimes does not provide a data bank, but it does suggest ways to resolve a problem—which must then be implemented by each human factors specialist dealing with an aspect of a specific system development problem."

Many H specialists attempt to explain the deficiencies in the applicability of human factors data in terms of the general reluctance of engineers to use the data:

"Probably the reason for this is that the leaders in industry have yet to be convinced that it is to their advantage to consider human factors data in the design and sale of their products. Fortunately, there are a few exceptions to this that I happen to know of in the agricultural implement industry, the automotive industry, the airlines industry and, to some extent, the computer industry." (*Christensen*)

There is a feeling also that "many problems which arise in a development program are system-specific" (*Fucigna*) for which research on general principles may not be satisfactory.

If we summarize the reasons adduced why human factors research fails to satisfy system development needs as completely as it should, the list looks like this (not necessarily in order of importance):

1. Improper format for presenting findings to engineers.
2. Insufficient time to perform needed research.
3. System development problems are too system-specific for research.
4. Reluctance of engineers and management to use available data.
5. Inability of research organizations to respond quickly enough to development problems.
6. Inability to translate research findings into practical solutions of problems.
7. Failure to disseminate information properly.

All of these reasons are valid in part, as is the contrary statement that a great deal of data is available but not being used. A basic consideration that was not expressed, however, is whether the human factors researcher is addressing himself to the significant behavioral problems of system development (if he does not, his research will be irrelevant and unsatisfactory), to which the inevitable reply is, what significant problems? (The author's ideas were presented in Chapter Four.)

It is important to note that the point was not raised by the panel. We might therefore infer that they are satisfied that research is attacking appropriate development problems. But is this actually the case?

If we examine the priority research areas indicated by the respondents to the previous question as representing problems of greatest importance, it is questionable whether most of these would actually satisfy the requirements of the engineer and specialist for design-relevant data. Of course, we could rebut this by asserting that not all human factors research should be directed at solving system development needs. Although this may be true,[5] the most immediate and pressing problem facing human factors research is the satisfaction of system development needs.

Question 4. What theoretical or methodological deficiencies can you see in human factors research as it is performed today and why? Are the methodological tools human factors research employs sufficiently precise or sophisticated or valid to perform the needed research? to supply us with the requisite data? If not, what do these tools lack?

Although some H specialists are satisfied with their methodological tools, at least half of our panel is disturbed about their adequacy. Those who are relatively satisfied see difficulties in the validation of theory, the absence of a comprehensive all-encompassing theory "to tie together the

[5] This assumes that the purposes of the human factors discipline extend beyond assistance to engineering and solution of system development problems. See the answers to question 10 for further comment on this point.

numerous facts available in the study of human factors" (*Weldon*), or in the application of human factors methods. A characteristic comment is that of *Fucigna:*

"The methodological tools for conducting human factors *research* are sufficiently precise, sophisticated and valid. The deficiencies lie in the application of the research data to system development efforts. Frankly, applied human factors is more of an art than a science.[6] This is not to say that the results are necessarily invalid, any more than one would say the art involved in practicing medicine is invalid. What is lacking may not be feasible to develop or for that matter appropriate, i.e., a precise tool for dealing with an imprecise element—the human element."

Fogel has "no quarrel with the tools used. Hopefully the more sophisticated statistical techniques are brought to bear as warranted." [7] Similarly *Bauer* approves of the methodology but "what is lacking is the ability to translate the results into meaningful practical terms of a usable form." He sees the primary methodological deficiency as one of "lack of utilization of the operational environment."

Orlansky feels that

"the primary limitation in current human factors research is how to deal with the problem of validity.[8] This limitation is due less to ignorance than to the lack of opportunity for adequate validation. A significant exception may be found in the general, but not yet complete, validation of human control dynamics in an aircraft simulator with actual flight. However, such validation is lacking, for example, in many aspects of manned space flight, and in large command and control systems. Opportunities for validation are limited more because of cost, complexity and, in some cases, lack of participation by key senior personnel than because of conceptual problems. One way of dealing with this well recognized problem

[6] This comment (H as art more than science) is one which the author has heard a number of times. Frankly, it sounds more like an apology than an explanation. The problem is that as long as it is accepted as a reasonable explanation, specialists will have little incentive to improve their discipline.

[7] The concept of statistical analysis as the solution to methodological problems ignores the fact that the problem must first be identified and then studied before statistics can be brought to bear on it. This kind of answer is often a cliche solution to serious problems. (Fogel indicates, however, that he was taking exception to the lack of statistical analysis which has often characterized human factors studies. He says he too is concerned "with structuring the problem . . . prior to the acquisition of data and the application of statistical interpretaton.")

[8] Validation-testing of the hypothesis in the operational environment. There is remarkably little of this throughout the discipline.

would be to develop a series of simulations which increase in their reality in order to understand the extent to which precise duplication of equipment and personnel is required for studies of validity. It is often proposed that models of man are required in order to undertake more adequate human factors research. This proposition is easy to make and difficult (I should say premature) to achieve in the near future.

"Limitations also exist, at present, in our ability to extrapolate from short term experiments to long term effects, e.g., extended space flight, continuous operation of command and control centers during crises and so forth. Another limitation, noted above, is the extent to which studies with lower echelon personnel, apply to senior personnel, such as college students for colonels, and colonels for generals."

Wissel says:

"To serve the demands of the applications human engineer, human factors research must be able to provide answers in a hurry or be able to anticipate the requirements of development and design . . . The tools of human factors research appear to be as sophisticated as we need at the moment. It may even be that our tools are too sophisticated. Very often we may need information on what the operator can accomplish in terms of pounds of force rather than being concerned about the significance of difference at the 1% level of ounces of force. Maybe we should be less concerned with the increase of vigilance resulting from adding the nth operator in the detection system than we should be with designing the equipment so we can be sure that the first operator will look at it more than ten percent of the time in the first place. With computers to assist in data analysis it should be possible to examine many more parameters in a much shorter time. I think we have the tools and techniques."

Christensen believes firmly that

"we generally still are unable to cast our data in such form that the engineer can use it sensibly to make design tradeoffs between human factors considerations and other considerations. Our scales need improvement; a ratio scale is a rarity in the human factors area and yet such scales are needed to make meaningful analyses of the relationships between human factors variables and the other variables that a design engineer must consider."

As far as theory is concerned, *Van Cott* feels that

"the biggest . . . problem is a lack of comprehensive theory and model of human performance that is sufficiently generalizable to be applicable to field performance, as well as laboratory situations, and which takes into account changing levels of performance as a result of learning and motivational states. In the area of methodology, the most glaring deficiency is with respect to the lack of standardized performance measurement batteries and measurement methods which can be used to assess human performance under a variety of environmental and task situations."

Woodson is hopeful:

"They (methods) are getting better all the time, I think. The biggest problem is providing money to develop methodology and keeping at it until we do meet our objectives. It's like the problem regarding performance measurement techniques, no one wants to pay for developing tools. It takes years to investigate the underlying relationships and causes of cancer, but we keep steadily at it. Not so with human performance measurement techniques and methods. We make a minor stab one year and then drop it for lack of funds or interest. Industry can't justify it, the government is tied up in rules and red tape, and the universities are too busy demonstrating."

A more balanced point of view is expressed by *Regan:*

"There is little in the way of theory which motivates human factors research. It tends to be pragmatic and problem oriented. In fact, there are those who might contend that the terms human factors and research are incompatible. However, human factors research as I know it seems to be characterized essentially by the same strengths and weaknesses, assets and liabilities that characterize psychological research generally. As with behavioral science research, human factors efforts need more precise and varied measures of human performance, more complex experimental tasks, much longer samples of behavior (longitudinal studies) and the incorporation into experiments of environmental factors which influence performance."

The lack of a unified theoretical structure for human factors is pointed out by *Lyman.* "There are many, many theories, e.g., signal detection theory, "man-in-the-loop", etc." *Weldon* is discouraged about this:

"As of this moment the task of getting a structure of theory for human factors seems unreasonably hopeless. For example if one accepts the proposition that a man is guided in his acts by the concepts he holds, then it is necessary to set up some inclusive taxonomy of concepts—in short, to describe the concept space of man.[9] . . . It seems to me . . .

[9] It is probably too much to expect of a young discipline like H that it have a comprehensive theory to integrate the varied phenomena it describes. Certainly, if

unreasonable to expect a theory structure to appear for the use of human factors researchers within a foreseeable time."

Like *Orlansky*, *Kraft* is concerned about the problem of validating human factors research:

"Empirical validations have their limitations as well. The cost of doing a good experiment is generally higher than the customer (in-house or government) will tolerate. One is forced to use small, non-representative samples, questionable experimental designs and techniques and equipment which haven't been properly validated. Our rationale for this is that some empirically derived data may be better than none and we all acknowledge that the engineer is more impressed with quantitative rather than qualitative findings. We can also blame the "real world" for placing us in the situation of doing applied research before sufficient basic research has been completed. Experimental replication, we rationalize, is not possible under industrially imposed constraints. Therefore, it is generally not possible for independent experimenters to get precisely the same results when they repeat an experiment.

"We also accept the premise that we cannot establish the necessary controls for many of our experiments to achieve the preciseness of results desired, but this does not stop us from going ahead.

"Thus . . . the tools employed by many human factors researchers are not precise, sophisticated nor valid. . . .

we look at psychology as a whole, such a unified descriptive theory is not to be found. It may be more realistic to think of microtheories dealing with parts or aspects of human factors. It may be possible, however, to develop some concepts to guide the research to be performed. This does not require a theory which integrates all known data about human factors, but it does require a set of concepts which suggest that certain human factors elements are more worthy of research than others. It may in fact be necessary to start with some such theoretical structure before we can proceed to meaningful research. (The author leaves it to the reader to decide whether a comparable theoretical structure of essential elements exists in psychology; personally he thinks it does.) Perhaps we need some conceptual examination of how human factors research should be performed. It sometimes seems to the author that there are too few people in the discipline who are concerned about the conceptual structure of human factors *as a whole*. There are obviously theories and theorists concerned with research subsets of the discipline, such as tracking, vigilance, reconnaissance; but is there anyone who considers H as a whole from a theoretical standpoint? Because of the problem orientation which the discipline has developed, H specialists tend to be doers rather than thinkers, at least they are not thinkers on a more or less global H scale. It is sobering thought that the scope and effectiveness of a discipline probably tends to correspond with the scope of its theoreticians. As Kraft says elsewhere, we are merchants of the expedient. Maybe we need to develop a branch of human factors which has leisure and inclination to develop a philosophy of human factors.

"These tools lack the history of standardization and validation necessary to make them precise and dependable. These weaknesses should not always reflect poorly on the human factors researcher who wants to and can do a better job—given sufficient money, time and an understanding customer. In the absence of these three elements, he does the best job he can with the tools he possesses—and sometimes at the expense of his profession. Despite the limitations of these tools for solid research, they do supply some of the requisite data needed by the human factors engineer to demonstrate a point to the design or project engineer.

"The human factors practitioner is not an experimental psychologist in the true sense. He is frequently a merchant of the expedient. He recognizes that the cost of experimental equipment and conducting experiments is ofttimes more than the customer or company will bear. Therefore, he limits his experiments to what he is able to do—which may not always be what he should do."

Question 5. Is human factors research performed in the laboratory being adequately applied to actual system development and/or operational problems? If not, why not? What are the problems in applying that research?

It should be apparent to the reader by now that human factors research is *not* being adequately applied to system development. The question then becomes, why is it not being adequately applied and what are the problems involved in that application? Some answers to this question follow:

"Some human factors research performed in the laboratory is not applied adequately to operational problems. The major reasons are the obvious ones. Some laboratory work is performed by individuals who are not aware of and, in some cases, are not interested in operating problems. This is matched, on the other hand, by operational personnel who are unaware of, and occasionally, not interested in applying relevant research findings. In the latter case, a convenient rationalization for stupidity or parochialism is to regard laboratory data as irrelevant to operational problems. For reasons associated with cost and experimental control, laboratory experiments often involve simplified rather than realistic equipment. Similarly, the subjects in laboratory experiments often are the well-known college sophomore or, his recent counterpart, the engineer who can be spared from his job. Such personnel are easily regarded as not typical of the trained expert on the job and it is clearly the researcher's responsibility to determine the extent to which this may or may not be true. Another problem is that so much time may be involved in building elaborate laboratory equipment that the operational equipment is built before the required data has been collected. Presumably, all of these problems can be reduced by better planning and communication between the research and operational communities which, after all, have many common interests." (*Orlansky*)

"We have a long way to go in devising laboratory situations that can be applied to system development and operation without a great deal of risk. We so often find that our laboratory research, in order to achieve scientific control and generality, is so contrived that it bears little relation to the real world situation." (*Van Cott*)

"One of the worst problems I've noted over the past 20 years is the lack of consistency in methods, approaches and metrics used by our laboratory friends. There is seldom any way to extrapolate conclusions because of the discrepancy between research and engineering metrics." (*Woodson*)

"In some cases, laboratory research is being adequately applied to system development and operational problems . . . The key to successful application is project engineering involvement. We see a problem which needs laboratory investigation, we interest project people in our approach, we secure their participation in the hardware development, and frequently we have them test the equipment before we run the test subjects. Then we conduct the experiments under their general surveillance. This type of involvement seems to be the key to successfully applying principles to design. I am unable to think of a single instance where we have used this approach and have failed to influence, in some way, the eventual design. However, we have undertaken some laboratory studies where we have failed to involve the appropriate engineers and we have found it difficult to convince them that design changes should be made on the basis of data alone. I believe this may be the heart of the problem. Our engineers are much less impressed with a high quality research report of the findings than they are by witnessing experiments and seeing some of the problems first hand. Once they have this exposure, then we can feed them the results of experiments and suggest appropriate design changes.

"Laboratory research performed by other organizations is less impressive to our engineers unless our human factors specialists serve as interpreters of the research reports and use them as a basis for selling recommendations. Part of the problem is that most engineers are unfamiliar with the human factors research which is being done and with the qualifications of the researchers. They are also unfamiliar with the reporting techniques, the methodology and the language used in these reports. Therefore, a trained middleman is needed to translate the results into a form which has meaning to the engineer." (*Kraft*)

"The problem is one of developing a category of scientist who has one foot in each of two worlds—research and application." (*Regan*)

"If the research being conducted is not broad enough to permit utilization of the results for system development or operational problems, then, by definition the research is not adequately performed. A basic problem relates to the fact that the development process is a very dynamic process, and it would be extremely difficult for human factors research activity to have the flexibility to compensate for every development change. Throughout the design and development process many concepts are modified so that the final hardware does not necessarily approximate the original hardware concept. . . . Over the development period it is quite likely that the state of the art might permit the elimination of certain manned functions and still permit the system to meet its requirements. At the same time, tests may reveal that the equipment cannot reliably meet other requirements and man must be introduced into the system to perform functions which were not originally anticipated. It is often the case that the original research planned to yield meaningful results for the operational system, can no longer be utilized because the operational system has changed radically." (*Wissel*)

It is interesting to note that Kraft sees the problem of research application as one of persuasion, not one of research "validity." His implication is that even when research is actually directed at a significant development problem and is done well, it may not be accepted without question by engineers unless they have somehow participated vicariously in the research process—a sort of engineering "community involvement." At first glance this is difficult to accept because it introduces an element of irrationality and emotionality into the research process; but second thoughts suggest that he may have a point.

Of course, Kraft is talking about research aimed at solving system-specific problems. A clear distinction must be drawn between research resulting from an immediate development problem and research directed at general problems of system development. The first type may well require engineering participation; the second need not—in fact cannot—require that participation.

Underlying the comments that have been made by the panel is the problem of validity, which the reader will note as having recurred a number of times during these discussions. Validity in the context of human factors research has two faces: the necessity of addressing research to "real" problems or issues and the necessity of simulating the real world environment during the study of that problem.

The "real" problem in the system-specific sense seems to be immediately apparent; it is whatever the engineers on a particular project decide is the question they want an answer to. The specialist may argue the correctness of the question asked, but at least it is relatively concrete.

Questions of a basic system development nature are something else again. Here there may be great dispute (among specialists, not engineers) about the nature of "genuine," "real," "important" development problems because, as was pointed out before, development has many aspects. An example of a "real" problem is the relationship between one or more equipment characteristics and the operator performance to be expected with these characteristics (see Chapter Four for other "real" —in the general sense—problems).

The other aspect of validity that human factors research must be concerned with is the reproduction of the operational environment in which an equipment/operator relationship is maintained. In a research situation it is impossible to reproduce that environment totally (i.e., the Heisenberg principle operates in behavioral research also), so we must select certain dimensions of that environment as being more representative of that environment than are others. Which dimensions are these? That too is an important subject for research, which has been largely untouched.

All of these obscurities, the illumination of which will be long in coming, need not retard research as long as the specialist analyzes these questions and derives the best answers to them that he can before he performs his research. What is disturbing is that so few researchers ask these questions.

Question 6. Is human factors research today attacking correct, that is, meaningful questions? What would you consider to be meaningful questions and why?

Weldon raises the point: "What is 'meaningfulness' in anything? This is about as profound and full of implications as any question one could ask. It leads straight into philosophy if pursued."

Meaningfulness can be viewed in several ways. If what the customer wants is interpreted as being meaningful, then responsiveness to customer needs, ability to satisfy these, represents meaningfulness; for example, *Wissel* would define meaningful questions as "those which can supply the correct answer to hardware systems." Meaningfulness may represent an opinion as to the way in which men should function in our technology. *Weldon* expresses this point of view quite well when he says:

"If Human Factors must raise its eyes above its immediate problems—its displays, controls, operator loadings, etc.—it will have to change its character. If it wants to have an opinion of its own about the meaningfulness of its work, it will have to create a professional opinion about the purpose of the systems it helps to develop, and thus also the goals of the society which uses those systems.

"So far, Human Factors has worked well within an established framework of accepted meaningfulness. It has not had to face profound problems of "what is worthwhile," of "which is better" or "what is best" for human beings. It has largely stuck to immediate problems and assumed either that those who initiated projects and paid for them knew what they wanted or what was meaningful, or assumed that it really didn't matter as long as the immediate problems were solved. After all Human Factors people are not paid to be philosophers, are not expected to be philosophers and probably would be on the "let-go" list if they tried to exercise the role of philosophers."

The financial dependency of human factors on the customer tends to reduce consideration of "meaningful" research. As *Orlansky* states:

"Human factors research today deals primarily with questions for which research support can be found. These questions may be more or less meaningful to our contemporary world. Contemporary human factors research can not be said to deal with transcendental problems: it goes where the money is. Up to now, this has led to concern with military problems. In recent years, there has been a healthy infusion of non-military problems, principally non-military manned space flight of interest to NASA. Some signs that human factors research can help improve society can be found in such projects as FAA support for research on air traffic control centers, company support for research on control of experimental automobiles, Post Office support for mail sorting machines, and so forth. The support for human factors research on problems of modern society such as rebuilding cities, devising automated hospitals and re-training adults will surely be available before too long."

We can also view the problem of meaningfulness in terms of the development of a unified research structure. *Van Cott* puts it well:

"Meaningful research can be viewed in two ways. One is its responsiveness to current system design problems. The other is in gradual development into a comprehensive body of generalizable research data that answers fundamental problems about the nature of man in systems. We have been better at responding to current design problems than in building a comprehensive data base about human performance. There are, of course, weaknesses in both. Often our approach to system design is based on expertizing, or just guessing. On the other hand, our approach to human performance theory has not kept pace with changes in the role of man in systems so that we can turn to what we know about human performance for relevant and applicable information."

There are then at least three ways of defining what is meaningful in human factors research:

1. Satisfying the customer's needs, whatever these are; here the user's opinion of the adequacy of the research is the criterion of meaningfulness.
2. Answering the general question, how do men function in a system environment; here there is no user criterion. The discipline itself or those who view the discipline from other sciences can help decide whether research is meaningful.
3. The application of human factors methodology to other than machine systems, that is, social systems; in this case the criterion of meaningfulness is the ability of the discipline to influence or modify these social systems.

At the present time human factors has not generally gotten beyond the first item, satisfying the customer's needs. However, some specialists like *Lyman* feel that

"certain problems, such as the understanding of the dynamics of human costs and values in a decision process and the factors in social organization at all levels are perhaps the most pressing problems of our time. We are on the verge of social engineering in the sense that, as a first step, large scale environmental design will become commonplace for the control of such matters as air, water, and noise pollution. This is almost sure to be followed by more emphasis on esthetic matters in city and community planning and in the development of the cultural aspects of society to meet the needs of the many, as well as the few."

For none of these three definitions of meaningfulness is human factors research considered entirely adequate. *Regan* considers that

"there is a tendency to study design features which account for small portions of the total variance associated with effective performance. In general, the conceptual (as compared with the perceptual/sensory) role of the operator in man-machine systems has received little attention. Many of the features of the human being's job need to be dealt with simultaneously. Particularly those which are uniquely human. Research should seek to identify, exploit and improve those features (e.g., problem solving) of human performance not easily dealt with by equipment system design."

However, *Kraft* is generally optimistic:

"Most of the research reports which come across my desk are addressed to problems which I consider to be meaningful to some potential *user.* (Italics those of the author) "By meaningful, I use the criteria of applicability. That is, can the data be applied to solving some design or operating problem, even if it is a small one."

However, like other specialists, he goes beyond the customer criterion.

"I also consider the research meaningful if it leads to a better understanding of human behavior and if it develops principles which can subsequently be applied to solutions of man-machine problems."

In a similar manner *Bauer,* who represents the industry interest in human factors goes well beyond the immediate man-machine interface in terms of what he considers to be meaningful questions.

"What would you consider to be meaningful questions, and why? Those that answer the current applied needs of systems development and operational environments and current applied cultural needs. Why are these questions meaningful? Because they are the current human factors needs of the society in which we live and human factors is an APPLIED discipline."

On the other hand, *Christensen* says:

"Although human factors research generally is directed toward meaningful questions, I am not nearly so sure that the answers are meaningful. Human engineers, for example, must attack questions regarding man's capabilities and limitations but attack them only if he can come out with results that the design engineer can use. Generally, this means that the abscissa in his experiment must represent a variable that the design engineer can manipulate. However, the design engineer does not need to know that 'a' is significantly greater than 'b'; but he needs to know how much greater and the consequences in terms of systems effectiveness of selecting various points along a dimensionalized 'a'."

This last point, which has been noted in previous chapters, is of crucial importance in any effort to influence system design.

A number of points should be made. Even if we apply the criterion that research which satisfies the customer's needs is meaningful, some judgment is required. There are after all two customers and two types of needs. The governmental agency soliciting research is the immediate customer for that research, but presumably it acts as agent for the engineer and specialist working on system development projects, who then become the ultimate customers.

Although it is true that research must satisfy the needs of system development, it was pointed out in the previous section that there are two types of system development needs, those of the immediate project and those of system development in general.

What is needed are research guidelines which are not merely a reflection of the researcher's or the customer's personal interest but are rather derived from some underlying theory of how humans function in a machine (i.e., system) context. Those factors which are hypothesized to have a primary effect on that human functioning should be the factors to be investigated first and with greatest effort. We come back then to the matter of priorities.

As problems become more molar (e.g., the application of human factors methods to social systems), the need for guidelines derived from a theoretical structure become even more important.

The relationships between some specific question posed by a new hardware system and the research required by that question are relatively obvious; such immediately apparent relationships are, however, not likely to be the case with social systems. It is inevitable that a theoretical structure to serve as a research framework will be developed if the discipline is to progress. The important thing is not that we drop everything else to develop such a theory immediately but that human factors specialists engage themselves in the effort to develop one ultimately.

Question 7. Are we making proper use of the real world, that is, operational environment, as a source of problems to solve? hypotheses to test? as a data source? If not, why aren't we, and what should be done?

It may be significant that without exception our experts explicitly or implicitly answered the question in the negative. The reasons for this are, however, by no means so unanimous. *Fucigna* claims that

"very few H personnel are supported to study H problems in an operational environment, and, to my knowledge, there is no formal system to provide feedback from the operational environment. Support of H efforts is primarily during the R&D phase of the system life cycle.

Until such support is extended to include phases beyond R&D, the real world problems will not be apparent."

Van Cott seems to feel that there is a learned tendency to avoid the "real world."

"Because of the origins of human factors in psychology there is a tendency to avoid the real world in favor of the laboratory. We do not make sufficient use of operational testing, using representative personnel and typical operational environments to develop concepts, test alternative designs, or check out the performance of a prototype man-machine system. Our laboratory tests all too often look at a single critical dimension of human performance (e.g., vigilance, tracking) and we seldom study these tasks in the context of the real-world man-machine relation."

Regan points out that

"the experimental tasks are not characterized by the richness, complexity, and uncertainty found in operational situations. One approach to achieving more representative tasks, problems, and environments would be to use these environments (e.g., a military situation such as a surface ship searching for a submarine) themselves in some controlled experimental way. To do this, the sponsors need to recognize that research is a mission in itself worthy of support and to include participation in research as a mission for operating forces. Product testing research is done in the market place. Human factors studies need a similar orientation. The reason this problem exists is identified in the first question. The sponsors need to see clearly the utility of the work. Some major formal effort should be inaugurated to identify past payoffs (both in theory and practice) resulting from human factors activity. In addition, a catalogue . . . of research and application issues facing human factors efforts should be prepared. These activities might help the management and human factors personnel to see just how and where contributions can be made and where work needs to be done."

Christensen considers this avoidance of the operational environment as

"one of the great deficiencies in our field. Bob Mills and I recently did a study in which we tried to address the problem of what operators actually do in systems once the systems are in the field. We were dismayed to find that there have been very few studies conducted under such conditions. Further, those that have been done suggest that once a system reaches the field it is used quite differently than was ever intended by those who conceived of it and designed it. I am strongly in favor of more of our human factors people going into the field to find out exactly how systems are used, under what conditions they are used and how the actual duties and activities of the operators (including maintenance men) differ from the intentions of the design engineer. I think it is an unusually rich field that has been barely scratched. I think the reason that we have not done more of this is that it is very difficult to do. The individual is always worried about whether or not he will be accepted in the field, whether or not he will be in the way, that he may be subjected to embarrassment when he tells the operators why he is there, etc. I have done a fair amount of this type of work, and I can assure you that if one has a clear and reasonable objective and conducts himself properly he will be welcomed and supported in most field situations. I find it a very rewarding area in which to work, and I am astonished that more of our young human factors people don't pursue this as their area of primary concern."

Woodson suspects that

"it is a matter of money more than anything. In addition, there is often the question of safety when you try to use the real world to provide the setting for an experiment. I personally think this is an area which should be investigated more fully in order to develop a long range plan for studies and development of techniques. After all we have very good examples of studies of heredity which are typical of this area. These take considerable time spans, but that is no excuse for not getting started."

Of course, there is the question how we define the real world, as *Kraft* points out.

"I do believe that much 'real world' experimentation is being done because in industry one cannot survive very long if one is unresponsive to the problems faced by the design engineers. However, there is a tendency on the part of some human factors researchers to shy away from 'real world' situations because the variables are difficult to manipulate and control or because they hesitate to become involved in engineering-oriented problem solving. Part of their resistance may be inherent in their academic psychological training, where they have been sheltered from or not have had access to 'real world' problems. Thus, when they move into a government or industrial laboratory, they find that much of what they have learned cannot be applied to the problems they must solve or that they are not equipped to handle the problem. What should be done to overcome this problem is what is no longer being done—that is, government agencies providing fairly sizable sustaining budgets to several academic laboratories and providing them access to operational problems."

Wissel feels that

"designing a valid study to be executed under operational conditions is quite difficult. With the availability of modern computers and simulators, it is tempting to carry out such studies in the laboratory. While it is true that variables can be better controlled under strict laboratory conditions, it is not impossible to obtain quite valid data from field studies. It is more difficult."

Bauer stresses the matter of personal interest.

"Too many human factors researchers are pursuing problems of personal academic interest that may not be real world critical problems (usually aren't) rather than pursuing those critical real world needs. What's even worse, they are encouraged and reinforced to behave in this manner by financial support from their empathetic counterparts in government. What should be done? Obviously, more rigid criteria of application should be required for the approval of a human factors research contract. Or it should be done in academic settings under the basic discipline."

Our respondents pointed out two main reasons why human factors researchers did not make sufficient use of the operational environment (OE): lack of money and a preference for the controlled conditions available in the laboratory. Although these are legitimate explanations, they are somewhat superficial because they ignore more fundamental questions. Must we use the OE? For what purpose? To what extent?

First of all, exactly what do researchers want to get from the OE as a test bed? The OE can be used in several ways. Obviously we must use the OE to validate laboratory findings because without such validation even extensively replicated research results remain only hypothetical.

The OE can also be used to suggest the variables which need investigation. When laboratory results are negated after being applied to the OE, it is usually because the variables which were meaningful in the laboratory are not meaningful in operations. Does this mean that all or most human factors research should be performed in the OE? Logically it seems so; but the practical difficulties of doing so are probably insuperable. What may be required is to make the laboratory resemble the OE more, and to make the OE more controllable. To do so, however, we must know precisely (not just generally) how the laboratory and the OE differ.

The crucial point is that we do not know what major characteristics distinguish the OE from the laboratory; whether it is, as Regan suggests, "richness, complexity or uncertainty" or some other dimensions, and that, until we determine what these differences are, we can make no real decisions on (a) what research should be performed in the laboratory or in the OE; (b) for what purposes (hypothesis-identification, hypothesis-testing, or result-verification); and (c) how operational research can be more carefully controlled to laboratory specifications or how laboratory test situations can be so structured that they more closely correspond to significant real world dimensions.

Question 8. Engineering involves three-dimensional physical parameters, human factors deals with multidimensional behavorial parameters. What kind of research should be performed to develop equations (mathematical or otherwise) which will translate the behavioral into their physical equivalents? Can you supply an example of research which was highly successful in providing these translation equations?

This question appeared to generate a certain amount of resistance from the panel, perhaps because it was phrased ambiguously. Its intent was to point out that in the process of designing man-machine systems the specialist implies certain relationships between the physical characteristics of the equipment and the performance he expects of the operator interacting with those characteristics. However, these relationships are only implied; for example, the use of the principles of control panel layout in MIL-STD 1472 is supposed to produce more efficient operator performance. Assuming that these effects occur (and to the extent that they do), the question is how do the physical characteristics exercise their effect? It seems reasonable to suppose that the physical characteristics of the equipment serve as stimuli which are perceived by the operator and that somehow these stimuli influence the operator's responses. But to say this is really to say only what is obvious. What we must know, for example, is which of the many characteristics of the equipment are *selected* by the operator as effective stimuli, and why, and how, once selected (consciously or unconsciously) by the operator, these stimuli cause the operator to vary his performance?

From the standpoint of the practitioner it may be sufficient that positive effects result when human engineering principles are utilized in design, but from a research standpoint an unexplained effect is an unanswered question; how physical design characteristics translate into behavioral effects is therefore perhaps the most fascinating problem the specialist encounters. Unearthing the relationship may also have more than a purely theoretical interest; knowledge of which equipment characteristics serve as positive stimuli for performance could determine how equipment should be designed to improve that performance. The questions implied by the relationships mentioned above underly the conceptual orientation described in Chapter Four.

At the moment few specialists seem to be concerned about the theoretical basis for the relationships between the physical and the behavioral perhaps because they are overwhelmed by more practical problems.

Christensen states that

"this is a very tough question. In one sense, the physical sciences have prospered because they have been able to identify constants. There is a real scarcity of constants in the human factors field. In order to improve in this area, it would appear to me that we have to do several things.

1. We must try to improve our scalar techniques, i.e., we must attempt to develop ratio scales wherever possible.
2. We must put our data in such a form that it can be compared with or played off against physical parameters, the so-called trade-off type of equation.

"With respect to the second part of your question, I suppose that the work on transfer functions has come about as close as anything that I can think of to putting data in a form useable by design engineers but even here we have barely progressed beyond the single dimensional tracking stage."

Fogel, however, feels that this is an unreal problem, as does *Van Cott:*

"I think this is a meaningless question. I would say, however, that some of the most successful research in bridging the gap between physical parameters is to be found in the application of information theory and information processing concepts by Fitts, Attneave, Garner, George Miller, Pollack and others.

"The real problem is not one of translation between human performance parameters and physical or engineering parameters, but defining parameters with respect to human performance that fully describe it. The limitations on the use of information theory, for example, are that it does not take into account the adaptive capabilities of man, nor the meaningfulness dimension that information inputs have. These are human properties and may not have a counterpart in the physical world. Defining these uniquely human parameters is the biggest problem that faces us today—and not that of physical to human performance translation."

Lyman indicates some hope that the mathematical model will help.

"It appears that the development of mathematical models has become a current hope for the solution to all man-conceived problems. With the advent of highly sophisticated information handling machinery, it is possible to conceive of simulating behavioral parameters in terms of inputs and outputs that are functionally equivalent to those observed in human behavior. Certain models have already been devised for particular aspects of human behavior such as human transfer functions, signal detection models, decision models, etc."

Kraft, however, has "a somewhat negative bias regarding the value of math modelling of human behavior. I am not aware of any examples of research which were successful in providing valid translation equations."

Weldon points out that the equations required to describe these relationships would be enormously complex. "They would be like nothing we have ever seen." He adds a very pertinent comment:

"Physics is no more physical than human factors, and no more real. Physics may have more appeal because its concepts are neater and more demonstrable. However, if engineering is going to include human factors within its scope then engineering will have to accept the unwelcome properties of human beings as they are, rather than insist upon the neat world of physics as the model to which all other areas must conform. Behavioral engineering, if it ever matures, must find its models in the world of organisms, not in the world of physics and chemistry."

If the answers provided to this question seem a bit unsatisfying, it may be because Human Factors as a discipline has not yet developed a distinctive way of looking at the behavioral/physical continuum. (The author is aware that he is opening himself to attack on this point. After all, psychology, with its relatively much longer history, is still bothered by body-mind dualism.) It seems to the author, however, that, simply because H̲ does straddle the world of the physical and the behavioral [10] (if we can distinguish them in this way—see Weldon's comment above), it must have a distinctive way of philosophically interrelating them. The fact that the question appeared to daunt our panel of experts suggests that H̲ is still somewhat philosophically immature.

Question 9. What is the influence of their psychological training on the performance of the people who perform human factors re-

[10] Even more immediately and directly than other behavioral sciences. It is possible to work around the body/mind relationship in psychology, for example, without any great loss in effective knowledge; but H̲ cannot because its efforts are specifically directed at modifying *physical systems in response to behavioral requirements* (and psychology's efforts are not).

> search? In terms of applying that research to actual engineering development problems? Is the influence of their training positive or negative for achieving human factors research goals—whatever these are?

The intent of this question was to find out whether psychology is actually as important to human factors as a psychologist might suppose. In formulating this question the author overlooked the fact that the group he was questioning had all had psychological training. The question was therefore a little naive. As we would expect, everyone felt that psychological training is indispensable for human factors work although there were some reservations about its being *sufficient*. The over-all impression we receive from the responses is that psychology is not enough, that human factors goes beyond psychology.

"Psychological training, in my estimation, is evidence of an interest in and provides sensitivity to requirements, capabilities and limitations of humans—a necessary but not the only ingredient in performing H̲ work effectively. For the *applied* H̲ specialists, training in psychology coupled with training and a strong interest in engineering and/or mathematics will increase the effectiveness of the individual considerably." (*Fucigna*)

"I don't think human factors as a discipline would have gotten started unless psychologists had gotten interested in man-machine systems. The training in scientific methodology and analysis received by the psychologist is a tremendous asset. But lack of an engineering background by many psychologists in human factors is a liability that must be corrected through changes in the human factors training curricula." (*Van Cott*)

"Unless the basic training has been extended into the specialty of engineering psychology, a psychologically trained person requires a great deal of a preliminary retraining in order to communicate with actual engineering development problems. Since I work principally with engineers who learn only enough psychology to build on their engineering background, the question of the amount of the psychology necessary for solving human factors problems remains. I believe that, generally speaking, psychological training is decidedly positive in its effect but without the physical science background it is not as effective as otherwise." (*Lyman*)

"Psychological training certainly is a beneficial influence on people who perform human factors research. Basic training in experimental psychology is fundamental to the field and undoubtedly benefits the design of studies for this complex endeavor. Some of the research that has been conducted and reported on by people without psychological training tends to oversimplify the various characteristics in order to describe such attributes in numerical terms. It is equally important that people trying to apply the results of human factors research have psychological training in order to evaluate the research findings." (*Wissel*)

We encounter also the feeling that there is a "real" world (of system development, presumably) which does not quite fit the relatively academic orientation which a psychologist receives in his training.

"I have been disturbed quite often by the tendency for human factors personnel (psychologists, physiologists and others) to slip into the same errors they accuse their engineering colleagues of, i.e., experting on the basis of spontaneous opinion. It's human nature I guess. However, I can't excuse a human engineer who can't even human engineer his own problem solving procedure. Training could be somewhat negative, I think. For example, a great number of former human engineering types give up and go back to teaching because they can't cope with the real world. I pose the question of whether they may not be passing their personal deficiency onto their students. I hope those who read this recognize that I am not referring to their technical integrity, but rather their refusal to accept the real world and the compromises which have to be met in it." (*Woodson*)

Coping with the real world may be a synonym for adjustment to the engineer and his foibles, as *Kraft* implies. (The reader may also recall *Weldon's* comment that, if engineering wants human factors, it must accept the unwelcome properties of the human.)

". . . The successful practitioner must realize that he must first learn to work with the engineers and physical scientists in an engineering environment and learn how to communicate with them and to think like them to a limited extent. The psychologist in industry must learn to work in the engineering world rather than vice versa. If he can't make this transition, even in part, he most likely will not be a successful human engineer. He should not regard this as selling out his 'birth right' but more as a condition of employment in industry.

"The 'hard core' experimental psychologist may encounter problems in applying his research to engineering problems or to an acceptance of the limitations placed on his research, but if he can learn to accept the constraints of his work situation, his training in psychology will be a distinct asset to the achievement of human factors research goals. The point I wish to make here is that the psychology training *per se* need not be a negative influence if the proper work attitude can be developed." (*Kraft*)

On the other hand, if we assume that the problems to be solved in human factors may not be quite the same as those of psychology (see the discussion in Chapter Four on differences in orientation between the two), it is possible that preoccupation with psychological variables may lead the specialist to overlook genuine human factors problems because the latter cannot be formulated in strictly psychological terms. Psychology is centered largely around the individual, whereas human factors requires an appreciation of the interplay between the individual and the system he operates.

"What is the influence of their psychological training on the performance of the people who perform human factors research?—in terms of applying that research to actual enigneering development problems? Superior performance in research; i.e., application, methods, etc. The fact that they are tackling the wrong problems is beside the point, as is the fact that their answers are academically oriented, rather than engineering-oriented.

"Is the influence of their training positive or negative for achieving human factors research goals—whatever these are? Their training is highly positive for the research goals that have been set for them—(incorrect though these goals may be)." (*Bauer*)

"I'm not sure that I completely understand the question. For example, is the influence of their training positive or negative? I believe that it depends to which aspects of their training you are referring. For example, I feel that the general content in psychology is positive. I believe that the psychologist brings to the human factors area considerable worthwhile knowledge about human capabilities. For example, consider the idea of absolute and differential thresholds and the limits of the senses. What happens to these senses under various conditions? I believe that the methods that we learned in psychology are not necessarily the best for the human factors research area. Specifically, although we are fairly good statisticians, we are not particularly good mathematicians. We are more concerned with the establishment of significant differences than we are with the generation of functions that will allow the engineer to make trade-off studies. On the other hand, we are reasonably skilled in the area of experimental design and generally have a fair understanding of *interaction* and its consequences in the operation and maintenance of systems. Incidentally, there are going to be many members of the human factors family that are going to feel, as many of them have in the past, that psychologists believe that psychology is the only science that contributes to human factors. Let's never forget the anthropologists, the physiologists, the bioacousticians, etc." (*Christensen*)

Christensen's point about the psychologist being concerned only about significant differences rather than absolute values is an especially important one, which has been pointed out before but which deserves repeated emphasis. The author has tried not to forget the other specialities within human factors, but there are so many it is impossible to give each his due.

Question 10. Should human factors be considered solely as a technology which supports engineering technology, or should it have any goals beyond this? Specifically, should the goal of human factors research be to answer specific questions raised by engineering problems; or should it be to develop principles describing how personnel perform in a man-machine context?

As indicated in an earlier question dealing with the meaningfulness of human factors research, it is possible to view H in several ways: as being *only* a support to engineering; as directed at describing how men function in a technological culture; [11] as a discipline capable of building social systems and solving problems in these systems; or as all of these. The industrial specialist is likely to see his discipline in its narrowest aspect because that is the "world view" with which he is most familiar. Can the human factors specialist aspire to anything more, considering the youth of the discipline and the severity of the technical and support (financial) problems it faces? It is encouraging to find that a majority of the respondents adopted a broader viewpoint. However, the following comments must be considered:

"Human factors information is a reasonably useful technology in the general field of engineering technology, important only when required for effective design" (*Orlansky*). *Fogel* indicated that he "for one, would be willing to restrict my attention to specific technical subject matter which arises within the man-machine context." *Kraft* made a very reasoned statement:

"Human factors is not ready to stand by itself as a technology, but it should aspire to this as it becomes better understood and accepted, and as the methodologies and data bases improve.

[11] In a sense parallel to sociology, which describes how human groups function in society, but concerned primarily with man–machine interactions.

"For the present, human factors should concentrate on answering specific engineering questions, but continue to push the frontiers of human knowledge.

"There is room for both engineering support and principle development at our stage of growth. We must be careful, however, not to claim expertise that we do not as yet possess. There is much more to be learned about the constituents of human behavior and performance before we are in a solid position to predict performance. We have pushed the field ahead at a fairly rapid rate and have earned the scorn of many medical people who frequently regard human factors specialists as charlatans. Therefore, it behooves all of us to be aggressively patient about the progress of our profession. If we believe in the value of what we are doing, if we make enough converts, and if we continue to generate valid new principles, we may eventually establish human factors as a technology.

"My feeling is that we should be honest about the limitations of human factors without being apologetic. We should take every opportunity to press the field forward and widely publicize our activities and successes, but we should avoid the implication that human factors is a panacea."

Weldon agrees with him:

"I do not see human factors as a technology. There are too many questions; they are too unreliably answered; there is such a quantity of information; there is a lack of sound theory to tie the information together. If a human factors man has information which gives him an answer or which points to a solution he needs, he is lucky. Usually he must use judgment and make a hard decision to get his answer, or do research that may or may not provide the answer sought. It may not be quite as bad as this, but it is too messy to be a technology.

"However, I am personally sold on the work of human factors. I think it is doing a very important job. It is opening a new area for the work of psychology and helping to bring psychology down to earth. It also is bringing a bit of sophistication about human beings to engineering."

Others, however, had perhaps a more optimistic attitude toward their discipline.

"I believe it (human factors) should have goals beyond supporting engineering technology. Many problems confronting or which should be confronting H personnel do not involve engineering. The manning and training area for systems void of hardware is an obvious one. Motivation, morale and safety are others which are appropriate for H personnel to consider independent of engineering technology. Health and welfare, while overlapping somewhat with other disciplines, are in some instances appropriate considerations for H personnel. H, in my opinion, goes beyond the man-machine context to include the man-man interface, man-environment interface, and the man-system (industrial, military and social) interface." (*Fucigna*)

"Human factors is an applied technology that supports system design. That is its main purpose and this system orientation distinguishes human factors from applied psychology. In order to support system design, human factors must not only be able to answer questions raised by engineering, but to anticipate these questions and identify other relevant areas of concern that relate to man-machine performance that might be overlooked by the engineer, and to develop a body of methods, principles and data that provide a foundation for system support.

"I tend, however, to take a broad view of systems. I include not only weapon and space systems, but also transportation, housing, educational devices, training equipment, and offices; in fact, all of those things which support the functioning of our social institutions are systems. In this sense, the breadth of human factors engineering is very broad and might even be considered, by some, to be synonymous with social engineering.

"I believe that among the applied sciences we occupy an unusual position: we understand systems, but our focus is on man. The same cannot be said of the engineering sciences nor of other applied social sciences (e.g., sociology, and social psychology), which may understand the problems but lack the methodological tools and systematic logic of exposure to the R&D process, which would permit them to develop policy and procedures for the humanization of life and of our social institutions." (*Van Cott*)

Christensen points out that we all agree with the proposition that H is a technology supporting engineering but goes on to say:

"Isn't the question really whether or not we agree that human factors should be *simply* a technology that supports only engineering technology? I would then answer this question this way. I think that human factors has a great and grave responsibility to support engineering technology. We are the people to whom they have a right to look for answers regarding how best to use man to meet specific systems objectives. I think, however, that we should not put ourselves in the position of allowing the engineering profession or any other profession to dictate what the objectives of a specific system should be. I think we should have a voice along with others such as the economist, the sociologist, etc.

in establishing the objectives of a system when it is conceived. In other words, I deplore the thought of a society designed only by engineers as much as I would deplore a Congress in which there were nothing but lawyers. I think as a profession that we must remain sensitive to what man's real objectives should be, what his aspirations are and/or should be, the brotherhood of man, the need for individual development and fulfillment, and such fine things. I don't consider them trite at all. In fact, I consider them so important that if we do not achieve those, then all the engineering devices in the world will have been designed in vain. I am even willing to take the position that in a *hardware* system, if the most important interactions are those involving man, then perhaps a competent human factors man, and not an engineer, should be in charge of the development of the system."

Woodson agrees:

"I tend to look at human factors on the broader scale which includes all human-related problems, even social. Let's face it, we are in a world full of viable, emotional people who do things with things and other people.

"Our engineering colleagues figure that human factors specialists know about people—'all about' people. We tell the engineer to take a global look at his system, and then we decide that our speciality, human factors, covers only those things we are interested in.

"As far as answering specific vs general problem questions, I don't think we can ever draw a line around the area of our responsibility. As long as it has to do with people, we had better get with it right now."

Lyman states that

"human factors research will, in my opinion, be most effective if it is directed both to specific questions raised by engineering problems and also to the broader questions concerning man's role with respect to his environment, other men, and the tasks he encounters while he carries out this individual life cycle."

These responses suggest a certain ambivalence on the part of the specialist. On the one hand, he feels it necessary to expand the horizons of his discipline; on the other, he is aware of its present limitations; until these are overcome, he has some difficulty accepting the concept of $\underline{H}$ as more than an adjunct to engineering.

If the specialist accepts the human factors *system orientation,* [12] however,

[12] Which assumes that the human factors analytic logic is appropriate to systems in general (large/small; simple/complex; man/man) as well as to man–machine systems in particular.

ever, it follows that H should be just as applicable to social systems (which in many aspects are largely manual) as to more mechanized systems. *Van Cott's* point that among applied sciences H occupies an unusual position is an understatement; it may better be phrased as an unique position. Logic therefore suggests that the human factors methodology (*once validated*) should be capable of being applied to the more complex social systems. Before this happens, however, that methodology must be refined and validated on the less complex man-machine systems to which it is presently being applied. The author has no patience with those who maintain that the human factors methodology *in its present form* (i.e., without significant improvement) can be applied successfully to any system. It should also be remembered that the application of H to social systems will require the adaptation of any theoretical structure presently available to H to the demands of the new systems.

APPENDIX A: RESPONDENT BIOGRAPHICAL DATA

Harold A. Bauer, BA, Psychology, 1948. Formerly Head, Human Factors Engineering, McDonnell-Douglas Corporation, Long Beach, California.

Julien M. Christensen, PhD, Psychology, 1959. Director, Human Engineering Division, Aerospace Medical Research Laboratory, Wright-Patterson AFB, Ohio. Former president of the Human Factors Society. Fellow, Division of Military Psychology, American Psychological Association, 1968. Fellow, Society of Engineering Psychologists, 1965. Air Force Decoration for Exception Civilian Service, 1966. National Science Foundation Fellow, 1957. Consultant to National Academy of Science. Lecturer, numerous universities. Inventor and author of numerous papers.

Joseph T. Fucigna, MA, Psychology, 1953. Executive Vice-President, Dunlap and Associates, Inc., Darien, Connecticut. 20 years experience with Dunlap in research involving systems analysis and planning, human performance research, training program development, equipment design, etc.

Lawrence J. Fogel, PhD, Biotechnology, 1964, MS, Electrical Engineering, 1952. President, Decision Science, Inc., San Diego, California. Specialist in bionics, information, and computer-based systems, author of *Biotechnology*, McGraw-Hill, 1964.

Jack A. Kraft, EdD, 1951. Formerly Assistant Manager, Biotechnology, Lockheed Missiles and Space Company, Sunnyvale, California. Former president of the Human Factors Society. Fellow of the American Psychological Association.

John Lyman, PhD, Psychology, 1951. Professor of Engineering and Psychology, Director, Biotechnology Laboratory, Department of Engineering, University of California, Los Angeles (one of the relatively few schools which provide strong biotechnological training). Former president of the Human Factors Society.

Jesse Orlansky, PhD, Psychology, 1940. Former vice-president of Dunlap and Associates, Inc. (1948–1960); since 1960 at the Institute for Defense Analyses, Arlington, Virginia. President of Division 21, Society of Engineering Psychologists, American Psychological Association.

James J. Regan, PhD, Psychology, 1957. Chief Psychologist and Head of the Human Factors Laboratory of the Naval Training Device Center, Orlando, Florida.

Harold P. Van Cott, PhD, Psychology, 1954. Director, Office of Communication Management and Development, American Psychological Association. Formerly Director of the Institute for Research on Human Performance, American Institute for Research, Washington, D.C. Consultant to the U.S. Army Research Office, the Ford Foundation and Temple University.

Roger J. Weldon, PhD, Psychology, 1952. Professor, Systems Engineering Department, University of Arizona, Tucson, Arizona, since 1952.

Joseph W. Wissel, PhD, Psychology, 1950. Senior Staff Engineer, Lockheed Missiles and Space Company, Sunnyvale, California. Responsible for direction of the Human Engineering and Maintainability Staff supporting development of the Poseidon and Polaris missiles.

Wesley E. Woodson, BA (Education), 1941. President, Man Factors, Inc., San Diego, California. Formerly with the Navy Electronics Laboratory and General Dynamics/Convair. Senior author of the most widely known human engineering guide for equipment design in America (if not the world).

APPENDIX B: THE OUTLOOK FOR H̲

Opportunities

The problem of how to end a book is just as difficult for an author as where to start it. Because the emphasis in this book has been largely on human factors in military systems and because previous sections have dealt with the *status quo* of H̲, so to speak, the author intends to conclude this chapter—and the book—by describing the opportunities which may arise for H̲ in nonmilitary applications.

If we look at the wide variety of systems and products in this country, there is obviously much potential outside of defense systems which is yet untapped by H. Many specialists conceptualize nonmilitary systems under a single rubric—commercial industry—and they believe that industry is the major area in which H can and must expand if only to provide a wider range of employment opportunities and more stability than defense work.

There are, however, opportunities beyond those of commercial industry. The range of potential nondefense H applications is briefly (and only grossly) indicated in Table 10-1. Included in the list are not only true commercial hardware systems (e.g., transportation, information processing and retrieval) but also social/governmental "software" systems (e.g., education, medicine, criminal justice). Also included are consumer products which are not true systems in the sense of complete man-machine interaction.

The author deliberately began the list in Table 10-1 with *educational systems*. This is because the opportunities in this area are tremendous, provided the specialist can put his research background to practical use. It is apparent from the recent and continuing ferment about public education that any discipline which can guarantee increased learning in shorter time can just about write its own ticket. When the need for trained industrial personnel is also considered, the economic resources available in education are even greater.

The need exists first and foremost in the research area. Present educational practices differ relatively little from those of a hundred years ago. If, for example, it were possible to apply to those practices the results of all the learning research which has been performed in the last 20 years, the results would be revolutionary, and the H specialist as training expert would be in tremendous demand. No less important is the need to develop new instructional technology (e.g., computer-assisted and self-instructional devices).

Of all the areas in which the H specialist might get involved, training is the one which is most closely related to his psychological background and experience and hence one in which he should find the greatest opportunities.

The only question is whether the specialist can actually supply the solutions to the surfeit of educational problems which we face. Within the military training area in which the specialist has been most involved, he has not been outstandingly successful; he has had difficulty applying his research to practical problems (see, for example, Mackie and Christensen, 1967).

More restricted opportunities exist in *medicine* (although see a some-

Table 10-1 Areas of Potential Nondefense H Applications

Social/Governmental

1. *Educational Systems*
   (a) research on and application of training principles to public and industrial education;
   (b) development of instructional technology (e.g., computer-assisted and self-instructional devices);
   (c) development of instructional software (i.e., training materials for [b]);
   (d) research and development of performance evaluation methods and standards;
   (e) development of curricula.

2. *Medicine*
   (a) human engineering design of medical instruments, equipment and hospital facilities (e.g., operating room);
   (b) improvements in medical training and information utilization.

3. *Criminal Justice*
   (a) research leading to improvements in police, court and correctional systems;
   (b) training programs for police personnel;
   (c) human engineering design of police, court and correctional equipment;
   (d) development of information retrieval and communications systems.

4. *Mail Processing*
   (a) human engineering design of mail processing equipment;
   (b) human engineering design of mail handling facilities (i.e., post offices);
   (c) design of improved mail sorting codes;
   (d) improved training of postal employees.

5. *Urban Environments*
   (a) human engineering assistance in housing design, street lighting, parks, etc.;
   (b) improvements in governmental systems providing human resources services (e.g., welfare, education—in other words, governmental management).

Commercial Systems

1. *Information Retrieval*
   (a) human engineering development of information retrieval hardware;
   (b) development of programs for inputting and retrieving information;
   (c) training of personnel to use computerized information handling systems.

2. *Transportation*
   (a) human engineering design of transport vehicles (automobiles, buses, trains, aircraft, ships);
   (b) human engineering design of passenger terminals and baggage handling facilities;
   (c) human engineering design of communications, reservations, and vehicle routing systems;
   (d) research on traffic safety, driver behavior, and road design (governmentally sponsored).

3. *Industrial Systems*
   (a) human engineering in development of automated warehouses, docks, and processing facilities;
   (b) human engineering in computer-aided consumer shopping facilities;
   (c) training of industrial personnel;
   (d) human engineering assistance in the design of manufacturing processes and production lines;
   (e) evaluation of industrial personnel performance and management procedures.

4. *Consumer Products*
   (a) human engineering assistance in product safety design;
   (b) human engineering in product "styling";
   (c) investigation of accidents resulting from product safety deficiencies.

5. *Recreation*
   (a) human engineering design of theatres, auditoria, and sports arenas;
   (b) human engineering assistance in environmental control, for example, lighting, noise;
   (c) human engineering design of reservations systems.

---

what optimistic viewpoint by Rappaport, 1970). A slight amount of work has already been done on human engineering of medical instruments, equipment, and hospital facilities (e.g., the design and layout of operating room equipment). This type of work requires comparatively little modification of basic H principles. The design of a more efficient operating room, for example, is not that much different (analytically, anyway) from the design of an aircraft cockpit, except for the specialized knowledge needed to make the human engineering inputs feasible. Requirements also exist for improved medical training and information systems.

In the *criminal justice* area the government has established a number of new federal agencies (particularly the Law Enforcement Assistance Administration) with considerable funding to update and improve the efficiency of local police, court and correctional organizations. Again no great modification of H principles is required to make major applications primarily because local police forces function in such a relatively primitive manner that even very simple H applications could make significant improvements. Areas of application include more systematic training of police personnel, the human engineering of police equipment (e.g., handcuffs, personal intercoms, the police car—for which see Clark and Ludwig, 1970), and the development of information retrieval and communications systems (e.g., for processing arrest records). The difficulty here, which is also found in medicine, is that the ingrained traditionalism of both specialities makes it difficult for the newcomer to secure an entree.

Although it is necessary to secure the cooperation of local police and medical organizations, the money for research and development comes from Washington, D.C., which is somewhat more receptive to new concepts.

Another governmental area in which H efforts could profitably be expended is *mail processing*. Although a small group of specialists within the Post Office's Department of Research and Engineering are presently working on the human factors aspects of system development and research, their number is pitiably small. The vast extent of the U.S. Post Office (handling 40 billion items per year) calls for vastly increased H assistance.

Ideally the specialist would like to apply his background to the improvement of *urban environments*. Because of the recent stress on "model cities" and the renovation of the "inner city," the national government has allocated substantial sums of money for these and similar programs. One obvious area of H application is the improved design of housing, which has been developed almost exclusively on a traditional basis. A small amount of H work is being performed in architectural design, but the effort is highly tentative. The specialist's background in facilities and work environment design could find useful application in the design of parks, street lighting, shopping centers, subways (see Raffle and Sell, 1969).

The great variety of governmental systems that do not produce hardware but exist to get things done cry out of course for the application of H techniques which presumably should be just as applicable to systems that do not develop hardware as to those which do. H techniques, emphasizing system analysis concepts, could be applied to governmental management.

Among commercial systems we may hope for expanded use of H assistance in *information retrieval* and utilization, if only because we can anticipate an increased "information explosion" in the future. H personnel have of course been making design inputs to the development of computerized information retrieval hardware (e.g., CRT displays, terminals). However, there is potentially a much greater area of H application in the development of programs for inputting and retrieving information. The specialist's background in information theory and communication would be of use here. Training the personnel who will make use of these computerized information handling systems represents a further area of potential H application.

Since much of the specialist's work on military systems has involved vehicles of one sort or another (e.g., aircraft, tanks, ships), it is reasonable to anticipate some application of his experience to civilian *trans-*

portation. Automobiles, buses, railroad equipment, aircraft, and ships possess significant needs for human engineering. We have seen (Chapter Two) the almost desperate need for the human engineering of automobiles; other types of transportation vehicles are in no better shape. Recent specialist experience in assisting in the design of new military transport aircraft (C-5A) and naval destroyers (DX series) indicates that substantial improvements in efficiency can be achieved through expanded human engineering design. The need for human engineering of passenger terminals and baggage handling facilities is apparent to anyone who has waited to board a plane or reclaim his baggage. Although a sizeable number of specialists have been working for some years in the areas of traffic safety, driver behavior, and road design, the need here is a continuing one.

The application of H̲ to *industrial systems* and *consumer products* has been meagre. Much more needs to be done to improve the design of manufacturing processes and production lines. Beyond that, the introduction of computerization, in a very tentative manner, to warehousing and consumer shopping, gives reason to hope that commercial industry will be more receptive than it has been to new ideas. Mention has already been made of the possibilities inherent in training of industrial personnel. There have been minor H̲ applications to the evaluation of personnel performance and management procedures (i.e., management consulting), but, considering the widespread inefficiency of industrial operations in general, the need for more H̲ activity in this type of work is apparent. Again the major factor here is the innate conservatism of industry, which, despite its presumed receptivity to new concepts, is still stubbornly traditional in its outlook.

Interest in design of consumer products for users and in product safety has expanded considerably in recent years as a consequence of people like Ralph Nader. It is natural then to think of the application of human engineering practices to product design since this would require no significant changes in methodology (as a representative example, see MacNeill, 1970). A few H̲ personnel have also become specialists in the investigation of accidents resulting from product safety deficiencies.

Practically no attention has been paid to one of the most important consumer areas, that of *recreation.* The application of human engineering design to theatres, auditoria, and sports arenas would be a natural extension of the specialist's work in facilities design.

Difficulties

Although the opportunities for H̲ expansion exist, the specialist has formidable difficulties to overcome. One of these difficulties lies in the nature of the specialist's system orientation which is not particularly adapted to work on commercial products.

A product differs from a system because it does not involve the latter's man-machine *interaction.* A system is defined by its closed loop characteristics: the operator manipulates the machine, the machine reacts; this reaction stimulates the operator who then responds by a further machine manipulation, etc.

The product (e.g., a can of soup) lacks the looping characteristic; it is the object of the operator's action, it does not react. Because of this lack of interaction it is usually much less sensitive to impinging stimuli than is the system. Its performance is not usually or greatly influenced by its environment or by alternate task uses. Moreover, the demands for capability which it makes of the user are an order of magnitude less than those which systems make of their operators.

When products become complex, however, they assume many of the attributes of the system. A prime example is the automobile; it makes heavy demands on its users and may even be injurious to their safety; its sensitivity to manipulation creates a form of man-machine interaction.

Since H̲ by its nature involves a system orientation, the call for its services in the development of products is somewhat muted. This is not to say that human factors principles (e.g., anthropometric constraints) cannot be highly useful to product development; for example, the design of most common appliances could be much improved if their designers made use of simple human engineering principles. However, the more complex analytic aspects of H̲ are not called into play in product design.

Another product characteristic which must be considered is that, to an extent much greater than is true of the system, it has or is or produces something to be *sold.* As a consequence, product developers view the user interaction with his equipment largely in terms of those factors which induce the user to *buy.* To them this interaction is essentially a matter of product "styling," which has so far been handled primarily by industrial designers.

As far as functional efficiency from a user standpoint is concerned, industry considers that these needs are largely satisfied by the designer. If the latter makes the equipment or the product work, this is assumed automatically to take account of user interfaces. We suspect that this attitude is very strongly entrenched in industrial product management, notwithstanding the volume of evidence which indicates that the design engineer does not take sufficient account of user needs.

In any event, as a consequence of these factors, the prospects for expansion of H̲ into product design do not look overly bright, and in fact we find few H specialists designing such products as automobiles, furniture, or appliances.

The opportunities for application of H to social/governmental systems are more hopeful but also limited. The hopeful aspect is that these systems have one factor in common: profit is not their determining criterion. Hence it is reasonable for these systems to provide money for human factors activity without thought of a commensurate financial return. Moreover, the emphasis on the "social" planning elements of government (e.g., welfare, pollution, law enforcement, transportation) make this type of system a natural opening for the H specialist.

When, however, H is put to work in these social systems (and this occurs more and more frequently), it is at the subsystem rather than the system level. For example, a police department may contract for a new, computer-controlled information retrieval system with graphic capabilities. Obviously the specialist, as part of the contractor team which designs such a system, participates in the design of the system; but he does not do so in a policy-making role, nor does his activity influence the over-all social system significantly.

Much publicity has been given, at least in the author's home state of California, to the adaptation of aerospace systems engineering methodology to analytic research of major social problems like water resources and law enforcement, and presumably a few H specialists play a role in such research efforts. The concept is very promising, but such projects are only in their infancy although we may expect to see more of them.

It is reasonable to suppose that many H contributions to social/governmental systems are and will continue to be in the traditional human engineering manner (e.g., work station layout, assistance in the design of equipment), rather than the design of the total system. To the extent that these new systems (new only in the sense that H has not been involved in their activities before) involve conventional man-machine interfaces, it is to be expected that specialists will be asked to work on them. Examples of such H applications in nonmilitary systems are the design of semi-automatic postal sorting equipment, research studies on the development of new ship navigation instruments, and a flurry of human factors research on varied transportation activities (e.g., trains, highway signs, etc.). This type of input will require no basic changes in H principles.

If H expands to new types of social/governmental systems, it will be those which, because they have serious implications for the user's health, safety, and/or comfort, create a demand for more efficient human factors design, the same demand which initiated H at the start of World War II. If H is, as we have suggested elsewhere, a means of controlling a technological culture which impacts harshly on the average citizen, its expansion will be to those systems and activities which have the explicit goal of minimizing that discomfort.

REFERENCES

Clark, G. E., and H. G. Ludwig, 1970. Police Patrol Vehicles, *Human Factors,* **12** (1), 69–74.

Mackie, R. R., and P. R. Christensen, 1967. *Translation and Application of Psychological Research.* Technical Report 716-1, Human Factors Research, Inc., Goleta, California, January.

MacNeill, R. F., 1970. Human Engineering Pleasure Boats, *J. Indust. Des. Soc. Amer.,* **3** (2), 35–39.

Meister, D. (Ed.), 1970. *The Status of Human Factors Theory, Research and Practice,* unpublished manuscript.

Raffle, A., and R. G. Sell, 1969. The Victoria Line–Passenger Considerations, *Appl. Ergonomics,* **1** (1), 4–11, December.

Rappaport, M., 1970. Human Factors Applications in Medicine, *Human Factors,* **12** (1), 25–35.

INDEX